Urban Mining for Waste Management and Resource Recovery

Urban Mining for Waste Management and Resource Recovery

Sustainable Approaches

Edited by

Pankaj Pathak and Prangya Ranjan Rout

CRC Press
Taylor & Francis Group
Boca Raton London New York

CRC Press is an imprint of the
Taylor & Francis Group, an **informa** business

First edition published 2022
by CRC Press
6000 Broken Sound Parkway NW, Suite 300, Boca Raton, FL 33487-2742

and by CRC Press
2 Park Square, Milton Park, Abingdon, Oxon, OX14 4RN

ISBN: 9781032061795 (hbk)
ISBN: 9781032061801 (pbk)
ISBN: 9781003201076 (ebk)

DOI: 10.1201/9781003201076

Typeset in Times LT Std
by KnowledgeWorks Global Ltd.

Contents

Preface

Rapid urbanization and incessant industrialization are causing major concerns for global development. In one way, conventional natural resources are depleting at a faster rate and in another way, huge amounts of solid and liquid wastes are being generated. This is posing several environmental, economic, and social threats to sustainable development. These threats are significant and alarming in developing countries due to the mismanagement of resources and improper waste processing practices. The major challenge occurs when solid and liquid wastes are not handled properly during collection, processing, and disposal, which causes deterioration of environmental quality and impairment to human health. However, with appropriate (scientific) management strategies, these anthropogenic wastes (human-made materials) can be explored as potential resources through the urban mining concept and could be a panacea for sustainable development.

In this book, Chapter 1 introduces the basics of urban mining and Chapter 2 describes GIS applications in solid waste management. Chapters 3 and 4 explore energy from food waste and biochar production. Moreover, Chapter 5 provides ideas for handling biomedical waste. Chapters 6 and 7 describe recycling techniques of plastics and reclamation of asphalt pavement materials. Chapter 8 presents details of resource recovery from agro-waste. Additionally, Chapters 9–12 explain the treatment of wastewater and resource recovery from municipal and industrial wastewater, whereas Chapters 13 and 14 describe advanced methods of resource recovery from wastewater.

Pankaj Pathak
Prangya Ranjan Rout

Editors

Dr. Pankaj Pathak is an Assistant Professor at SRM University, Andhra Pradesh, India, with a keen interest in sustainable waste management, green energy resources, and geochemistry, including sustainable handling of hazardous waste and associated environmental impacts. She has been involved in various research projects viz., waste to resource, nuclear waste management and characterization of buffer materials, and e-waste management. She has also gained research experience in dealing with hazardous solid and liquid waste treatment technology, along with hydrometallurgical recovery of metals from waste streams, their safe disposal, and remediation techniques.

Dr. Prangya Ranjan Rout is presently serving as an Assistant Professor in the Department of Biotechnology, Thapar Institute of Engineering and Technology, Patiala, Punjab, India. He also worked as a researcher at INHA University, Incheon, Republic of Korea from 2018–2020. He holds an MTech degree in biotechnology and a PhD in environmental engineering. His research interest lies in the domain of anaerobic digestion, bioconversion of wastes to wealth, emerging contaminant removal, membrane technology, resource recovery and reuse, and wastewater treatment. He has authored over 50 publications, including refereed journal articles, book chapters, national and international conference presentations, technical notes, and a published patent. Some of the awards he has received include the Odisha Young Scientist Award 2017, Best Practice Oriented Paper 2019 from ASCE-EWRI, and Outstanding Reviewer 2019 from ASCE. He is an Associate Editor of the *ASCE Journal of Hazardous, Toxic, and Radioactive Wastes* and has served as a guest editor of a special issue of the journal. He is also actively involved in editing contributed book volumes for internationally renowned publishers, such as CRC Press/Taylor & Francis, John Wiley & Sons, and ASCE.

1 Basic Concepts, Potentials, and Challenges of Urban Mining

Sasmita Chand,[1] Prangya Ranjan Rout,[2]
and Pankaj Pathak[3]
[1]Center of Sustainable Built Environment, Manipal
Academy of Higher Education, Manipal School
of Architecture and Planning, Manipal, India
[2]Department of Biotechnology, Thapar Institute
of Engineering and Technology, Patiala, India
[3]Department of Environmental Science,
SRM University, Andhra Pradesh, India

CONTENTS

1.1 INTRODUCTION TO URBAN MINING

Population explosion mediated surge in demand has led to massive consumption of natural resources (Mohanty et al., 2021). In addition, rapid urbanization, industrialization, economic growth, and enhancement of people's living standards have led to considerable buildup of natural resources in products, infrastructures, and in the waste deposits (Krook and Baas, 2013). As a consequence, the rate of waste

DOI: 10.1201/9781003201076-1

generation has risen drastically and natural resource reserves are on a depleting trend (Lee et al., 2021). Therefore, in the current scenario, material resource crunch and suitable waste disposal are emerging as critical global challenges. Contextually, recycling-induced resource reclamation from wastes is a smart approach to realize the dual benefits of recovering valuable recyclable materials and protecting the environment by streamlining the waste disposal issues (Gutberlet, 2015; Simoni et al., 2015). The simultaneous resource recovery and waste management can be achieved through a growing remedial practice named urban mining (Park et al., 2017).

For the first time, the phrase urban mining was presented in the 1980s to describe metal recovery from waste manufactured goods in an urban setting (Nanjo, 1988). Professor Nanjo of Japan coined the term to promote recycling and reuse. The term has evolved since then and nowadays, it is being used to describe all the processes related to reclaiming varieties of resources from diversified anthropogenic wastes. So, the concept of urban mining emphasizes securing raw materials from anthropogenic sources, particularly in cities. In contrast, conventional mining signifies the extraction of raw materials from natural resources, as depicted in Figure 1.1 and compared in Table 1.1 (Copper Alliance, 2020). Another simplified description of urban mining is the recuperation of resources/materials in the anthroposphere/technosphere. All the man-made things and their interactions with the four spheres of Earth (atmosphere, hydrosphere, biosphere, and lithosphere) constitute the anthroposphere or the technosphere (Kuhn and Heckelei, 2010; Zalasiewicz, 2018). As the anthroposphere contains vast amounts of a wide variety of materials, urban mining targets to utilize these materials as a source of raw material supply to produce a variety of new products. Urban mining aims to utilize the waste of today and emphasizes capturing the value contained in the waste of tomorrow. By allowing the collection of wastes/discarded products and ensuring their return to the material cycle as secondary raw materials, urban mining appears to be a key element of the circular economy. Through the supply of secondary resources, urban mining offers a degree of independence from natural resources and it complements conventional mining in meeting the higher demand for resources, thereby ensuring enhanced supply security (Tercero et al., 2020).

Urban mining is the practice of managing resources. It plays a significant role in recycling and environmental safeguarding via the extraction of valuable metals, materials, and energy from anthropogenic stocks or waste streams

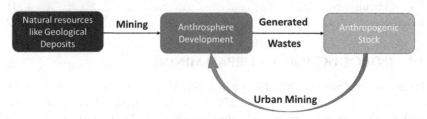

FIGURE 1.1   Resource flow from natural deposits to the anthroposphere and their recycle-reuse in anthrosphere through urban mining.

TABLE 1.1

Comparison between Conventional and Urban Mining

Si. No.	Conventional Mining	Urban Mining
1.	Extraction of valuable minerals from natural sources like geological deposits.	Recovering/recycling valuable materials from anthropogenic stocks in urban areas.
2.	The predominant source of the majority of metals used in the anthroposphere.	For some of the rare metals, urban mining is gradually becoming the sole source.
3.	Can independently meet the increasing demand.	Can complement conventional mining in meeting the increasing demand.
4.	Alone it cannot meet the rising demand for electrical and electronic appliances.	In cities across the globe, millions of appliances are yet to be recovered.
5.	Process of extracting metals is complex and cost incurring.	Raw material recovery from e-wastes can be done in simple and cost-effective ways.
6.	Tailing management and disposal issues.	Disposal issues of residues after recovery.
7.	Resource availability in designated areas.	Concentrated in urban areas.
8.	Geo-political issues in classical mining.	No such issues in urban mining.
9.	Risk in accessing and replacing reserves.	No such risks involved.
10.	Difficult to access community support and social license to operate.	Easy access to community and social support to operate.
11.	No direct involvement of public in operation.	Public involvement in waste collection.
12.	Classical or conventional mining has a significant environmental impact.	Avoids negative impacts on human beings and the environment to a greater extent.

(Zeng et al., 2018). Urban mines comprise different composite streams and are identified in urban areas as potential sources of resources. In the context of urban mining in India, urban mines are majorly composed of waste dumps and landfills, electronic waste (e-waste), building stock, old vehicles, municipal solid waste (MSW), etc. Moreover, in an international framework, the composition of urban mines consists of two categories: (a) short-term urban mines, where we basically find commodity materials that include waste electrical and electronic equipment (WEEE), waste generated during industrial production processes, from consumer goods, biomass, and packaging wastes; and (b) long-term urban mines constitute of industrial, commercial, and residential buildings, MSW, as well as industrial landfills, infrastructures such as canals, bridges, and streets, and construction and demolition (C&D) waste (Arora et al., 2017; Pathak and Chabhadiya, 2021). Electronic waste (e-waste) is amounting out of these stocks uncontrollably and emerging as a potential global problem (Pathak et al., 2017). Thus, taking into environmental consideration, recently urban mining, including e-waste, has established an important concern owing to its business opportunity, profitable prospects, livelihood sources, and eventually achieving the Sustainable Development Goals (SDGs), 2030 agenda (Arya and Kumar, 2020). It has also been reported that urban mining is being called a pioneering battleground of sustainability, which can play an important role in environmental sustainability in developing countries especially, in India (Rout et al., 2020; BW Businessworld, 2021; Indiatimes, 2021).

1.2 POTENTIAL SOURCES OF URBAN MINING

In India, natural resource depletion is very rapid because of myriads of interlinked aspects like exponential population outbursts, rising incomes, swift urbanization, and industrialization. The overall resource consumption is very high, but the per capita material consumption compared to other developed and developing nations is still low. Out of the total material consumption in India, approximately 97% are extracted and only 3% are imported. If the trends are allowed to continue, by the year 2050, the requirement of materials is expected to touch 25 billion tons, with a major surge expected to be seen in minerals, metals, and fossil fuels (GIZ-IGEP, 2013). The metropolises in developing and developed countries contain a huge amount of materials, buildings, infrastructure, and landfills (Ghosh, 2020). To accomplish these anthropogenic stocks, the foremost aims of urban mining anticipated for environmental protection, resource conservation as a major pillar in economic benefits. Therefore, urban mining is being sought worldwide to recover resources from anthropogenic stocks, and these stocks can significantly contribute to the resource economy, as shown in Figure 1.2.

These secondary resources potentially safeguard the environment and bring about resource and financial benefits in a developing country like India. Thus, the

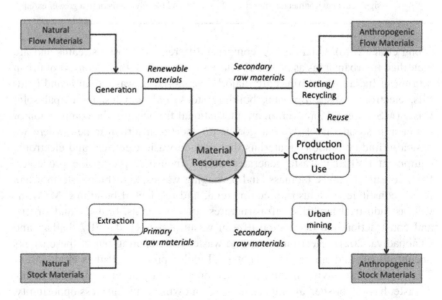

FIGURE 1.2 Scheme for describing the material flows among different material resources, as natural vs. anthropogenic. (Adopted from Cossu and Williams, 2015.)

concept and practice of urban mining are very relevant to urban waste sources like end-of-life vehicles (ELVs), C&D waste, energy efficient lighting, and e-wastes. The usage of secondary resources is meant for both the environment and economy of sustainable cities, significantly reducing CO_2 emissions. These cities would not distinguish between resources and wastes and would contemplate innovative ideas to foster the utilization of secondary resources (Arora et al., 2017).

The resource stock categories can be differentiated in many ways and can be anticipated as potential sources of valuable resources in the Indian context as:

1.2.1 LANDFILLS

Garbage disposal in Indian cities is very haphazard, which creates a lot of pollution and nuisance. However, a change has been seen in a few cities in which they are dumping the wastes through a scientific and engineered approach called sanitary landfills. The composition of these waste streams have been observed like combustibles as leather (5–35%), metals (<1%), rubber, plastic, textile, soil (40–68%), and rocks and inert wastes (18–30%). The quantity of precious recoverable materials like metals from the ultimate waste disposal is pretty low because of the high diversion rate of valuables through the informal sector. Additionally, the combustible fraction of the waste can be utilized to generate energy, inert debris and stones could be utilized for the purpose of construction, and the final soil fraction could be used as soil amendments for non-edible crops or as filling material with tested for contamination (Arora et al., 2017).

Enhanced landfill mining (ELFM) is the process of excavation and valorization of landfilled wastes to value-added materials and energy resources using state-of-the-art conversion technologies (Jones et al., 2013). ELFM approach facilitates valorization of landfilled wastes to both waste-to-energy (WtE) and waste-to-material (WtM) through recycling and incineration processes, thereby ensuring improved reuse rates, better recycling, and boosted energy valorization (Van Passel et al., 2013). ELFM considers landfills as transitory storage facilities rather than a permanent solution from which the landfilled wastes are eventually extracted for further valorization. The resource recovery helps reduce the use of virgin materials and landfill footprint, where the reclaimed space from landfill closure can have other societally beneficial usages. ELFM also aims to sequester a significant portion of the CO_2 that arises during the energy valorization process. ELFM technology looks promising, though it is still in its initial phase of development. Therefore, ELFM approach can reach its full potential by combined incentives of energy utilization, material recycling, and nature restoration along with the requirement of strategic policy decisions (Rout et al., 2018).

1.2.2 CONSTRUCTION AND DEMOLITION (C&D) WASTE

C&D waste is generated when demolition activities occur in the subway, flyover, bridges, roads, buildings, and remodeling. C&D waste mostly consists of non-biodegradable and inert materials (plastic, wood, metal, plaster, concrete, etc.).

The C&D waste is growing progressively in India. As the Technology Information Forecasting and Assessment Council (TIFAC) estimates, the C&D waste produced was around 15 million tons per year in 2001. As per the estimation by the Centre for Science and Environment (CSE), around 500 million tons of C&D waste was generated (TIFAC, 2001; CSE, 2014; NITI, 2018), but a general perception around was that these were gross underestimations and outdated. As per TIFAC (2001), the composition of C&D waste comprises bricks and masonry (31%), soil, sand and gravel (36%), ceramics, plastics, wood, metals (10%), concrete (23%), etc. Based on these characterizations and composition, C&D wastes can be used in wooden frames, metal fittings, the construction sector, etc. The left-over gravel, sand, and soil could be consumed as landfill cover. C&D waste appearing in the form of mortar, bricks, and concrete can be processed well and could be transformed into aggregates, which then could be directly utilized in building materials as paving block, bricks, concrete, etc. (Behera et al., 2014; Arora et al., 2017).

The C&D wastes can either be conventionally dumped in landfills or reused to produce natural materials in new concrete production. The huge amount of generated C&D wastes has resulted in space restraint in landfills. Therefore, recycling of C&D wastes can be adjudged as one of the superior alternatives to overcome the landfill space occupancy issues and reduce the overall quantities of C&D wastes (Tam, 2008). The cost-benefit comparison of the conventional practice of landfill dumping and recycling of the C&D wastes as aggregates for fresh concrete making revealed that the recycling process receives an annual net benefit of around $30,916,000, whereas the conventional technique receives an annual negative net benefit of approximately −$44,076,000 (Tam, 2008). Therefore, the modern C&D wastes recycling approach, as depicted in Figure 1.3, is an economical method,

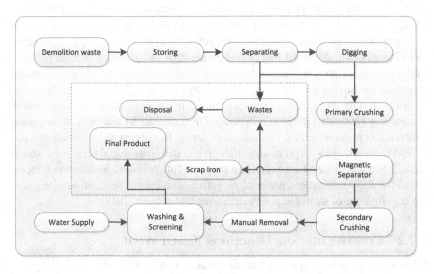

FIGURE 1.3 Sustainable construction and demolition waste recycling scheme. (Adopted from Rout et al., 2016.)

which can also be helpful in achieving construction sustainability and protecting the environment simultaneously (Rout et al., 2016).

1.2.3 END-OF-LIFE VEHICLES

The rate of growth of the Indian auto industry has been phenomenal. It has been one of the fastest growing industries across the country. In total, 23.36 million units of vehicles were produced in 2014–2015 (Society of Indian Automobile Manufacturers, 2015). The requirement for two and three wheelers has increased in the Indian marketplace as they offer inexpensive ways for transportation for a significant portion of the population. The annual domestic sales in three consecutive years 2012–2015 were 17.7, 18.4, and 19.9 million cars, respectively. The total number of registered vehicles in the year 2015 was over 200 million (Society of Indian Automobile Manufacturers, 2015). By 2015, an estimated 8.7 million vehicles are expected to reach ELV status, 83% of which were from the two-wheeler segment. An estimated 22 million vehicles and 80% of two-wheelers are expected to attain ELV status by 2025 (Akolkar et al., 2016). According to the central pollution control board (CPCB), the semiformal sector involving scrap dealers and dismantlers is mainly involved in reusing, recycling, or disposing of discarded vehicles. Approximately, 75% of the weight of a vehicle comprises metals. These are recycled and recovered through secondary metal processing units.

1.2.4 MUNICIPAL SOLID WASTE (MSW)

India is facing an uphill challenge of ever-increasing MSW generation. Much of the litter can be owed to improved income, fast urbanization, changing lifestyle, and economic trends. MSW generation has seen a steep rise from 34 million tons to 80 million tons during 2000 to 2015 and an expected 200 million tons by 2030. MSW in India mostly comprises of a large organic fraction (40–60%), ash and fine earth/soil constitute 30–40% of the waste rest is comprised of recyclables like metals, glass, and plastic (about 10%) by weight (Kaushal et al., 2012; Nandy et al., 2015; Pujara et al., 2020). The waste composition is diverted by an informal network comprising of scrap dealers, sorters, and door-to-door collectors. Additionally, the organic waste from MSW can be converted to composting in most of the cities, as shown in Figure 1.4.

Nearly 84–90% of MSW is dumped in landfills and the biodegradation of the organic fractions of MSW under anaerobic conditions inside the landfills produces landfill gas (LFG) (Shan et al., 2010). The LFG can be collected to produce energy, thereby minimizing greenhouse gas emissions to the atmosphere and generating revenues for the municipality corporations. As per an estimation by the USEPA, during the years 1995–2001, 253 Teragrams (Tg) of CO_2 equivalents of methane were used in projects aiming to convert LFG to energy and 350.3 Tg of CO_2 equivalents of methane was flared (USEPA, 2004). The recovered LFG can directly be used as a fuel or to up-concentrate bio-methane through

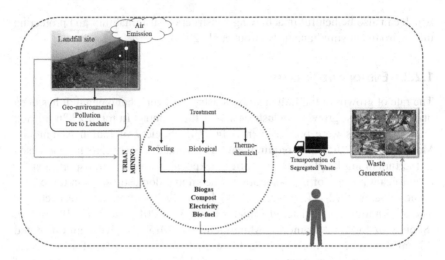

FIGURE 1.4   Process of urban mining for managing municipal solid waste. (From Pujara et al., 2019.)

the removal of CO_2 from LFG. LFG can also be utilized in the production of synthetic natural gas, liquid hydrogen, synthetic hydrocarbons, and methanol (Muradov and Smith, 2008).

1.2.5 ELECTRONIC WASTE (E-WASTE)

Among the above anthropogenic stocks, e-waste is the fastest growing secondary source for valuable and precious metals. As per the United Nations (UN) Global e-Waste Monitor (2020), globally 53.6 million metric tons of electronic waste were generated in 2019. The global annual e-waste volume has seen a 20% huge jump with 44.7 Mt in 2016 and 52.2 million tons or 6.8 kg per inhabitant by 2021 (Livemint, 2021). It has also been predicted the global e-wastes, in particular batteries or plugs, will reach 74 Mt by 2030 (www.ewastemonitor.info). It has also been reported that with an exponential growth rate of a population of 1.21 billion (Census, 2011) and an estimated population to be 1.39 billion (upcoming census 2021), a rapid generation of e-waste in India is expected. The data for e-waste generation for top countries have been reported that India generates e-waste annually about 3 Mt and third position after China and the United States and it might rise to 5 million tons by 2021 (Census of India, 2011; Arya and Kumar, 2020; Census of India, 2020; Mongabay-India, 2021). Thus, for the proper management of this urban mining stock like e-waste, a value chain must be explored as a sustainable approach. Several methodologies are being taken into consideration for the urban mining potential of e-waste as an awareness of accountability. Additionally, the importance is also being placed on upstream solutions through the recycling processes (The Diplomat, 2021). In terms of e-waste management, in Southern Asia, since 2011, India has e-waste legislation rules, making it compulsory that only authorized recyclers and dismantlers can only collect e-waste. The total number of authorized recyclers who comply with the e-waste (Management) Rules, 2016, stands at 312. Compulsory responsibilities and collection targets are transferred to the manufacturers through extended producer responsibility (EPR) (Mongabay-India, 2021). Therefore, for recycling, we have to implement various effective constructions and integrated approaches for the whole e-waste value chain to meet reprocessing targets in an eco-friendly manner like eco-friendly electrical and electronic equipment (EEE) product design, life cycle assessment, reduce, recycle, reuse, and recover (4R) approach, EPR, circular resource exchange and management, development of eco-friendly techniques (refurbishing, remanufacturing, etc.) to manufacture value-added products from e-wastes, etc. (Arya and Kumar, 2020; DownToEarth, 2021).

Therefore, the overall objectives of urban mining mediated e-waste handling are environmental protection and economic benefits through conservation of material resources, which is a substantial pillar in economic activities. Furthermore, urban mining can be taken as part of green circular economy, product life cycle, zero waste, smart city application, sharing cities, and economy (Rau, 2019). Raw materials can be processed and transformed into new products and consumer goods. The International Society of Waste Management, Air and Water (ISWMAW) and International Solid Waste Association (ISWA) have recognized that the recycling, reuse, and repair approaches by informal sectors like microenterprise achieve significant recycling and save local authorities in large cities (Ghosh, 2020).

1.2.6 INDUSTRIAL WASTES

The exponential growth rate of populations and speedy industrialization has given rise to an enormous amount of liquid and solid wastes generated by industries like tanneries, sugar, food processing, distilleries, pulp and paper, dairies, slaughterhouses, starch, poultries, etc. Industrial wastes could broadly be categorized into hazardous (radioactive materials, heavy metals, trace organics, pesticides, etc.) and non-hazardous (paper, cardboard, wood, textiles, packaging, etc.) industrial wastes. Usually, these wastes are dumped randomly without adequate treatment on land and discharged into aquatic bodies. This becomes an important source of environmental pollution and degradation. Thus, the huge amount of wastes generated from various industries has to be managed in an integrated manner and it should be essential to obtain approval for their treatment and disposal from corresponding State Pollution Control Boards (SPCBs) by appropriate rules. In recent times, hazardous wastes are used as an alternate fuel, and non-hazardous wastes are used for biogas production and digestate, fertilizer, ethanol, and power generations (EAI, 2021). Based on the Ministry of New and Renewable Energy data, it has been reported that the total assessed energy generation prospective from urban and industrial sectors in India is about 5690 MW (Ministry of New and Renewable Energy, 2021).

1.2.7 BIOMEDICAL WASTES (BMWS)

The wastes that are infectious, hazardous, corrosive, and radioactive and generated from medical activities like diagnosis, treatment, and immunization are known as biomedical waste (BMW). The main sources of BMW are characterized as primary like hospitals, nursing homes, dispensaries, research labs, immunization centers, animal research centers, dialysis units, blood banks, and industries and secondary like medical clinic, home treatment, slaughter houses, funeral services, and educational institutes (Tiwari and Kadu, 2013). With an increase in population, which is anticipated to reach 1.65 billion by 2030, the number of health care, hospitals, and laboratories has increased in Indian cities, and BMW constitutes nearly 50–60% of the total solid waste. The bulk of BMW produced in India has been reported as 530 metric tons per day and the bulk of the BMW related to COVID-19 alone was 101 metric tons per day in 2020 (Statista, 2021). Therefore, proper management needs to be implemented. If the management is inappropriate, this may result in numerous problems comprising spreading contagious diseases and various other environmental contamination (Rai et al., 2020; Goswami et al., 2021). Several treatment methodologies have been adopted for BMW treatment, such as chemical treatment, incineration, microwave irradiation, dry and wet thermal treatment, landfill disposal, and incineration (Kulkarni, 2020). So far BMW management in India is concerned, BMW (Handling and Management) Rules, 1998 was notified under the provision of Environment (Protection) Act 1986. Again, it was revised in 2016 and amended in 2018 to increase the appropriate management procedures (BMWM Rules, 2018).

1.3 CHALLENGES ASSOCIATED WITH URBAN MINING

Urban mining is a significant tool to fulfill the demand for raw materials in various industrial activities and its further expansion. Urban mining is a new substitute for developing economies and is structured by the concept of circular economy (Arora et al., 2017). However, the urban mining sector does not consider a formal recycling system for recovering materials and matter and poses severe effects on health and the environment, as most of the activities are handled by the informal sectors. Based on the study by several researchers, 18 barriers were identified for urban mining and are shown in Figure 1.5 (Kazançoglu et al., 2020).

To understand the different challenges associated with urban mining, Kazançoglu et al. (2020) have provided 6 different dimensions and 18 barriers for the evolving economy, as shown in Figure 1.5. It indicates that each dimension has more than one barrier and causing huge impacts not only on the environment but also on the economy, business, and government rules and legislations. Therefore, formal recycling should be done to sustainably utilize the urban mining tools for emerging economy.

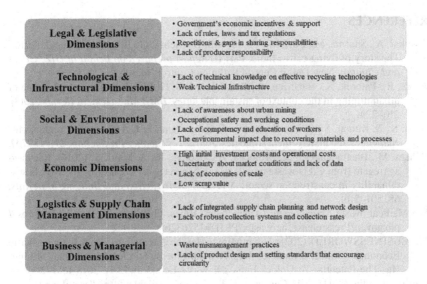

Legal & Legislative Dimensions	• Government's economic incentives & support • Lack of rules, laws and tax regulations • Repetitions & gaps in sharing responsibilities • Lack of producer responsibility
Technological & Infrastructural Dimensions	• Lack of technical knowledge on effective recycling technologies • Weak Technical Infrastructure
Social & Environmental Dimensions	• Lack of awareness about urban mining • Occupational safety and working conditions • Lack of competency and education of workers • The environmental impact due to recovering materials and processes
Economic Dimensions	• High initial investment costs and operational costs • Uncertainty about market conditions and lack of data • Lack of economies of scale • Low scrap value
Logistics & Supply Chain Management Dimensions	• Lack of integrated supply chain planning and network design • Lack of robust collection systems and collection rates
Business & Managerial Dimensions	• Waste mismanagement practices • Lack of product design and setting standards that encourage circularity

FIGURE 1.5   Six dimensions and 18 barriers of urban mining . (From Kazançoglu et al., 2020.)

1.4   CONCLUSIONS

Urban mining represents an opportunity to recover raw materials from anthropogenic sources, particularly in urban areas/cities. The mainstream material hibernating stocks are located in the urban areas/cities in the form of complex waste streams like C&D wastes, BMW, e-waste, MSW, industrial wastes, and ELVs. The cities constantly face the waste disposal issues due to these wide spread complex waste streams and also there are loss of critical resources through the waste streams. Urban mining is a holistic approach composed of advanced recycling techniques to improve resource recovery from the anthroposphere, reallocate valuable materials to a second life in the form of active products, and safeguard the environment by minimizing the contaminant emission levels. However, in practice, the recycling and recovery processes are not very effective due to the large quantities of the generated wastes and heterogeneous characteristics of the wastes. Some of the key challenges faced by urban mining include lack of advanced techniques to mine specific obsolete reserve stocks in an inexpensive and correct way, failing to choose an appropriate technology that can transform inert anthropospheric stocks into marketable materials and energy carriers, dearth of understanding of policy instruments and their role in regulating societal impacts of urban mining, and non-integration of resource recovery aspects in city planning as well as landfill transformation processes. In a nutshell, the state-of-the-art urban mining technology is still largely theoretical; the development of applicable technologies improved recycling standards, better city planning, and conducive environmental, as well as social factors will be key elements in demonstrating the feasibility and application of urban mining concepts in practice.

REFERENCES

Akolkar, A., Sharma, M., Puri, M., Chaturvedi, B., Mehra, G., Bhardwaj, S., et al. (2016). The story of dying car in India. Part II. *Report prepared on behalf of GIZ, CPCB and Chintan*. New Delhi: Central Pollution Control Board.

Arora, R., Paterok, K., Banerjee, A., & Saluja, M. S. (2017). Potential and relevance of urban mining in the context of sustainable cities. *IIMB Management Review*, 29(3), 210–224.

Arya, S., & Kumar, S. (2020). E-waste in India at a glance: Current trends, regulations, challenges and management strategies. *Journal of Cleaner Production*, 271, 122707.

Behera, M., Bhattacharyya, S., Minocha, A., Deoliya, R., & Maiti, S. (2014). Recycled aggregate from C&D waste and its use in concrete – A breakthrough towards sustainability in construction sector: A review. *Construction and Building Materials*, 68, 501–516.

Bio-Medical Waste Management (Amendment) Rules. (2018). Online available: https://pib.gov.in/Pressreleaseshare.aspx?PRID=1526326.

BW BUSINESSWORLD. (2021). Urban Mining Has the Potential to Unlock India's e-Waste Economy. Online available: http://www.businessworld.in/article/Urban-Mining-Has-The-Potential-To-Unlock-India-s-E-Waste-Economy/15-12-2018-165172/

Census of India. (2011). 2011 Census of India. Online available: https://en.wikipedia.org/wiki/2011_Census_of_India

Census of India. (2020). 2021 Census of India. Online available: https://en.wikipedia.org/wiki/2021_Census_of_India#:~:text=India's%20population%20was%20estimated%20to%20be%201%2C393%2C409%2C038%20as%20per%202021%20upcoming%20census.

Copper Alliance. (2020). The promise and limits of urban mining. Online available: https://copperalliance.org/2020/11/17/the-promise-and-limits-of-urban-mining/.

Cossu, R., & Williams, D. (2015). Urban mining: Concepts, terminology, challenges, *Waste Management*, 45, 1–3.

CSE. (2014). *Construction and Demolition Waste*. New Delhi: Centre for Science and Environment. Online available: http://www.cseindia.org/userfiles/Construction-and%20-demolition-waste.pdf.

DownToEarth. (2021). International e-Waste Day: Why India needs to step up its act on recycling. Online available: https://www.downtoearth.org.in/blog/waste/international-e-waste-day-why-india-needs-to-step-up-its-act-on-recycling-73786

EAI. (2021). Industrial Wastes in India. Online available: http://www.eai.in/ref/ae/wte/typ/clas/india_industrial_wastes.html.

Ghosh, S. K. (2020). *Urban Mining and Sustainable Waste Management*. Berlin, Heidelberg: Springer.

GIZ-IGEP. (2013). India's future needs for resources: Dimensions, challenges and possible solutions. *Indo-German Environment Partnership*. New Delhi: GIZ-India.

Goswami, M., Goswami, P. J., Nautiyal, S., & Prakash, S. (2021). Challenges and actions to the environmental management of bio-medical waste during COVID-19 pandemic in India. *Heliyon*, 7(3), e06313.

Gutberlet, J. (2015). Cooperative urban mining in Brazil: Collective practices in selective household waste collection and recycling. *Waste Management*, 45, 22–31.

Indiatimes. (2021). Urban mining is the key to sustain the cities of future. Online available: https://timesofindia.indiatimes.com/blogs/da-one-speaks/urban-mining-is-the-key-to-sustain-the-cities-of-future/

Jones, P. T., Geysen, D., Tielemans, Y., Van Passel, S., Pontikes, Y., Blanpain, B., Quaghebeur, M., & Hoekstra, N. (2013). Enhanced landfill mining in view of multiple resource recovery: A critical review. *Journal of Cleaner Production*, 55, 45–55.

Kaushal, R., Varghese, G., & Chabukdhara, M. (2012). Municipal solid waste management in India—Current state and future challenges: A review. *International Journal of Engineering Science and Technology*, 4(4), 1473–1489.

Kazançoglu, Y., Ada, E., Ozturkoglu, Y., & Ozbiltekin, M. (2020). Analysis of the barriers to urban mining for resource melioration in emerging economies. *Resources Policy*, 68, 101768.

Krook, J., & Baas, L. (2013). Getting serious about mining the technosphere: A review of recent landfill mining and urban mining research. *Journal of Cleaner Production*, 55, 1–9.

Kuhn, A., & Heckelei, T. (2010). Anthroposphere. In: P. Speth, M. Christoph, & B. Diekkrüger (Eds.), *Impacts of Global Change on the Hydrological Cycle in West and Northwest Africa* (pp. 282–341). Berlin, Heidelberg: Springer.

Kulkarni, S. J. (2020). Biomedical waste scenario in India – Regulations, initiatives and awareness. *Biomedical Engineering International*, 2(2), 0086–0092.

Lee, E., Rout, P. R., & Bae, J. (2021). The applicability of anaerobically treated domestic wastewater as a nutrient medium in hydroponic lettuce cultivation: Nitrogen toxicity and health risk assessment. *Science of the Total Environment*, 780, 146482.

Livemint. (2021). India among top 5 nations in e-waste generation: Report. Online available: https://www.livemint.com/Politics/DKDWemxZwRmWddDpfcfo1H/India-among-top-5-nations-in-ewaste-generation-Report.html.

Ministry of New and Renewable Energy (MNRE). (2021). Waste to energy. Online available: https://mnre.gov.in/waste-to-energy/current-status.

Mohanty, A., Rout, P. R., Dubey, B., Meena, S. S., Pal, P., & Goel, M. (2021). A critical review on biogas production from edible and non-edible oil cakes. *Biomass Conversion and Biorefinery*, 1–18.

Mongabay-India. (2021). The why and how of disposing electronic waste. Online available: https://india.mongabay.com/2020/08/explainer-the-why-and-how-of-disposing-electronic-waste/

Muradov, N., & Smith, F. (2008). Thermocatalytic conversion of landfill gas and biogas to alternative transportation fuels. *Energy Fuels*, 22, 2053–2060.

Nandy, B., Sharma, G., Garg, S., Kumari, S., George, T., Sunanda, Y., et al. (2015). Recovery of consumer waste in India: A mass flow analysis for paper, plastic and glass and the contribution of households and the informal sector. *Resources, Conservation and Recycling*, 101, 167–181.

Nanjo, M. (1988). Urban mine, new resources for the year 2,000 and beyond. *Tohoku Daigaku Senko Seiren Kenkyusho Iho*, 43(2), 239–251.

NITI. (2018). Strategy for promoting processing of construction and demolition (C&D) waste and utilization of recycled products. Ministry of Housing and Urban Affairs, Government of India. Online available: https://niti.gov.in/writereaddata/files/CDW_Strategy_Draft%20Final_011118.pdf.

Park, J. K., Clark, T., Krueger, N., & Mahoney, J. (2017). A review of urban mining in the past, present and future. *Advances in Recycling and Waste Management*, 2(2), 127.

Pathak, P., & Chabhadiya, K. (2021) Recycling of rechargeable batteries: A sustainable tool for urban mining. In: Baskar C., Ramakrishna S., Baskar S., Sharma R., Chinnappan A., Sehrawat R. (Eds.) *Handbook of Solid Waste Management*. Singapore: Springer. https://doi.org/10.1007/978-981-15-7525-9_74-1.

Pathak, P., Srivastava, R. R., & Ojasvi (2017). Assessment of legislation and practices for the sustainable management of WEEE in India. *Renewable & Sustainable Energy Reviews*, 78, 220–232.

Pujara, Y., Govani, J., Patel, H. T., Chabhadiya, K., Vaishnav, K., & Pathak, P. (2020). Waste-to-energy: Suitable approaches for developing countries. In: *Alternative Energy Resources – The Way to a Sustainable Modern Society, The Handbook of Environmental Chemistry*. Berlin, Heidelberg: Springer.

Pujara, Y., Pathak, P., Sharma, A., & Govani, J. (2019). Review on Indian municipal solid waste management practices for reduction of environmental impacts to achieve sustainable development goals. *Journal of Environmental Management*, 248, 109238.

Rai, A., Kothari, R., & Singh, D. P. (2020). Assessment of available technologies for hospital waste management: A need for society. In: *Waste Management: Concepts, Methodologies, Tools, and Applications* (pp. 860–876). Hershey, PA: IGI Global.

Rau, S. (2019). Options for Urban Mining and Integration with a Potential Green Circular Economy in the People's Republic of China.

Rout, P. R., Bhunia, P., Ramakrishnan, A., Surampalli, R. Y., Zhang, T. C., & Tyagi, R. D. (2016). Sustainable hazardous waste management/treatment: Framework and adjustments to meet grand challenges. In: *Sustainable Solid Waste Management* (pp. 319–364). Reston, VA: American Society of Civil Engineers.

Rout, P. R., Verma, A. K., Bhunia, P., Surampalli, R. Y., Zhang, T. C., Tyagi, R. D., Brar, S. K., & Goyal, M. K. (2020). Introduction to sustainability and sustainable development. In: *Sustainability: Fundamentals and Applications*. New York, NY: John Wiley & Sons, Ltd.

Rout, P. R., Bhunia, P., & Rao, Y. S. (2018). Landfill technologies and recent innovations. In: *Hand Book of Environmental Engineering* (pp. 293–302). New York, NY: McGraw-Hill International Publishing.

Shan, G. B., Zhang, T. C., & Surampalli, R. Y. (2010). Bioenergy from landfills. In: S. K. Khanal, R. Y. Surampalli, T. C. Zhang, B. P. Lamsal., R. D. Tyagi, & C. M. Kao (Eds.), *Bioenergy and Biofuel from Biowastes and Biomass*, Reston, VA: ASCE.

Simoni, M., Kuhn, E. P., Morf, L. S., Kuendig, R., & Adam, F. (2015). Urban mining as a contribution to the resource strategy of the Canton of Zurich. *Waste Management*, 45, 10–21.

Statista (2021). Volume of the coronavirus (COVID-19) biomedical waste generated across India as of July 2020, by state (in metric tons per day). Online available: https://www.statista.com/statistics/1140324/india-covid19-biomedical-waste-generated-by-state/

Society of Indian Automobile Manufacturers (SIAM). (2015). 55th Annual Convention. Online available: https://www.siam.in/event-overview.aspx?mpgid=30&pgidtrail=30&eid=165

Tam, V. W. Y. (2008). Economic comparison of concrete recycling: A case study approach. *Resources, Conservation, and Recycling*, 52, 821–828.

Tercero, L., Rostek, L., Loibl, A., & Stijepic, D. (2020). The promise and limits of urban mining.

The Global e-Waste Monitor. (2020). Online available: http://ewastemonitor.info/

The Diplomat. (2021). The Potential of Urban Mining. Online available: https://thediplomat.com/2013/11/the-potential-of-urban-mining/

TIFAC. (2001). Utilization of waste from construction industry. New Delhi: Technology Information, Forecasting and Assessment Council.

Tiwari, A. V., & Kadu, P. A. (2013). Biomedical waste management practices in India-a review. *International Journal of Current Engineering and Technology*, 3(5), 2030–2034.

U.S. Environmental Protection Agency (USEPA). (2004). *Inventory of U.S. Greenhouse Gas Emissions and Sinks: 1990–2002.* Washington, DC: Office of Atmospheric Programs, U.S. Environmental Protection Agency, Government Printing Office.

Van Passel, S., Dubois, M., Eyckmans, J., De Gheldere, S., Ang, F., Jones, P. T., Van Acker, K. (2013). The economics of enhanced landfill mining: Private and societal performance drivers. *Journal of Cleaner Production*, 55, 92–102.

Zalasiewicz, J. (2018). The unbearable burden of the technosphere. *The UNESCO Courier*, 71, 15–17.

Zeng, X., Mathews, J. A., & Li, J. (2018). Urban mining of e-waste is becoming more cost-effective than virgin mining. *Environmental Science & Technology*, 52(8), 4835–4841.

2 Current Trends and Future Challenges for Solid Waste Management

Generation, Characteristics, and Application of GIS in Mapping and Optimizing Transportation Routes

Danush Jaisankar,[1] Sailesh N. Behera,[1,2] Mudit Yadav,[1] and Hitesh Upreti[3]
[1]Environmental Engineering Laboratory, Department of Civil Engineering, Shiv Nadar University, Greater Noida, Gautam Buddha Nagar, Uttar Pradesh, India
[2]Centre for Environmental Sciences and Engineering, Shiv Nadar University, Greater Noida, Gautam Buddha Nagar, Uttar Pradesh, India
[3]Fluid Mechanics Laboratory, Department of Civil Engineering, Shiv Nadar University, Greater Noida, Gautam Buddha Nagar, Uttar Pradesh, India

CONTENTS

DOI: 10.1201/9781003201076-2

2.1   INTRODUCTION

With rapid increase in urbanization and sound economic expansion with improved standard of living of humans, the quantity and complexity in properties of municipal solid waste (MSW) in urban areas of India have changed with substantial increase in per capita per day generation of waste (Sharholy et al., 2008; Srivastava et al., 2015; Mandal, 2019). The population of India is increasing fast with 1.3 billion in base year of 2015, and it is predicted that the population will reach 1.5 billion by the year 2030 with a growth rate in the range from 1.2% to 1.6%. Based on the trends of past studies and reports, it is quite clear that there is substantial increase in solid waste generation quantity or per capita generation in India due to rise in economic status of people (Singh et al., 2011; Suthar and Singh, 2015). Looking at the waste management scenarios in Indian cities, it has been observed that these aspects are not completely handled smoothly, the way in which it is being managed by any developed countries across the globe. This is because of non-availability of sufficient resources, infrastructure, and suitable services to dispose solid wastes from most of the households for MSW management in Indian cities. The statistics reveals that MSW generated in India is about 188,500 ton/day and per capita waste generation in cities varies from 0.2 kg/person/day to 0.8 kg/person/day (Gupta et al., 2015; Khan et al., 2016; Dutta and Jinsart, 2020).

In the perspective of the cost of management, it has been seen that collection and transport of MSW share a large portion of total waste management cost in the range from 70% to 100% in Indian cities. This cost share for these components of waste management system in Indian cities are higher than typical values for modern waste management system practiced in cities of developed countries. The reason for such trends may be due to less allocation of funds for proper disposal of solid waste with unawareness of pretreatment for materials and practicing illegal dumping of waste on ground (Chattopadhyay et al., 2009; Singh et al., 2011). Among cities of India, Delhi (the national capital territory) is a special case and is the most densely populated with urbanized patterns throughout. The annual growth rate in population of Delhi during the last decade (2001–2011) was 2.2% almost 1.6 times the national average (Census of India, 2011; http://census2011. co.in). Delhi is also a commercial hub, providing employment opportunities and accelerating the pace of urbanization, resulting in increase in MSW generation. Improper and non-organized management of solid waste creates undesirable consequences such as disease transmissions, odor spreading, nuisance creation, atmospheric pollution, land contamination, water pollution, fire hazards, esthetical nuisance, and economic losses. The main challenge is managing the waste generated through rapid growth, more specifically in developing countries that lack the public service infrastructure to manage municipal waste. In recent years, we witnessed the trend of use and throw (disposable) consumables and this matter has contributed significantly to solid waste generation, and this unwanted matter has heavily impacted the environment, public health, and produce socioeconomic problems. This precarious behavior needs to be thoroughly examined, and changed, so that a sustainable way of living can happen.

In cities of developed countries, the cost of transportation of solid waste constitutes 50% of the total expenditure, whereas, in cities of developing countries, the same is 85–90% (Ghose et al., 2006; Lu et al., 2006; Karak et al., 2012). During this decade, it has been a challenge for the administrative authorities to optimize the resources meant for management of wastes in a city. Therefore, necessary improvements in the existing practice with better tracking system are required to provide an efficient waste management system that can optimize the routine work of MSW management to save the cost of expenditure and time of operation. Several studies in the past modeled the transportation paths of solid waste management to propose the routes of transfer and transport of waste from collection points from user ends to processing/recovery or disposal sites that can reduce the cost of hauling distance of transport vehicles (Apaydin and Gonullu, 2007; Karadimas et al., 2007; Liu, 2009; Aremu, 2013; Singh and Behera, 2019).

The geographic information system (GIS) is a software-based platform, which allows its technical users to capture, store, manage, and analyze a large volume of spatially or geographical reference data collected from various sources with systematic and secured digitized archives (Ghose et al. 2006; Behera et al., 2011a,b). The GIS enables the reader and management personnel to visualize and interpret the data for better understanding of relationships and patterns with graphical representations of results in a user-friendly manner (Liu, 2009; Sanjeevi and Shahabudeen, 2016). ArcGIS is a powerful tool that allows user to create spatial maps and interpret results based on activity level data or different thematic layer information. To manage such huge data and information with details of various locations meant for better solid waste management, ArcGIS serves the purpose to develop, store and manage bigger database, and produce various thematic layers of digitized maps. Therefore, this software is becoming popular among researchers to present complicated and large database in many ways.

This present work is a first study of its kind that provides various scenarios of solid waste management in Delhi, a megacity and national capital of India. The specific objectives of this book chapter are provided as follows: (1) assessment on various scenarios in existing trends in solid waste management including generation and composition, storage and collection, transportation between collection points to disposal sites, processing and recovery, and landfill sites for disposal solid waste; (2) presentation of methods for development of spatially resolved waste generation maps using GIS, waste generation maps with respect to land-use patterns, and waste generation maps for projected years; and (3) presentation of methods on finding proposed transportation route for reduction of hauling distance of transport vehicles using Network Analyst tool in GIS.

2.2 METHODOLOGY WITH DISTINCT APPROACHES

2.2.1 CHARACTERISTICS OF THE STUDY DOMAIN

We have selected the study domain as the National Capital Territory of Delhi, India, to conduct our research, present MSW status and issues, apply our methodology for generation of GIS-based maps, find out the shortest transportation routes, and suggest future way forward for better practice of MSW. This northern Indian city lies between the latitude of 28°24'17" to 28°53'00" N and longitude of 76°50'24" to 77°20'37" E. Figure 2.1 shows the location of Delhi in the map of India and presents the boundary of the study domain, Delhi. The population of Delhi in 1901 was 17.4 lakh and increased to 40.0 lakh in 1951 and 94.2 lakh in 1991. A sharp increase of population was recorded in the last two decades with the population being 138 and 168 lakh in 2001 and 2011, respectively, having an annual growth rate of 2.2% during this time span, and it was predicted that the growth rate would reach 3.3% during 2019–2020 (Census of India, 2011; http://census2011.co.in; Chandramouli and General, 2011). Areas of the states of Haryana and Uttar Pradesh surround the border of Delhi. Emerging and developed cities of Gurugram, Noida, Greater Noida, Faridabad, and Ghaziabad (come under National Capital Region – NCR) are situated close to Delhi. The trends and patterns of population growth, urbanization, transportation growth, and industrial clusters in NCR are similar to Delhi. The river Yamuna is the one of the largest rivers in northern India, and it flows through the region of Delhi. The political area of Delhi is around 1,484 km² (1,115 km² as urban and 369 km² designated as rural), and the city has more plantation areas compared to other cities in the country (Mohan et al., 2011; Pradhan and Kumar, 2014). From an environmental pollution point of view, there are severe problems of air pollution, pollution of water bodies in the region, and solid waste handling and management issues prevailing in the city. Delhi has been known as one of the most polluted cities in the world with annual average concentration of PM_{10} (particulate matter with aerodynamic diameter ≤ 10 μm) around 190 μg/m³ (Gupta et al., 2017; Yadav et al., 2020).

Three municipal entities are contained within the study domain and these three entities are Municipal Corporation of Delhi (MCD), New Delhi Municipal Council (NDMC), and Delhi Cantonment Board (DCB) (Kennedy et al., 2009; Lata et al., 2020). The area covered by MCD is 1,399 km² (94% of total area) and

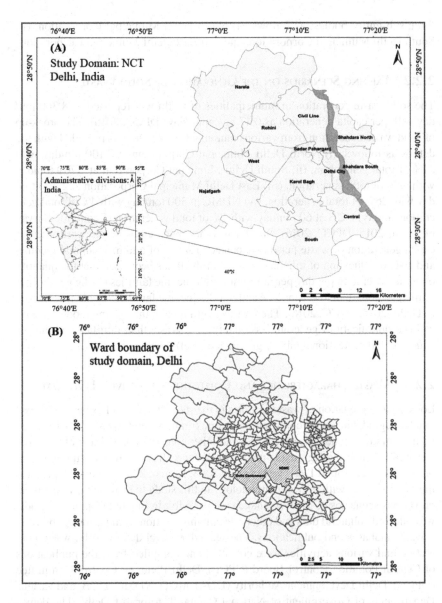

FIGURE 2.1 Study domain: (A) Location of Delhi on map of India and boundary of study domain, Delhi. (B) Map showing boundary of wards of Delhi used for further processes in GIS.

population of about 97% of total population. The remaining 6% area and 3% population are covered by both NDMC and DCB. There are 12 districts in Delhi, and a total of around 278 wards in all municipalities of Delhi (Figure 2.2). Figure 2.1 with Delhi map having digitized ward boundaries was used for generation of further maps in GIS. It should be noted that the MCD entity is again divided into three sub-entity bodies, as follows: (i) North Delhi Municipal Corporation, (ii) South Delhi Municipal Corporation, and (iii) East Delhi Municipal Corporation.

2.2.2 EXISTING SCENARIOS ON THE GENERATION OF SOLID WASTE

The solid waste generation in municipalities of Delhi was reported as 8,360 ton/ day with per capita generation as 0.47 kg/capita/day (DPCC, 2016). The break-up of solid waste generation from various municipalities of Delhi is provided with the details as follows: (i) North Delhi Municipal Corporation at 3,100 ton/day with 37% of total generation, (ii) South Delhi Municipal Corporation at 2,700 ton/day with 32% of total generation, (iii) East Delhi Municipal Corporation at 2,200 ton/ day with 26% of total generation, (iv) NDMC at 300 ton/day with 4% of total generation, and (v) DCB at 60 ton/day with 1% of total generation (Economic Survey of Delhi, 2015; DPCC, 2016). Similar to other mega cities of the word, the per capita generation of waste per day depends on status of a family (annual income) and level of education of inhabitants in a family. It has been seen that the quantity of waste generation per day per family for high-income families (>2 kg per day per family) is more than lower-income and middle-income families (Economic Survey of Delhi, 2015; DPCC, 2016). The families with inhabitants having moderate educational qualifications generate less quantities of waste than families with higher educated and professional (above graduation level) inhabitants.

2.2.3 WASTE CHARACTERISTICS AND COLLECTION OF ACTIVITY LEVEL DATA

Looking at segregation of waste samples in the study domain, it has been observed that biodegradable matter contributing the maximum weight (%) at >50% followed by inert matter (stones, brocks, ashes, and other construction and demolition matter) at >30%, non-biodegradable matter >13%, and all remaining including paper, plastic, metal at around 12% (Talyan et al., 2008; Kumar, 2013). Figure 2.2 presents the steps adopted with overall methodology in this study to provide the assessment on various scenarios of MSW management in Delhi. In the first step of this study, we collected collateral data and other relevant information from various public and private databases and publications. The secondary level data on solid waste generation and various activities were collected and compiled from the publications of Central Pollution Control Board (CPCB), Delhi Pollution Control Committee (DPCC), Delhi Development Authority (DDA), MCD, NDMC, DCB, and various Departments of Government of National Capital Territory of Delhi. The demographic and ward-wise population was collected from MCD, NMDC, DCB, and website of Indiastat (Indiastat, 2020; https://www.indiastat.com/). The geo-information data were obtained from Google Earth maps, DDA maps, and the Bhuvan website, which is the Indian Geo-Platform of Indian Space Research Organization (ISRO) (Bhuvan, 2020; https://bhuvan.nrsc.gov.in/bhuvan_links.php).

The maps from OpenStreetMap and Maps of India were used for further geo-referencing and digitization for our relevant purposes and uses on the Platform of ArcGIS. The waste segregation and properties data were collected from private

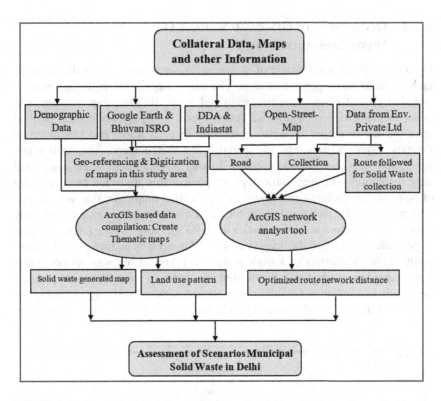

FIGURE 2.2 Overall methodology with steps for data and information collection on assessment of various scenarios for MSW in Delhi.

organization like AG Enviro Infra Projects Pvt. Ltd. The data for waste collection points on dhalaos, dustbins, and open sites were collected from the MCD. In addition to collection and compilation of the aforementioned secondary data those were used in inventory process of solid waste generation, we conducted a limited number of *in situ* field surveys in a locality under MCD. Under this scope of work, we conducted *in situ* field surveys in two wards (around 50 households in each of the wards) of east Delhi to conduct questionnaire and compile data with different income group families to decide on waste generation rate in kg/person/day. Three categories of households were considered for primary surveys with income group localities criteria of DDA having high-income group and middle-income group flats. The resettlement clusters were surveyed for obtaining relevant information on low-income group families. The methodology of this study was segregated with two major methods, as follows: (1) development of spatially resolved waste generation maps using the ArcMap tool of ArcGIS software, and (2) finding the optimized route network distances for transportation of waste from collection points to disposal sites using Network Analyst tool of ArcGIS software. The waste generation map with respect to land-use pattern was also generated using ArcGIS software.

2.2.4 Development of GIS-Based Spatially Distributed Waste Generation Maps

Figure 2.3 shows a schematic of sequences of steps followed in the development of solid waste generation map. The major steps involved in this process included geo-referencing the base map and connecting the map with attribute table, incorporation of ward-wise activity level and population data, method selection for inventory of waste generation in each ward, and generation of thematic layers with ward-wise spatially distributed maps. The solid waste generation map exercise was conducted using ArcMap with small-scale ward wise resolution. The important steps adopted in this study for the generation of maps are enumerated as follows: (i) for creating maps, ArcMap was opened with a specific file name and saved for further use; (ii) a base map of the study domain with distinct ward boundaries of Delhi was obtained for further process of geo-referencing; (iii) the image was geo-referenced by adding ground control points with coordinates of known points; (iv) ArcCatalog was opened and a new shapefile was created in the desired location by a right click with a feature type such as a polygon; (v) Editor was opened with start editing and using the create feature tab, the shapefile was selected followed by clicking on polygon,

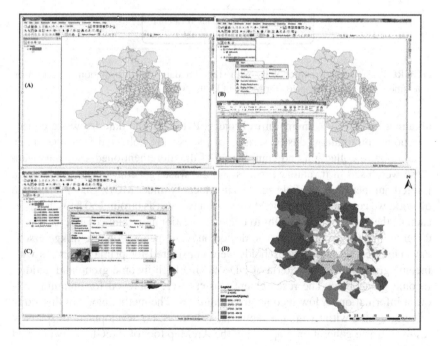

FIGURE 2.3 Schematic of sequences in generation of solid waste map: (A) geo-referencing and digitization of a base map, (B) incorporation of ward-wise activity data as an attribute, (C) method for map development, and (D) generation thematic layer with spatial-distributed solid waste.

the image file was digitized; (vi) required population data were entered in an Excel file and a multiplier of per capita waste generated per day was used to find the quantity of solid waste generation; (vii) the Excel sheet was imported as a layer using conversion tools in the toolbox, and with a right click on the existing layer, the join option was selected and related with respect to FID joining the layer; (viii) after joining the layer, and with a right click on the layer, properties were selected, and using the symbol tab, quantities and graduated colors were selected, and finally the desired value in the field tab was chosen; and (ix) after selecting the required color, title, legend, north arrow and scale, the solid waste generated map was created.

Similar to the above process in development of ward-wise solid waste generated map, land-use pattern wise map was generated to assess the quantity of waste generated in various land-use patterns of the study area. The important steps for map generation are summarized as follows: (i) ArcMap was opened for this project and named with specific file for further use; (ii) relevant data and information for the study area were taken from Google Earth (GE), DDA maps and Bhuvan (Indian Geo-Platform of ISRO); (iii) the images were geo-referenced with the similar ways as explained in the last paragraph; (iv) based on the classification of land-use area types requirement, respective areas were marked by drawing polygons in GE and then the .kml files were imported into ArcMap, which were converted to layer (ArcGIS-supported layer file) by using the conversion tools in the toolbox; and (v) by comparing the maps from various data sources, land-use and land-cover (LU-LC) maps were created. The base year for this study was 2015 and the activity-level information were projected for the past and future years of 2011 and 2021, respectively, for generating GIS maps of waste generation with spatial resolution ward wise.

2.2.5 TRANSPORTATION ROUTE OPTIMIZATION TECHNIQUES

The objective of finding the shortest route between collection points and transfer/ processing unit or disposal site was conducted in the ArcGIS inbuilt Network Analyst Tool, which was based on Dijkstra's algorithm (Dijkstra's, 1959; Chang et al., 1997; Sanjeevi and Shahabudeen, 2016). With this concept, the shortest distance between a single source/point to rest of sources/points in a well-defined network was estimated. This algorithm has been practiced for identification of an optimized route from a specific source vertex to every vertex in a complex routing network to decide on routing and sub-routing paths. The details of the basic concepts and methods for iteration processes to derive such network analysis are described somewhere else (Singh and Behera, 2019). In brief, the optimized distance in a network was found from the inbuilt algorithm in ArcGIS, with the following steps, described as: (i) the starting node, "s" was assigned with zero distance value; (ii) the starting node was labeled as a permanent node, and the distance value of infinite was assigned with other temporary node, marked as "d"; (iii) the remaining distance values were updated in a set of temporary nodes, labeled as "j," which could be reached from the existing node; and (iv) the new distance value of nodes was updated for every link in a network, expressed as Equation (2.1).

$$N_d = Min\left\{D_j,(D_i + C_{ij}\right\}$$ (2.1)

where N_d represents new updated distance, D_j represents the distance of nearest node before updating, D_i represents current distance of node, and C_{ij} represents distance of existing link in which "i" represents the index current node and "j" represents the index of temporary node.

The new distances were updated in a set of temporary nodes using expression of Equation (2.1). Such process of iteration was continued till all nodes reached from starting node to temporary node, and the updated process was then completed. The required primary data for this network analysis were thematic map layers of different utilities of the study area including road network, boundary map, location information of collection points, and transfer/processing station or disposal site. In this study, optimal route analysis was conducted between locations of 20 collection points or Dhalaos to the transfer station located in Vijay Ghat to find the shortest hauling distance of waste transport vehicles. Figure 2.4 shows the schematic of steps in Network Analysis using ArcMap tool in ArcGIS that included geo-referencing and digitization of road network in the study area, identification of

FIGURE 2.4 Schematic of steps in Network Analysis in ArcMap: (A) Geo-referencing and digitization of road network, (B) identification of collection points, (C) tracing of paths from collection points to disposal locations, and (D) determination of shortest hauling distance of vehicles.

waste collection points in specific area, tracing the paths of transport by transfer vehicles from collection points to disposal locations, and finally determination of the shortest hauling distance of vehicles from collection points to disposal site.

2.3 RESULTS AND DISCUSSION

2.3.1 Scenarios in Current Trends in Solid Waste Management

2.3.1.1 Generation and Composition of Solid Waste

Although the data of population growth trends were shown till 2011 in methodology section with growth rate of 2.2% per year, the MCD data show a 3.3% growth rate annually in 2017 with respect to population of 2016. From that perspective, it is quite likely that the rate of increase in solid waste generation per capita per day would be higher. Specifically, it was seen that total generation rate per day for the entire study area 8,360 ton/day with a population of 168 lakh living in around 36 lakh household during 2011. This statistical data MSW generated showed a rate of 0.49 kg/capita/day. In the same way, it was predicted by the Planning Department of Delhi that population would grow at a rate of 3.1% per year by 2021, whereas the waste generation per capita per day would grow by around 3.8% annually (total waste generation ~15,000 ton/day). In this regard, the per capita generation increase was more compared to population growth rate showing that the change in lifestyle of people and dependency on packed food and more disposable materials could be the reason for such spike in generation of solid waste on unit basis.

Looking at the source of generation of waste in Delhi for our study period (base year 2015) at city level, it was observed that residential waste contributed the highest at 54% to total generation of waste followed by commercial waste (generated from shopping complexes) at 18%, vegetables and fruit markets at 10%, industries at 9%, 7% from construction and demolition sites, and remaining 2% from hospitals and medical facilities (DPCC, 2016). Looking at the trends in composition of solid waste from 1982 to 2015, it was observed as follows: (i) maximum contribution was from biodegradable matter at 58% followed by inert materials (stones, bricks, ashes) at 29%, paper at 6%, and non-biodegradable at 5% during 1982, (ii) maximum contribution was from biodegradable matter at 38% followed by inert materials (stones, bricks, ashes) at 35%, non-biodegradable at 14%, paper at 6%, and plastics at 6% during 1995, and similar trend was observed during 2002, and (iii) maximum contribution was from biodegradable matter at 40% followed by inert materials (stones, bricks, ashes) at 35%, non-biodegradable at 11%, plastics at 7%, and paper at 6% during 2015.

The trends showed that the plastics and paper are becoming one of the significant sources in due course of time, specifically during the last decade in the study area. With increase in economic status of people in the urban region, a substantial rise in use of plastics, paper, and their packing materials might be the cause for such trend. One more trend was observed that there was no change in contributions from paper showed in the last two decades showed that recycling of paper has been in its maximum capacity in Delhi. The composition proportions of plastics and non-biodegradables have been increasing substantially, which showed a reflection on improvement of living standards and consumerist attitude of residents. In addition, the contribution of inert materials (primarily composed of construction and demolition waste) in the composition of total waste increased substantially in the last two decades, indicating that substantial increase in construction and demotion sites in Delhi in the last decade.

2.3.1.2 Storage and Collection Solid Waste

For executing the tasks of primary collection of solid waste to disposal process in the municipality area of Delhi, the third-party private contractors are being employed in some zones. These third-party contractors have their own manpower force under various categories including sweeping staff, waste collectors, and drivers to operate vehicles from the collection points to respective disposal sites. It was observed that the total employed sweeping staffs for waste management work was in ratio of 1:216 persons and 1:326 persons for areas under entities of MCD and NDMC, respectively (DPCC, 2016). These values were better than recommended criteria of 1:500 set by Central Public Health and Environmental Engineer Organization (CPHEEO). The entire area of Delhi is divided into 14 zones by the regulatory and executive bodies, and every zone is comprised of different receptacles, where the wastes from various corners are collected. The private contract operators normally keep two set of waste bins with distinct colors in blue and green for collection of non-biodegrade/recyclable waste, and bio-degradable waste, respectively.

Based on the information from MCD and DPCC reports, these designated waste bins are emptied on daily basis into vehicles separately with same color of blue and green (DPCC, 2016). The operators tried to do segregation of these two major categories of solid waste before collected into different vehicles. In addition, the operators at NDMC use open containers with 900 community bins made-up of masonry structure and 1,000 metallic skips with capacity of 1m³. Under ideal scenario of MSW management, the waste is normally stored at the source of generating locations until further collection to disposal sites. However, such ideal scenario does not happen in practical sense in the streets with higher population density of in Delhi. With unavailability of system of storage of waste at sources, the wastes are thrown on road sides of the street. To avoid such mismanagement, the operators installed dhalaos, which are made up of concrete structures with dimensions in the range from 4.5 m × 3 m to 13.5 m × 9 m with the capacity to hold of 4–16 tons of solid waste. The requirements for such storage space depend on solid waste generation per capita per day, population counts, and frequency of collection. Overall, installed dhalaos and dustbins act as transfer stations from which collection vehicles collect and unload at disposal sites.

Out of 2,403 dhalaos installed at various location in Delhi, 2,021 located in MCD zone, 221 in NDMC zone, and 161 in DCB zone. As there were no standard norms on maintaining the distance between receptacles (dhalaos) or no of such facilities per unit area, people face difficulty to deposit wastes directly into dhalaos. A number of complains have been received from the residents in irregularity in collection of waste, roaming of animals around these collection points, and bad maintenance of these dhalaos. All these factors helped in development of odor and nuisance causing unhygienic conditions with stored wastes. Because of all such disadvantages, MCD has been replacing these older practices by putting trolleys and smaller bins with higher frequency of collection. Although, improvement is taking place in the collection system of MSW management, the problems are still there in high-density population streets in Delhi. There is need of more robust methods and infrastructure with better tracking systems to improve the collection system in Delhi.

2.3.1.3   Transportation of Solid Waste

The transportation system has the pivotal role in solid waste management in any city, and hence it requires more organized structure to provide well-defined, cost-effective paths for transportation routes of collection vehicles. This component of management links between collection of wastes from storage locations and transfer to processing unit or disposal/landfill site. A large fleet of vehicles are being operated and maintained by MCD, NDMC, and DCB for transportation and secondary collection of solid wastes from various waste receptacles to the processing or disposal/landfill sites.

The trend showed that around 75% of waste has been collected for transfer through transportation to processing or disposal/landfill sites, and the problems of mismanagement happen because of non-availability of adequate vehicles and constant breakdown and maintenance issues of existing operating vehicles. Looking at the categories of vehicles for the purpose of transportation of wastes in Delhi, it was observed that 722 vehicles were being employed by MCD with a break-up of 585 refuse removal trucks and 137 loaders. Around 722 vehicles with category of auto tippers were being used for primary collection of wastes from houses in MCD. Eight zones under MCD were privatized under the Public Private Partnership (PPP) scheme with some better equipped vehicles with 305 larger hydraulic vehicles and short range tippers or dumpers for transportation of wastes to disposal sites. There were around 85 and 15 trucks operated and maintained under NDMC and DCB, respectively, for the purpose of transportation of waste. Overall, it has been observed that unequal distribution of transportation vehicles among zones in Delhi was noticed.

To manage with insufficient transportation vehicles in areas under MCD, the transportation vehicles were operated in two shifts at a rate of one trip per shift based on the distance between the collection points and disposal site. There have been cases of under-utilization of transportation vehicles due to several factors including irregular, poor and improper maintenance, unsystematic parking of vehicles, loss of time while waiting in the queue for fueling or moving from one station to another for fueling, and many others. Compared to MCD, operational efficiency in NDMC and DCB was better.

2.3.1.4   Processing and Recovery of Solid Waste

The scenario of processing and recovery of solid waste at city level are summarized as follows: (1) with North Delhi Municipal Corporation (under MCD), out of

3,100 ton/day waste generated, 1,500 ton/day sent to Bhawana waste processing site, 400 ton/day sent to waste to energy plant located at Okhla, and 1,200 ton/day sent to Bhalswa landfill site; (2) with South Delhi Municipal Corporation (under MCD), out of 2,700 ton/day waste generated, 1,250 ton/day sent to waste to energy plant located at Okhla, 800 ton/day sent to Bhalswa landfill site, 500 ton/day sent to Okhla landfill site, and 150 ton/day sent to compost plant located at Okhla; (3) with East Delhi Municipal Corporation (under MCD), all 2,200 ton/day waste generated were sent to Ghazipur landfill site; (4) under NDMC, all 300 ton/day waste generated were sent to waste to energy plant located at Okhla; and (5) under DCB, all 60 ton/day waste generated were sent to Okhla landfill site. Overall, the trends in processing of waste showed that more than 50% waste generated were not processed. Specifically, 3,700 ton/day out of 8,360 ton/day of waste generated was processed. Therefore, the number of processing units should be increased in terms of waste to energy, getting recyclable recovered matter, and compost processing.

2.3.1.5   Landfill Sites for Disposal of Solid Waste

The landfill is an engineered designed deposit location with pits or trenches, where solid wastes are finally disposed of. Compared to ordinary landfill, sanitary landfill has better advantages in terms of controlling the environmental risks associated with the disposal of wastes, tracking the process of disposal and emissions, making the land available to subsequent to disposal and other purposes. In case of Indian cities, proper engineered sanitary landfills are hardly found, as the waste generation load is higher compared to the allocated land and financial funds available to develop, operate, and maintain such facilities. Figure 2.5 shows

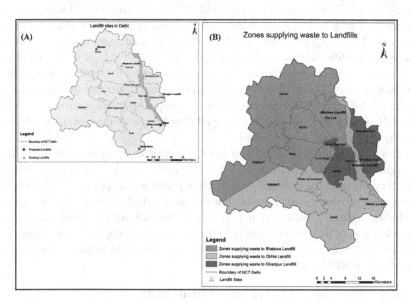

FIGURE 2.5   Landfill sites in Delhi: (A) Locations of existing and proposed landfill sites, and (B) zones supplying waste to specific landfill sites.

location of existing and proposed landfill sites in Delhi, and zones or areas that supply waste to the specific landfills. During the study period, it was seen that Delhi has three major land sites such as Bhalswa (in north Delhi with area of 21 ha), Ghazipur (in east Delhi with area of 29 ha), and Okhla (in south Delhi with area of 16 ha). Ideally, there was less space available for further dumping of waste, as the capacities of these landfills were exceeded due to large quantity of waste generated and disposed. However, due to demand to resolve such issues of handling of generated waste, waste dumping was allowed at these sites. As a result, the shape of these sites looked like artificial mountains. MCD is in the planning process to develop more landfill sites to meet the demand of disposal of wastes, and several specific areas have been identified by MCD for selection of sites fit for landfill. Under such initiatives, MCD has proposed three more landfill sites to be commissioned in the localities of Jaitpur, Bawana, and Bhatti mines.

2.3.2 Waste Generation Maps in the Base Study Year

Figure 2.6 shows the map generated from GIS with ward-wise spatial distribution of solid waste generated during the base year of this study (i.e., 2015). The legend displays different colors with five ranges of waste generation in the unit of kg/day varied from 7,794 to 74,295 kg/day in the wards of MCD. From Figure 2.6, it was observed that the index with "black" color are the regions where the waste generated was maximum relative to the other wards, and the trend kept on decreasing with corresponding display colors varying from "black" to "dark gray" traversing the descending order through "darker gray," "white," and "light gray." Around 272 wards in the city were considered for mapping work in this study, and the thematic map was created based on the population in these wards individually.

The following results were obtained in terms of percentage of distributions of solid waste generation among wards based on decreasing order of magnitude with display of specific color ranges in the legend of the map, as: (1) black [46,418–74,295 kg/day] constituting 12.2% areas having important locations of Hastal, Matiala, and Jaidpur; (2) darker gray [34,728–46,418 kg/day] constituting 18.4% areas; (3) white [27,332–34,728 kg/day] constituting 31.1% areas having important locations of Khayala, Ramnagar, and Dabri; (4) light gray [21,972–27,332 kg/day] constituting 29.8% areas; and (5) dark gray [7,794–21,972 kg/day] constituting 8.5% areas having important locations of Kashmiri gate and Daryaganj.

From the trends of above results, it was clearly assessed that majority of the wards were producing less quantity of waste [light gray, and dark gray] at 40.8% wards, and next segment [white] with 31% of wards were producing medium level of solid waste generation. On overall assessment, it was found that the maximum contribution to total generation at city level were made from medium and low level regions [white, light gray, and dark gray] constituting 69.4% of total waste generated in the city. The total generation of solid waste in the city was estimated at 11,300 ton/day with an average of per capita generation of waste at 0.52 kg/person/day.

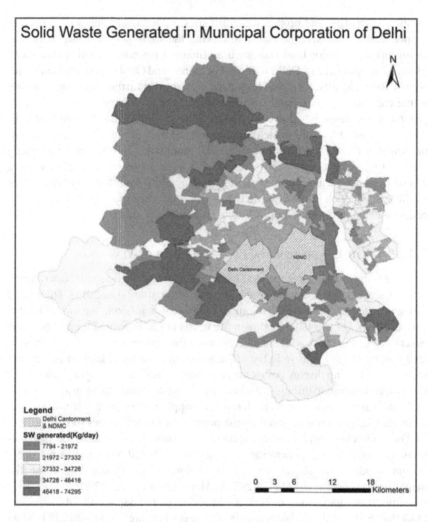

FIGURE 2.6 Spatially distributed solid waste generation map during the base study year (2015).

Figure 2.7 shows the map generated from GIS on spatial distribution of per capita per day waste generation in various land-use pattern areas of the city. This map was created for the base year of this study (2015) to classify waste generation based type of land-use areas such as agricultural, forest, restricted, residential, institutional, industrial, commercial, and others. The areas covered by individual land-use pattern as a percentage of total area of Delhi are provided as follows: agricultural with 28.2%, forest with 3.1%, restricted with 2.1%, residential with 24.6%, institutional with 3.9%, industrial with 2.8%, commercial with 3.5%, others with 31.1%, and river with 0.68%. Hence, a maximum percentage (31.1% for others) of the land area is being used up by barren lands, greeneries and to

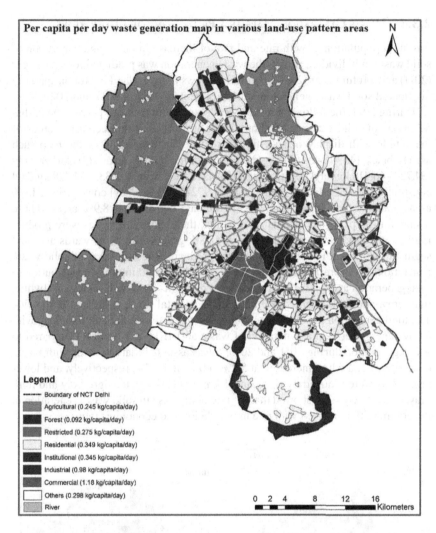

FIGURE 2.7   Spatial distribution of per capita waste generation in various land-use pattern areas.

construct roads followed by agricultural and residential land usage which correspond to 28.2% and 24.6%, respectively. The per capita per day for solid waste generation values do not follow the same trend of distribution with industrial and commercial possessing values at 0.98 and 1.18 kg/person/day, respectively, albeit sharing only 2.7–3.5% of the total land space in Delhi. On the contrary, space occupied by the majority of the land such as agricultural, residential, and other activities contribute very less in solid waste generation with their per capita generation values varied between 0.24 and 0.34 kg/person/day. The river contributed negligible wastes and hence, we did not consider in this analysis.

2.3.3 Waste Generation Maps as a Prediction for Other Years

Based on population growth rate and rate of increase in per capita generation of solid waste in individual wards, the waste generation was predicted for a past year (2011) and a future year (2020). Figure 2.8 shows maps created in GIS on spatially distributed solid waste generation during projected years of 2011 and 2020.

During 2011, the following results were obtained in terms of percentage of distributions of solid waste generation among wards based on decreasing order of magnitude with display of specific color ranges in the legend of the map such as: (1) black [46,418–69,938 kg/day] constituting 11.6% areas, (2) darker gray [34,728–46,418 kg/day] constituting 17.7% areas, (3) white [27,332–34,728 kg/day] constituting 30.0% areas, (4) light gray [21,972–27,332 kg/day] constituting 31.9% areas, and (5) dark gray [7,337–21,972 kg/day] constituting 8.9% areas. These results for the past year of 2011 showed that the majority of wards were producing less quantity of waste [light gray, and dark gray] at 40.8% of wards and next segment [white] with 30% of wards were producing medium level of solid waste generation. On overall assessment, it was found that maximum of contribution to total generation at city level was made from medium- and low-level regions [white, light gray, and dark gray] constituting 70.8% of total waste generated in the city. The total generation of solid waste in the city was estimated at 8,560 ton/day with an average of per capita generation of waste at 0.41 kg/person/day. Compared to waste generation during 2011 and 2015, it was assessed that the areas with black, darker gray, and white increased at 5.2%, 4.0%, and 3.7%, respectively, and lower level areas were reduced by 6.6% and 4.5%, respectively, for light gray and dark gray, respectively. This showed that the rate of increase in solid waste generation in the community increased with rise in population and economic status.

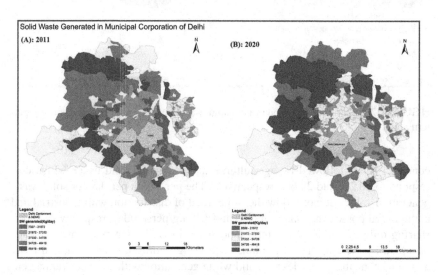

FIGURE 2.8 Spatially distributed solid waste generation map during projected years: (A) 2011 and (B) 2020.

For the future prediction of the year 2020, the obtained results in terms of percentage of distributions of solid waste generation among wards showed the following trends: (1) black [46,418–81,558 kg/day] constituting 13.1% areas, (2) darker gray [34,728–46,418 kg/day] constituting 19.8% areas, (3) white [27,332–34,728 kg/day] constituting 32.9% areas, (4) light gray [21,972–27,332 kg/day] constituting 25.9% areas, and (5) dark gray [8,556–21,972 kg/day] constituting 8.3% areas. Similar to trends for the base year (2015) and the past year (2011), the results of future year (2020) showed that the majority of the wards were producing less quantity of waste [light gray, and dark gray] at 34.2% of wards, and next segment [white] with 32.9% of wards were producing medium level of solid waste generation. It was seen that the maximum of contribution to total generation at city level was made from medium- and low-level regions [white, light gray, and dark gray] constituting 67.1% of total waste generated in the city. The higher level regions black, darker gray, and white) contributed the remaining 32.9% to total waste generation of the city. The total generation of solid waste in the waste predicted for future year (2020) at 15,600 ton/day with an average of per capita generation of waste at 0.64 kg/person/day. With comparative assessment for years (present – 2015 and future – 2020) on estimation of solid waste generation and its spatial distribution, it was observed that the areas with black, darker gray, and white increased at 7.4%, 7.6%, and 5.8%, respectively, and lower level areas were reduced by 13.1% and 2.4%, respectively for light gray and dark gray, respectively in future year (2020) with respect to present year (2015). Overall assessment on waste generation trends in the aspects of spatial distribution and per capita generation during 2011, 2015, and 2020 showed that rise in quantity of solid waste generation occurred with a consistent pattern, and the reason for such trends might be due to improvement in economic status and lifestyle of residents, and their dependence on more disposable matter for packing and parcel purposes on a day-to-day basis.

2.3.4 PROPOSED TRANSPORTATION ROUTES AND REDUCTION IN HAULING DISTANCE

Following the procedures as explained in the methodology section, ArcGIS Network Analyst tool was used, choosing 20 collection locations or Dhalaos as the starting nodes and the transfer station location in Vijay Ghat as the destination node. Table 2.1 presents the detailed information of collection points and ward number, transfer station as Vijay Ghat. The inbuilt algorithm in GIS identified the shortest path and presented graphically on a GIS-generated map along with route distance for each of the desired links. We compared the proposed and optimized route distance generated from this approach in GIS with the existing (current) route distance that is being practiced as the distance between individual collection point and transfer station. The detailed results from this approach of Network analysis are presented in Table 2.1 containing the current distance, optimized distance, distance saved, and percentage of saving in hauling distance of transport vehicle. Some representative GIS-generated maps are displayed in Figure 2.9 with four distinct routes that are marked in Table 2.1, i.e., the routes between IARI

TABLE 2.1

Results from Network Analysis on Transportation Routes with Comparison between Current Distance (km) versus Optimal Distance (km)*

Collection Point with Area Name	Ward Name	Ward No.	Transfer Station	Current Distance (km)	Optimized Distance (km)	Savings in Hauling Distance (km)	% of Savings (Hauling Distance)
IARI	Rajinder Nagar	149	Vijay Ghat	10.2	9.5	0.7	6.9
Jaypee Siddharth Hotel	East Patel Nagar	95	Vijay Ghat	9.5	8.8	0.7	7.4
Dev Nagar	Dev Nagar	92	Vijay Ghat	9	7.7	1.3	14.4
Sharai Rohilla	Karol Bagh	91	Vijay Ghat	8.3	7.9	0.4	4.8
S.P. Mukherjee Market	Model Basti	90	Vijay Ghat	7.9	6.4	1.5	18.9
Faiz Road	Model Basti	90	Vijay Ghat	8.4	6.2	2.2	26.2
Cycle Market	Model Basti	90	Vijay Ghat	7.3	5.9	1.4	19.2
Manak Pura	Model Basti	90	Vijay Ghat	7.2	6.1	1.1	15.3
Double Storey Quarters Motia Khan	Model Basti	90	Vijay Ghat	7.2	5.6	1.6	22.2
Chitragupta Road	Paharganj	89	Vijay Ghat	7.4	5.3	2.1	28.4
Bharat Nagar	Paharganj	89	Vijay Ghat	7.1	5.3	1.8	25.4
Ratan Market	Paharganj	89	Vijay Ghat	5.5	4.5	1	18.2
Gali no: 6 Multani Dhanda	Paharganj	89	Vijay Ghat	6.5	5	1.5	23.1
Qutab Road	Ram Nagar	87	Vijay Ghat	5.6	4.9	0.7	12.5
Farash Khana	Idgah Road	85	Vijay Ghat	4.3	4	0.3	6.9
Rani Jhansi Road	Deputy Ganj	76	Vijay Ghat	9	5.2	3.8	42.2
Bahadurgarh Road	Deputy Ganj	76	Vijay Ghat	6.9	4.7	2.2	31.9
Azad Park	Chandni Chowk	80	Vijay Ghat	3.7	2.8	0.9	24.3
A-2 Block Petrol Bunk Shastri Nagar	Kashmir gate	77	Vijay Ghat	6.7	5.8	0.9	13.4
Ahata Kedara	Qasabpura	88	Vijay Ghat	7.7	5.2	2.5	32.5

* Maps are displayed in Figure 2.9 for the Area's Italicized with existing path and optimized (proposed) path.

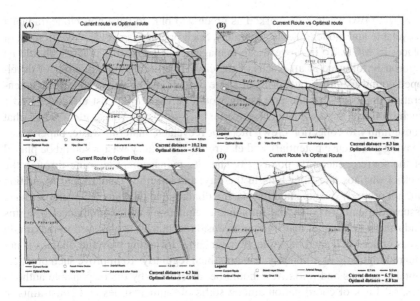

FIGURE 2.9 Maps showing the details of transport route with comparative assessment as current versus optimized between collection point and transfer station: (A) IARI and Vijay Ghat, (B) Sharai Rohilla and Vijay Ghat, (C) Farash Khana and Vijay Ghat, and (D) Shastri Nagar and Vijay Ghat. For more details, refer Table 2.1.

and Vijay Ghat, Sharai Rohilla and Vijay Ghat, Farash Khana and Vijay Ghat, and Shastri Nagar and Vijay Ghat. These four distinct paths showed the following results: (1) current distance = 10.2 km, and optimized distance = 9.5 km for the route between IARI and Vijay Ghat, (2) current distance = 8.3 km, and optimized distance = 7.9 km for the route between Sharai Rohilla and Vijay Ghat, (3) current distance = 4.3 km, and optimized distance = 4.0 km for the route between Farash Khana and Vijay Ghat, and (4) current distance = 6.7 km, and optimized distance = 5.8 km for the route between Shastri Nagar and Vijay Ghat. The optimal route distances were generated from GIS, and from comparative analysis between current and optimized route distances, the results showed an average saving of 1.43 km or 19.7±9.8% of total hauling distance in this specific network. The range for percentage of saving in hauling distance of transport vehicles varied from 4.8% to 42.2%.

2.3.5 FUTURE CHALLENGES AND THE WAY FORWARD FOR FUTURE SCOPES OF STUDY

From this preliminary investigation with primary and secondary data, and information compiled in this study, we find the following issues prevailed in the study area, which need to be improved to have an efficient and productive solid waste management system. These points are enumerated as follows: (1) Currently the executive bodies of the city practice source segregation methods in two ways by deploying two waste bins marked as green and blue to have non-biodegradable/ recyclable and biodegradable separately. Such practice is not effective in most of places in MCD. To improve waste management system, this practice should be rigorously followed at all places of the city. (2) Similar to the practice of developed countries, more categories of waste bins should be deployed at source generation points so that the processing would be easier and cost effective. Under such initiative, separate categories waste bins for plastic waste, glass waste (that cause injury to collection crew), metal or recyclable material waste, paper waste (having good reuse potential), and electronic and battery wastes should be placed at the source generation points. (3) The city management is going through the lack of dhalaos and litter bins in the city, and specific standard and regulation are not properly documented for placing dhalaos and dustbins. To overcome such issues, the authorities should implement regulations with compliance to norms and rules. (4) Unequal distribution of waste receptacles is observed in maximum areas under MCD, and the collection container including bins and dhalaos is not cleared regularly or efficiently or maintained regularly. To improve such practice, suitable actions and initiatives should be undertaken by the responsible authorities. (5) Lack of coordination among staffs of municipalities and constraints in financial budget that restrict the management system to provide efficient and advanced components similar to developed countries. To fix such issues, more PPP initiatives should be promoted by the government. (6) Number of regular operators required for different components of management is much lesser than required. Most of the requirements are fulfilled by employing temporary staff personnel that put some problems to management system obstructing smooth functioning. (7) Transfer and transport vehicles are poorly maintained due to inadequate workshop facilities and maintenance procedures. Frequent breakdown and vehicles used sometime are out of service for long periods. The bureaucratic system and procurement process are sometime cumbersome and slower that obstruct non-availability of spare parts for vehicles. For better management system, well-equipped and advanced vehicles with more hydraulic fittings with short range tipper and dumpers should be included in the fleet meant for transfer and transportation. Equal distribution of waste transfer and transport vehicles should happen in all zones of the municipalities. The overhauling vehicles engaged in transportation purposes should be replaced frequently. (8) The existing number of landfills are not adequate to manage waste disposal properly in the study area, as these are overloaded and exhausted. More landfills should be planned and constructed on priority basis. During planning and construction of landfills, they should be designed as sanitary landfills so that it would be easier for the management to track the systems efficiently. (9) The incineration plants and composting facilities are not exploited properly to their full capacities in getting maximum recovery from solid waste through the principle of waste-to-energy. (10) The authorities should declare some zones as ideal zones, where the best practice of management of international standards should be implemented. There should be encouragement through awards and recognition practices from the authorities to motivate and promote the best practices.

2.4 CONCLUSIONS

The National Capital Territory of Delhi, India was selected as the study domain to present MSW status and issues, apply our methodology for generation of GIS-based maps, finding out the shortest transportation routes, and suggest future way forward for better practice of solid waste management. The total generation of solid waste in the city was estimated at 11,300 ton/day with an average of per capita generation of waste at 0.52 kg/person/day during 2015. The assessment on waste generation trends in the aspects of spatial distribution and per capita generation during the years 2011, 2015, and 2020 showed that a rise in quantity of solid waste generation occurred with a consistent pattern, and the reason for such trends might be due to improvement in economic status and lifestyle of residents, and their dependence on more disposable matter for packing and parcel purposes on a day-to-day basis. From the route optimization study, it was found that optimal route distances can be saved with an average saving of 1.43 km or 19.7±9.8% of total distance savings in 20 locations of a specific network. The range for percentage of saving in hauling distance of transport vehicles varied from 4.8% to 42.2%.

REFERENCES

Apaydin, O., Gonullu, M.T. 2007. Route optimization for solid waste collection: Trabzon (Turkey) case study. *Global NEST Journal*, 9(1): 6–11.

Aremu, A.S. 2013. In-town tour optimization of conventional mode for municipal solid waste collection. *Nigerian Journal of Technology*, 32(3): 443–449.

Behera, S.N., Sharma, M., Dikshit, O., Shukla, S.P. 2011a. GIS-based emission inventory, dispersion modeling, and assessment for source contributions of particulate matter in an urban environment. *Water, Air, & Soil Pollution*, 218(1): 423–436.

Behera, S.N., Sharma, M., Dikshit, O., Shukla, S.P. 2011b. Development of GIS-aided emission inventory of air pollutants for an urban environment. *Advanced Air Pollution*, 279–294.

Bhuvan. 2020. Indian Geo-Platform of Indian Space Research Organization (ISRO). https://bhuvan.nrsc.gov.in/bhuvan_links.php. Last accessed in December 2020.

Chandramouli, C., General, R., 2011. Census of India 2011. *Provisional Population Totals*. Government of India, New Delhi. http://census2011.co.in.

Chang, N.B., Lu, H.Y., Wei, Y.L. 1997. GIS technology for vehicle routing and scheduling in solid waste collection systems. *Journal of Environmental Engineering*, 123(9): 901–910.

Chattopadhyay, S., Dutta, A., Ray, S. 2009. Municipal solid waste management in Kolkata, India – A review. *Waste Management*, 29(4): 1449–1458.

Dijkstra, E.W. 1959. A note on two problems in connexion with graphs. *Numerische Mathematik*, 1(1): 269–271.

DPCC. 2016. Annual Report of Solid Waste Management in Delhi, Delhi Pollution Control Committee. https://www.dpcc.delhigovt.nic.in. Last accessed in December 2020.

Dutta, A., Jinsart, W. 2020. Waste generation and management status in the fast-expanding Indian cities: A review. *Journal of the Air & Waste Management Association*, 70(5): 491–503.

Economic Survey of Delhi. 2015. Environmental Concerns, Chapter 8, pp. 114–152. http:// delhiplanning.nic.in/. Last accessed in December 2020.

Ghose, M.K., Dikshit, A.K., Sharma, S.K. 2006. A GIS based transportation model for solid waste disposal – A case study on Asansol municipality. *Waste Management*, 26(11): 1287–1293.

Gupta, N., Yadav, K.K., Kumar, V. 2015. A review on current status of municipal solid waste management in India. *Journal of Environmental Sciences*, 37: 206–217.

Gupta, S., Gadi, R., Mandal, T.K., Sharma, S.K. 2017. Seasonal variations and source profile of n-alkanes in particulate matter (PM10) at a heavy traffic site, Delhi. *Environmental Monitoring and Assessment*, 189(1), 43.

Indiastat. 2020. Indiastat: Socio-Economic Statistical Data and Facts about India. https:// www.indiastat.com. Last accessed in December 2020.

Karadimas, N.V., Papatzelou, K., Loumos, V.G. 2007. Optimal solid waste collection routes identified by the ant colony system algorithm. *Waste Management and Research*, 25(2): 139–147.

Karak, T., Bhagat, R.M., Bhattacharyya, P. 2012. Municipal solid waste generation, composition, and management: The world scenario. *Critical Reviews in Environmental Science and Technology*, 42(15): 1509–1630.

Kennedy, L., Duggal, R., Lama-Rewal, S.T. 2009. Assessing urban governance through the prism of healthcare services in Delhi, Hyderabad and Mumbai. In *Governing India's Metropolises: Case Studies of Four Cities*. Routledge.

Khan, D., Kumar, A., Samadder, S.R. 2016. Impact of socioeconomic status on municipal solid waste generation rate. *Waste Management*, 49: 15–25.

Kumar, A. (2013). Existing situation of municipal solid waste management in NCT of Delhi, India. *International Journal of Social Sciences*, 1(1): 6–17.

Lata, K., Kumar, P., Banerjee, A. 2020. Geospatial intelligence for smart living – Case of New Delhi. In *Smart Living for Smart Cities* (pp. 145–181). Springer, Singapore.

Liu, J. 2009. A GIS-based tool for modelling large-scale crop-water relations. *Environmental Modelling and Software*, 24(3): 411–422.

Lu, L.T., Hsiao, T.Y., Shang, N.C., Yu, Y.H., Ma, H.W. 2006. MSW management for waste minimization in Taiwan: The last two decades. *Waste Management*, 26(6): 661–667.

Mandal, K. 2019. Review on evolution of municipal solid waste management in India: Practices, challenges and policy implications. *Journal of Material Cycles and Waste Management*, 21(6): 1263–1279.

Mohan, M., Pathan, S.K., Narendrareddy, K., Kandya, A., Pandey, S. 2011. Dynamics of urbanization and its impact on land-use/land-cover: A case study of megacity Delhi. *Journal of Environmental Protection*, 2(09): 1274.

Pradhan, J.K., Kumar, S. 2014. Informal e-waste recycling: Environmental risk assessment of heavy metal contamination in Mandoli industrial area, Delhi, India. *Environmental Science and Pollution Research*, 21(13): 7913–7928.

Sanjeevi, V., Shahabudeen, P. 2016. Optimal routing for efficient municipal solid waste transportation by using ArcGIS application in Chennai, India. *Waste Management & Research*, 34(1): 11–21.

Sharholy, M., Ahmad, K., Mahmood, G., Trivedi, R.C. 2008. Municipal solid waste management in Indian cities – A review. *Waste Management*, 28(2): 459–467.

Singh, R.P., Tyagi, V.V., Allen, T., Ibrahim, M.H., Kothari, R. 2011. An overview for exploring the possibilities of energy generation from municipal solid waste (MSW) in Indian scenario. *Renewable and Sustainable Energy Reviews*, 15(9): 4797–4808.

Singh, S., Behera, S.N. 2019. Development of GIS-based optimization method for selection of transportation routes in municipal solid waste management. In *Advances in Waste Management* (pp. 319–331). Springer, Singapore. https://doi.org/10.1007/978-981-13-0215-2_22

Srivastava, V., Ismail, S.A., Singh, P., Singh, R.P. 2015. Urban solid waste management in the developing world with emphasis on India: Challenges and opportunities. *Reviews in Environmental Science and Bio/Technology*, 14(2): 317–337.

Suthar, S., & Singh, P. (2015). Household solid waste generation and composition in different family size and socio-economic groups: A case study. *Sustainable Cities and Society*, 14, 56–63.

Talyan, V., Dahiya, R.P., Sreekrishnan, T.R. 2008. State of municipal solid waste management in Delhi, the capital of India. *Waste Management*, 28(7): 1276–1287.

Yadav, A., Behera, S.N., Nagar, P.K., Sharma, M. 2020. Spatio-seasonal concentrations, source apportionment and assessment of associated human health risks of PM2.5-bound polycyclic aromatic hydrocarbons in Delhi, India. *Aerosol and Air Quality Research*, 20: 2805–2825.

3 Food Waste to Energy through Advanced Pyrolysis

Shukla Neha and Neelancherry Remya
School of Infrastructure, Indian Institute of
Technology Bhubaneswar, Argul, Odisha, India

CONTENTS

3.1 INTRODUCTION

Food waste (FW) is the part of food whose edibility cannot be recovered. FW is generated throughout the food system, including production, processing, distribution, and consumption. FW generated from households, food processing industries, and hospitality sector accounted to around 1.3 billion tons of annual FW generation, which is one-third of all the food produced to feed human (Schanes, Dobernig, and Gözet 2018). Amounts of FW generation vary as per the demographic location, financial condition, availability of agricultural lands, etc. Australia generates 361 kg of FW per capita per year followed by U.S. (278 kg/capita/year), Sweden (200 kg/capita/year), Canada (123 kg/capita/year) U.K. (74.7 kg/capita/year), India (51 kg/capita/year), China, and Greece (44 kg/capita/year) (FAO 2017b). Due to economic and population growth, this FW generation will be double folded by 2050 (FAO 2017a). FW contributes 8% of the global human-made greenhouse gas emission (FAO 2019). Also, FW occupies ~30% of the total available land which is 1.4 billion hectares (Karabulut et al. 2018).

DOI: 10.1201/9781003201076-3

3.2 CONVENTIONAL FW DISPOSAL TECHNIQUES

The disposal of FW is a major environmental problem. Even though several orga-
nizations around the globe are working to deliver the unused food to the needed
ones, around 40% of the FW ends up drains or sewers, which leads to block-
ages (Azzurra, Massimiliano, and Angela 2019). On the other hand, landfilling of
municipal solid waste (MSW) with a substantial amount of FW results in several
issues like soil and groundwater pollution by leachate and air pollution due to
landfill gases (Nandan et al. 2017; FAO 2019; Fung et al. 2019; EPA 2020).

Composting and anaerobic digestion are the biological treatment method for
FW. Aerobic decomposition of FW by composting converts FW into a dark,
earthy-smelling and crumbly end product called compost, which can be used
to improve soil quality by retaining the maximum nutrients such as nitrogen,
phosphorous, and potassium (Agegnehu et al. 2016). High moisture content
of FW (74–90%) makes it suitable for composting (Cerda et al. 2018). In the
anaerobic digestion process, FW is converted to methane and carbon dioxide.
Several constraints like high organic to ash ratio, unsystematic collection system (inert material like glass/plastic collected with FW), high sensitivity to
operating conditions like pH, carbon to nitrogen ratio, moisture content and
aeration rate, large area requirements, odor issues, high energy consumption,
long active duration (4–12 weeks); the release of toxic gases such as ammonia,
methane, hydrogen sulfide, nitrous oxide which are allied with the presence
of anoxic/anaerobic zones in the solid matrix and volatile organic compounds
and transportation constraints, etc., make biological treatments unfavorable for
FW management on a large scale (Kosseva 2011; Papargyropoulou et al. 2014;
Cerda et al. 2018).

3.3 THERMAL TREATMENT TECHNIQUES FOR FW

FW contains 80% of volatile matter, 15% of fixed carbon, and 5% of ash content
(Choe et al. 2017; Jo et al. 2017; Elkhalifa et al. 2019). Hence, thermal treatment such as incineration, gasification, hydrothermal carbonization (HTC), and
pyrolysis causes high volume reduction of FW within less treatment time with
less dependency on environmental conditions. Even though incineration of FW
resulted in high volume reduction irrespective of the property of FW, the process
resulted in the release of dioxins and furans, which are a great environmental threat leading to air (Girotto, Alibardi, and Cossu 2015; Kibler et al. 2018).
Also, incineration obstructs the recovery of valuable chemical compounds and
nutrients from FW (Gao et al. 2017). Gasification thermally converts FW into
gases at high temperatures (typically >700°C) under limited/controlled amounts
of oxygen. These gases, known as syngas or producer gas, composed of nitrogen,
hydrogen, carbon monoxide, and carbon dioxide are used as an energy source for
electricity generation and production of high-value chemical products (Begum,
Rasul, and Akbar 2012). The low heating value of FW owing to high moisture
content (~70%) is the major bottleneck in the gasification process (Pham et al.

2015). On the other hand, FW is treated through hot and compressed water (180–300°C) for 5 min to 8 h in HTC (Qiao et al. 2011; Li, Diederick, Flora, and Berge 2013). The process converts wet FW into a hydrophobic, homogenized and carbon-rich solid product, which is known as hydrochar (Pavlovič, Knez, and Škerget 2013). HTC of FW generated hydrochar with high heating value (HHV) of 20–45 MJ/kg having 50–70% of carbon (Saqib, Baroutian, and Sarmah 2018; McGaughy and Toufiq Reza 2018). On the other hand, one of the major reported shortcoming of the HTC process is the constant heat capacity of water. Also, carbonization process conditions (e.g., temperature and reaction time) are the major factor to achieve maximum energy recovery by HTC (Brown 2011). Pyrolysis is the thermal degradation/breakdown process of the feedstock in the absence of oxygen at an elevated temperature of 300–700°C (Rajasekhar et al. 2015). It is a simple and flexible process with ease of control of operating parameters. A wide variety of feedstock such as garden waste, tire, plastic, wood chips, sewage sludge, animal waste, FW, etc. can be easily pyrolyzed to yield useful end products, i.e., non-condensable gases (or pygas) that can be used in turbine and stream engines, solid biochar, which can be used as adsorbent or soil amendment, and bio-oil, which can be considered as a petroleum blend or as a high-grade fuel (Rajasekhar et al. 2015).

3.4 DIFFERENT HEATING MODES OF PYROLYSIS

Pyrolysis employs heating of FW with the electrical system at varied pyrolysis conditions such as temperature, heating rate, particulate size, residence time, and inert gas flow rate. Figure 3.1 illustrates the schematic diagram of a pyrolysis reactor. Slow, intermediate, fast, and flash pyrolyses are considered as the different heating modes of pyrolysis, which can be achieved by stated pyrolysis

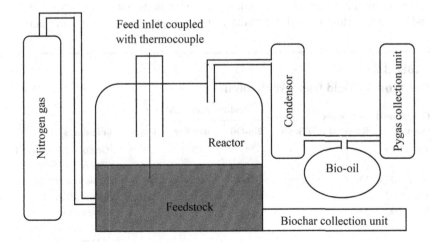

FIGURE 3.1 Schematic diagram of pyrolysis reactor.

TABLE 3.1

Different Heating Modes of Pyrolysis

Pyrolysis	Heating Rate (°C/min)	Temperature (°C)	Residence Time (s)
Slow	<60	300–500	1,800
Intermediate	60–75	400–600	<1,800
Fast	>75	500–600	10
Flash	>6,000	700–1,200	<2

parameters. Table 3.1 represents the different heating modes of pyrolysis. In slow pyrolysis heating rate of FW maintained <60°C/min with moderate reaction temperatures (300–500°C) and long residence time (~1,800 s) (Lombardi, Carnevale, and Corti 2015). On the other hand, a heating rate of 60–75°C/min, reaction temperature of 400–600°C and residence time <1,800 s is maintained in intermediate pyrolysis (Waluyo, Makertihartha, and Susanto 2018). The following conditions have to be maintained for fast pyrolysis: heating rate >75°C/min, the temperature of 500–600°C, and residence time of 0.5–10 (Budarin et al. 2009). Conversely, flash pyrolysis is carried out at very high heating rate (>6,000°C/min), high temperature (700–1,200°C) and low residence time (<2 s) (Xiu and Shahbazi 2012; Li et al. 2013).

Table 3.2 summarizes the experimental outcomes from the pyrolysis of FW having different heating modes. It was observed that slow pyrolysis is used to generate high yield of biochar, whereas intermediate pyrolysis yields maximum bio-oil (40–60%) with 20–30% of non-condensable gases and 15–25% of biochar (S. Wang et al. 2017). Fast pyrolysis not only suppresses secondary reactions but also polymerization of volatile intermediate compounds of feedstock to obtain high bio-oil yield in a shorter span of residence time. Furthermore, flash pyrolysis generates high amounts of non-condensable gases along with diminished bio-oil and biochar yields. Bio-oil generated from flash pyrolysis is extremely acidic,

TABLE 3.2

End-Product Yield from FW Pyrolysis

Operational Parameters (Sample Wt./Temperature /Time)	End-Product Yield (%)			References
	Bio-Oil	Biochar	Pygas	
100 g, 800°C	54	25	21	(Grycová, Koutník, and Pryszcz 2016)
500 g, 400°C, 1 h	25.6	32.2	42.2	(Liang et al. 2015)
1 g, 700°C	42	30	18	(Park et al. 2020)
150 g, 400°C, 2 h	8	5	87	(Soongprasit, Sricharoenchaikul, and Atong 2019)

reactive, very viscous, comprise minor amounts of residue solids and high aqueous substances (Horne and Williams 1996; Kan, Strezov, and Evans 2016; Daful and R Chandraratne 2020).

3.5 ADVANCE PYROLYSIS OF FW

Non-conventional/advanced pyrolysis methods use the advancement of conventional pyrolysis techniques for the improvisation of pyrolytic end-product yield and quality. Advanced pyrolysis techniques comprise several approaches such as co-pyrolysis, catalytic and microwave pyrolysis in different heating modes which are discussed in upcoming sections. Table 3.3 includes different forms of advanced pyrolysis with their operation conditions and end-product yields.

3.5.1 CO-PYROLYSIS

Co-pyrolysis involves pyrolysis of a feedstock mixture comprising two or more biomass, plastic, or other waste material. Co-pyrolysis improves the quantity and quality of pyrolysis end products due to the synergistic effects and interactions between the components in the feedstock mixture. Conventional pyrolysis can be easily converted to co-pyrolysis by blending and mixing the feedstock prior to pyrolysis. Technically, co-pyrolysis can be implemented without the need for any solvents or catalysts, which signifies that the co-pyrolysis requires lower energy input. Pyrolysis of peanut crisps and waste cereal yielded 34% and 62% of bio-oil respectively, whereas, co-pyrolysis of peanut crisps and waste cereal resulted in 46% of bio-oil yield (Grycová, Koutník, and Pryszcz 2016). Similarly, co-pyrolysis of soybean and polyvinyl chloride (PVC) improved bio-oil yield with up to 69% reduction in the tar content in bio-oil compared to the pyrolysis

TABLE 3.3
Advance Pyrolysis of FW

Type of Pyrolysis	Operational Parameters (Sample Wt./Temperature /Time)	End-Product Yield (%)			References
		Bio-Oil	Biochar	Pygas	
Co-pyrolysis	100g, 800°C	57	23	20	(Grycová, Koutník, and Pryszcz 2016)
Catalytic pyrolysis	20 g, Jordanian volcanic tuff as catalyst (0.1 catalyst to feed ratio (wt./wt.), 500°C	30	23	47	(Aljbour 2018)
MW pyrolysis	50 g, 550 W, 15 min	49	47	4	(Parth, Akash, and Remya 2020)
MW-catalytic pyrolysis	30 g of sample with 5 % of catalyst (MgO, Fe_2O_3, MnO_2, $CuCl_2$ and NaCl) 400 W	20	40	40	(Liu et al. 2014)

of soybean and PVC separately (Tang et al. 2018). The co-pyrolysis reduced oxygenates in bio-oil by depressing the concentrations of acids, phenolic compounds, ketones, and aldehydes. Desirable hydrocarbons, esters and alcohols were also presented in the bio-oil obtained from the co-pyrolysis of FW and wood bark. Also, co-pyrolysis of the FW and wood bar improved the bio-oil HHV up to 12 MJ/kg which was initially 2.9 MJ/kg obtained from pyrolysis of FW only (Park et al. 2020).

3.5.2 Catalytic Pyrolysis

Catalytic pyrolysis employs different catalysts such as acids, bases, metals, zeolite, or carbon during FW pyrolysis to improve the properties of the end products. Catalytic pyrolysis of FW evidenced improved heating rate, catalytic cracking, and deoxygenation reactaion. The process yielded bio-oil with less oxygenates and nitrogenates due to the conversion of oxygenates to CO_2, CO, and H_2O and nitrogenates to N_2 (Czajczyńska et al. 2017; Kim et al. 2020).

Catalytic pyrolysis is classified as *in situ* and *ex situ* process based on the mode of catalyst addition. In *in situ* catalytic pyrolysis or catalyst mixing method, FW is blended with the catalyst within the pyrolysis reactor (Figure 3.2a). *In situ* catalytic pyrolysis of FW inhibits tar formation and promotes the production of bio-oil with naphthalene and aromatic compounds along with non-condensable gases such as carbon mono-oxide and carbon dioxide (Hu and Gholizadeh 2019). Single reactor configuration demands very less equipment space and makes the process economical. However, ineffective solid/solid interaction and simultaneous heating of incompatible FW/catalyst system with different activation temperatures, often lead to poor deoxygenation (Sankaranarayanan et al. 2015). *Ex situ* catalytic pyrolysis or catalyst-bed system employs a modified reactor setup, which separates FW supported on quartz wool at the upper section the reactor from the catalyst secure at the bottom section (Figure 3.2b). The heating process starts with the catalyst activation, which is further followed by the thermal degradation of FW with the help of a heat carrier. This concept provides smooth

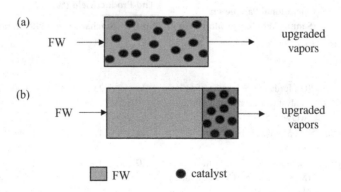

FIGURE 3.2   Catalytic pyrolysis process (a) *in situ*, and (b) *ex situ*.

control on pyrolysis, promotes catalytic activities during the reaction, and gives better optimization process without influencing each other. Also, this configuration allows longer vapor residence time, which stimulates secondary cracking of the volatiles for the production of non-condensable gases (Xiong et al. 2017). Thus, *ex situ* pyrolysis promotes production of non-condensable gases containing monocyclic aromatics and reduces catalyst contamination by minerals such as K, Na, Ca, Mg, etc. in FW. The process solves the issue of biochar-catalyst separation and recovery, which improves the process efficiency with reduced tar production. However, coke formation on the surface of the catalyst may cause catalyst deactivation (Praveen Kumar and Srinivas 2020).

3.5.3 MICROWAVE PYROLYSIS

Microwave (MW) technology utilizes MW irradiation as the heating source for pyrolysis of FW. MW causes molecular motion in the feedstock by the rotation of dipoles or by the migration of ionic species present in it, which resulted in internal heat due to friction (Remya and Lin 2011). MW pyrolysis resulted in the rapid pyrolytic conversion of FW, rice husk, and other lignocellulose materials to end products (Okumura, Hanaoka, and Sakanishi 2009; Lam and Chase 2012; Shukla, Sahoo, and Remya 2019; Sahoo and Remya 2020). Molecular-level volumetric heating of FW and heat transfer from FW to surface caused differential heating and release of volatiles due to the depolymerization and fragmentation of the FW (Nomanbhay et al. 2017). Table 3.3 enlisted various end-product yields obtained from MW pyrolysis of FW. End-product yield from MW pyrolysis of FW was attributed to different operating parameters like sample weight, feed characteristics, reaction temperature, reaction time, residence time, etc. (Miura et al. 2004; Chen et al. 2008; Huang et al. 2008; X. Wang et al. 2008; Budarin et al. 2009; Wan et al. 2009; Liu et al. 2014; Suriapparao and Vinu 2015a). High temperature with long residence time resulted in higher yield of non-condensable gas, low temperature with high residence time resulted in biochar yield, and higher temperature with less residence time improved bio-oil yield (Yadav et al. 2019). MW pyrolysis of FW showed low emission of polycyclic aromatic hydrocarbons (PAHs) compared to conventional pyrolysis (Domínguez et al. 2008). Biochar from MW pyrolysis of FW showed improved fixed carbon content and surface area ($877 \ m^2/g$) (Gronnow et al. 2013). Bio-oil yield up to ~35% was reported in several studies conducted on MW pyrolysis of FW (Grycová, Koutník, and Pryszcz 2016; Kadlimatti, Raj Mohan, and Saidutta 2019). MW pyrolysis of FW alone resulted in the production of oxygenated end products such as sugars, aldehydes, ketones, acids, and phenols, which resulted in bio-oil with reduced HHV, thermal instability, and corrosiveness (Abnisa, Wan Daud, and Sahu 2011). On the other hand, MW co-pyrolysis FW with other hydrogen-rich materials such as plastics improved the yield of hydrogen-enriched bio-oil. Besides having numerous advantages over the conventional pyrolysis process, bio-oil recovery from MW pyrolysis is still limited due to high dependency on material properties and operating parameters (Himmel et al. 2007; Arshad et al. 2017; Salman, Nomanbhay, and Salema 2019).

3.6 ASSESSMENT OF THE PYROLYSIS PROCESS VIA END-PRODUCT QUALITY, ENERGY YIELD, AND ENERGY EFFICIENCY

Practical implementation of the pyrolysis process depends on the quality of the resultant end products. FW pyrolysis reported the efficient conversion of FW into bio-oil and biochar with HHV. Bio-oil consisted of acids, sugars, alcohols, ketones, aldehydes, phenols and their derivatives, furans, and other mixed oxygenates (Grycová, Koutník, and Pryszcz 2016; Kadlimatti, Raj Mohan, and Saidutta 2019). Based on the feedstock, operating conditions and type of pyrolysis HHV of bio-oil varied from 2.9 to 12 MJ/kg (Dai et al. 2018; McGaughy and Toufiq Reza 2018). Co-pyrolysis of FW produced yellow colored bio-oil with typical petroleum hydrocarbons odor and less water content whereas pyrolysis of FW alone resulted in reddish-brown bio-oil with an irritable odor (Brebu et al. 2010). Conversely, biochar from FW pyrolysis was a carbon-rich matrix with some inorganic compounds with an HHV of 22.7–23.8 MJ/kg, which is comparable with typical coal (Rehrah et al. 2014; Kim et al. 2020; Park et al. 2020).

Analysis of the process in terms of energy yield and energy efficiency would enable the assessment of feasibility and applicability of the pyrolysis process. The efficiency of the FW conversion techniques was calculated for different yields of the process, their HHV, energy yield (EY), and net energy efficiency using Equations (3.1), (3.2), and (3.3) and reported in Table 3.4.

$$EY_{bio\text{-}oil} = \frac{HHV_{bio\text{-}oil} \ x \ Y_{bio\text{-}oil}}{HHV_{feedstock}} \qquad (3.1)$$

$$EY_{bio\text{-}char} = \frac{HHV_{bio\text{-}char} \ x \ Y_{bio\text{-}char}}{HHV_{feedstock}} \qquad (3.2)$$

Net energy efficiency $(\%)$

$$= \frac{HHV_{bio\text{-}oil} \ x \ W_{bio\text{-}oil} + HHV_{bio\text{-}char} \ x \ W_{bio\text{-}char}}{HHV_{feedstock} \ x \ W_{feedstock}} x \ 100 \qquad (3.3)$$

where $HHV_{bio\text{-}oil}$, $HHV_{biochar}$, and $HHV_{feedstock}$ are the higher heating value of bio-oil, biochar, and feedstock (MJ/kg), respectively, and $Y_{bio\text{-}oil}$ and $Y_{biochar}$ are the yield percentage of bio-oil and biochar, respectively.

FW pyrolysis showed 52% of bio-oil yield with 30% of biochar yield with HHV of ~13 MJ/kg for both end products (refer Table 3.4). MW pyrolysis of the same resulted in the decreased yield of bio-oil (by 42.3%) and slightly increased

TABLE 3.4

Energy Yield and Net Energy Efficiency of the Different Pyrolysis Processes for FW

Pyrolysis Conditions	Yield (Wt.%)		HHV (MJ/kg)			Energy Yield		Net Energy Efficiency (%)	Reference
	Bio-Oil	Biochar	Feedstock	Bio-Oil	Biochar	Bio-Oil	Biochar		
Pyrolysis (500°C, 240 min)	52	30	20	13	13	34	39	73	(Opatokun et al. 2016)
Microwave pyrolysis (450 W, 20 min)	30	34	10	4	20	12	68	80	(Kadlimatti, Raj Mohan, and Saidutta 2019)
Catalytic pyrolysis of olive cake with Jordanian volcanic tuff (500°C)	30	23	22	15	19	20	20	40	(Aljbour 2018)
Microwave catalytic pyrolysis of FW with silica bead (450 W for 25 min)	35	38	20	8	21	14	40	54	(Suriapparao and Vinu 2015b)
Co-pyrolysis of peanuts crisps and waste cereals (800°C, 30 min)	46	25	25	14	19	26	19	45	(Grycová et al. 2016)
Microwave co-pyrolysis of FW with LDPE (550°C, 800 W, 15 min)	49	47	15	5	25	19	78	97	(Parth, Akash, and Remya 2020)

yield of biochar (by 13.3%) with the increased yield of pygas. MW pyrolysis improvised the biochar HHV by ~54%, which enhanced the net energy efficiency by ~10% as compared to pyrolysis. Catalytic pyrolysis promotes the secondary cracking of volatiles and resulted in the highest pygas yield (47%). Even though the resulted bio-oil and biochar constitute significant HHV, lower yield of both end products leads to low net energy efficiency (40%). Furthermore, MW catalytic pyrolysis increased the bio-oil and biochar yield by ~17% and 65% respectively. Low HHV of bio-oil (8 MJ/kg) and moderate HHV of biochar (21 MJ/kg) resulted in 35% of improvised net energy efficiency of MW catalytic pyrolysis as compared to catalytic pyrolysis. Co-pyrolysis resulted in a significant yield of bio-oil (46%) and biochar (25%) with the comparable HHV (~17 MJ/kg). Hence, 45% of net energy efficiency was achieved (Table 3.4). On the other hand, MW co-pyrolysis provided 49% of bio-oil and 47% of biochar having the HHV of 5 MJ/kg and 25 MJ/kg, respectively. Conversely, MW intensified the carbonation of biochar and improved the energy density, resulting in a high HHV compared to that of feedstock (Kadlimatti, Raj Mohan, and Saidutta 2019). Thus, MW co-pyrolysis showed significant improvement in net energy efficiency (97%) compared to that of other forms of pyrolysis.

3.7 CONCLUSIONS

FW is a major contributor to MSW generation, which accounts for 1.3 billion FW generation per annum. Despite having the availability of numerous disposal methods, thermal techniques are the most commonly adopted method because of the recovery of useful end products. Among incineration, gasification, HTC, and pyrolysis, the pyrolysis of FW considered beneficial and useful due to the production of varieties of value-added end products like bio-oil and biochar, which holds great energy potential. A significant amount of bio-oil can be easily achieved by intermediate or fast pyrolysis; whereas for maximum production of biochar, slow pyrolysis is considered beneficial and to recover the highest yield of pygas, flash pyrolysis is highly recommendable. Based on the desirable end-product yield, a suitable advance pyrolysis technique can be selected. To achieve maximum bio-oil yield (57%), pyrolysis of FW can be considered, whereas higher biochar yield (47%) was reported with MW pyrolysis. Also, maximum pygas can be achieved with catalytic pyrolysis (47%). To improve the end-product yield and quality in terms of HHV of bio-oil up to 12 MJ/kg with the lesser presence of oxygenates and furans and HHV of biochar up to 23.8 MJ/kg and surface area of 877 m^2/g, advanced pyrolysis techniques such as co-pyrolysis and MW pyrolysis can be considered as a promising technique. Besides, considering the highest yield of bio-oil (49%) and biochar (47%) and their respective HHV (5 and 25 MJ/kg), MW co-pyrolysis process found the best technique by providing the highest percentage of net energy efficiency (96.65%) when compared to other aid-in pyrolysis techniques. Solving the technical challenges related with the pyrolysis of FW for the production of value-added products will aid and strengthen its application in waste management and the production of numerous chemicals and valuable energy sources that can be used directly or further processed into other forms of end products.

REFERENCES

Abnisa, Faisal, W. M. A. Wan Daud, J. N. Sahu. 2011. Optimization and Characterization Studies on Bio-Oil Production from Palm Shell by Pyrolysis Using Response Surface Methodology. *Biomass and Bioenergy*. doi:10.1016/j.biombioe.2011.05.011.

Agegnehu, Getachew, Adrian M. Bass, Paul N. Nelson, Michael I. Bird. 2016. Benefits of Biochar, Compost and Biochar-Compost for Soil Quality, Maize Yield and Greenhouse Gas Emissions in a Tropical Agricultural Soil. *Science of the Total Environment*. doi:10.1016/j.scitotenv.2015.11.054.

Aljbour, Salah H. 2018. Catalytic Pyrolysis of Olive Cake and Domestic Waste for Biofuel Production. *Energy Sources, Part A: Recovery, Utilization and Environmental Effects*. doi:10.1080/15567036.2018.1511649.

Arshad, Haroon, Shaharin A Sulaiman, Zahid Hussain, Yasin Naz, Firdaus Basrawi. 2017. Microwave Assisted Pyrolysis of Plastic Waste for Production of Fuels: A Review. *MATEC Web of Conferences* 131 (02005). doi:10.1051/matecconf/201713102005.

Azzurra, Annunziata, Agovino Massimiliano, Mariani Angela. 2019. Measuring Sustainable Food Consumption: A Case Study on Organic Food. *Sustainable Production and Consumption*. doi:10.1016/j.spc.2018.09.007.

Begum, Sharmina, M G Rasul, Delwar Akbar. 2012. An Investigation on Thermo Chemical Conversions of Solid Waste for Energy Recovery. *International Journal of Environmental and Ecological Engineering* 6 (2): 624–30.

Brebu, Mihai, Suat Ucar, Cornelia Vasile, Jale Yanik. 2010. Co-Pyrolysis of Pine Cone with Synthetic Polymers. *Fuel*. doi:10.1016/j.fuel.2010.01.029.

Brown, Robert C. 2011. Thermochemical Processing of Biomass: Conversion into Fuels, Chemicals and Power. *Thermochemical Processing of Biomass: Conversion into Fuels, Chemicals and Power*. doi:10.1002/9781119990840.

Budarin, Vitaly L., James H. Clark, Brigid A. Lanigan, Peter Shuttleworth, Simon W. Breeden, Ashley J. Wilson, Duncan J. Macquarrie, et al. 2009. The Preparation of High-Grade Bio-Oils through the Controlled, Low Temperature Microwave Activation of Wheat Straw. *Bioresource Technology* 100 (23): 6064–68. doi:10.1016/j.biortech.2009.06.068.

Cerda, Alejandra, Adriana Artola, Xavier Font, Raquel Barrena, Teresa Gea, Antoni Sánchez. 2018. Composting of Food Wastes: Status and Challenges. *Bioresource Technology*. doi:10.1016/j.biortech.2017.06.133.

Chen, Ming qiang, Jun Wang, Ming xu Zhang, Ming gong Chen, Xi feng Zhu, Fan fei Min, Zhi cheng Tan. 2008. Catalytic Effects of Eight Inorganic Additives on Pyrolysis of Pine Wood Sawdust by Microwave Heating. *Journal of Analytical and Applied Pyrolysis* 82 (1): 145–50. doi:10.1016/j.jaap.2008.03.001.

Choe, Jeehwan, Knowledge M. Moyo, Kibum Park, Jeongho Jeong, Haeun Kim, Yungsun Ryu, Jonggun Kim, Jun Mo Kim, Sanghoon Lee, Gwang Woong Go. 2017. Meat Quality Traits of Pigs Finished on Food Waste. *Korean Journal for Food Science of Animal Resources*. doi:10.5851/kosfa.2017.37.5.690.

Czajczyńska, Dina, Theodora Nannou, Lorna Anguilano, Renata Krzyzyńska, Heba Ghazal, Nik Spencer, Hussam Jouhara. 2017. Potentials of Pyrolysis Processes in the Waste Management Sector. *Energy Procedia*. doi:10.1016/j.egypro.2017.07.275.

Daful, Asfaw G, Meegalla R Chandraratne. 2020. Biochar Production From Biomass Waste-Derived Material. In *Encyclopedia of Renewable and Sustainable Materials*. doi:10.1016/b978-0-12-803581-8.11249-4.

Dai, Minquan, Hao Xu, Zhaosheng Yu, Shiwen Fang, Lin Chen, Wenlu Gu, Xiaoqian Ma. 2018. Microwave-Assisted Fast Co-Pyrolysis Behaviors and Products between Microalgae and Polyvinyl Chloride. *Applied Thermal Engineering*. doi:10.1016/j. applthermaleng.2018.02.102.

Domínguez, A., Y. Fernández, B. Fidalgo, J. J. Pis, J. A. Menéndez. 2008. Bio-Syngas Production with Low Concentrations of CO_2 and CH_4 from Microwave-Induced Pyrolysis of Wet and Dried Sewage Sludge. *Chemosphere*. doi:10.1016/j. chemosphere.2007.06.075.

Elkhalifa, Samar, Tareq Al-Ansari, Hamish R. Mackey, Gordon McKay. 2019. Food Waste to Biochars through Pyrolysis: A Review. *Resources, Conservation and Recycling*. doi:10.1016/j.resconrec.2019.01.024.

EPA. 2020. *Food Recovery Hierarchy – Sustainable Management of Food – US EPA*. EPA.

FAO. 2017a. The State of Food and Agriculture 2017. Leveraging Food Systems for Inclusive Rural Transformation. Population and Development Review.

———. 2017b. *The Future of Food and Agriculture: Trends and Challenges*. Food and Agriculture Organization of the United Nations. doi:10.4161/chan.4.6.12871.

———. 2019. *The State of Food and Agriculture, 2019: Moving Forward on Food Loss and Waste Reduction*. Lancet.

Fung, Leonard, Pedro E. Urriola, Lawrence Baker, Gerald C. Shurson. 2019. Estimated Energy and Nutrient Composition of Different Sources of Food Waste and Their Potential for Use in Sustainable Swine Feeding Programs. *Translational Animal Science*. doi:10.1093/tas/txy099.

Gao, Anqi, Zhenyu Tian, Ziyi Wang, Ronald Wennersten, Qie Sun. 2017. Comparison between the Technologies for Food Waste Treatment. *Energy Procedia*. doi:10.1016/j. egypro.2017.03.811.

Girotto, Francesca, Luca Alibardi, Raffaello Cossu. 2015. Food Waste Generation and Industrial Uses: A Review. *Waste Management*. doi:10.1016/j.wasman.2015.06.008.

Gronnow, Mark J., Vitaliy L. Budarin, Ondřej Mašek, Kyle N. Crombie, Peter A. Brownsort, Peter S. Shuttleworth, Peter R. Hurst, James H. Clark. 2013. Torrefaction/ Biochar Production by Microwave and Conventional Slow Pyrolysis – Comparison of Energy Properties. *GCB Bioenergy*. doi:10.1111/gcbb.12021.

Grycová, Barbora, Ivan Koutník, Adrian Pryszcz. 2016. Pyrolysis Process for the Treatment of Food Waste. *Bioresource Technology*. doi:10.1016/j.biortech.2016.07.064.

Grycová, Barbora, Ivan Koutník, Adrian Pryszcz, Kateřina Chamrádová. 2016. Pyrolysis Processing of Waste Peanuts Crisps. *GeoScience Engineering*. doi:10.1515/ gse-2015-0024.

Himmel, Michael E., Shi You Ding, David K. Johnson, William S. Adney, Mark R. Nimlos, John W. Brady, Thomas D. Foust. 2007. Biomass Recalcitrance: Engineering Plants and Enzymes for Biofuels Production. *Science*. doi:10.1126/science.1137016.

Horne, Patrick A., Paul T. Williams. 1996. Influence of Temperature on the Products from the Flash Pyrolysis of Biomass. *Fuel*. doi:10.1016/0016-2361(96)00081-6.

Hu, Xun, Mortaza Gholizadeh. 2019. Biomass Pyrolysis: A Review of the Process Development and Challenges from Initial Researches up to the Commercialisation Stage. *Journal of Energy Chemistry*. doi:10.1016/j.jechem.2019.01.024.

Huang, Y. F., W. H. Kuan, S. L. Lo, C. F. Lin. 2008. Total Recovery of Resources and Energy from Rice Straw Using Microwave-Induced Pyrolysis. *Bioresource Technology* 99 (17): 8252–58. doi:10.1016/j.biortech.2008.03.026.

Jo, Jun-Ho, Seung-Soo Kim, Jae-Wook Shin, Ye-Eun Lee, Yeong-Seok Yoo. 2017. Pyrolysis Characteristics and Kinetics of Food Wastes. *Energies* 10 (8): 1191. doi:10.3390/en10081191.

Kadlimatti, H. M., B. Raj Mohan, M. B. Saidutta. 2019. Bio-Oil from Microwave Assisted Pyrolysis of Food Waste-Optimization Using Response Surface Methodology. *Biomass and Bioenergy.* doi:10.1016/j.biombioe.2019.01.014.

Kan, Tao, Vladimir Strezov, Tim J. Evans. 2016. Lignocellulosic Biomass Pyrolysis: A Review of Product Properties and Effects of Pyrolysis Parameters. *Renewable and Sustainable Energy Reviews.* doi:10.1016/j.rser.2015.12.185.

Karabulut, Armağan Aloe, Eleonora Crenna, Serenella Sala, Angel Udias. 2018. A Proposal for Integration of the Ecosystem-Water-Food-Land-Energy (EWFLE) Nexus Concept into Life Cycle Assessment: A Synthesis Matrix System for Food Security. *Journal of Cleaner Production.* doi:10.1016/j.jclepro.2017.05.092.

Kibler, Kelly M., Debra Reinhart, Christopher Hawkins, Amir Mohaghegh Motlagh, James Wright. 2018. Food Waste and the Food-Energy-Water Nexus: A Review of Food Waste Management Alternatives. *Waste Management.* doi:10.1016/j.wasman.2018.01.014.

Kim, Soosan, Younghyun Lee, Kun Yi Andrew Lin, Eunmi Hong, Eilhann E. Kwon, Jechan Lee. 2020. The Valorization of Food Waste via Pyrolysis. *Journal of Cleaner Production.* doi:10.1016/j.jclepro.2020.120816.

Kosseva, M. R. 2011. Management and Processing of Food Wastes. In *Comprehensive Biotechnology*, Second Edition. doi:10.1016/B978-0-08-088504-9.00393-7.

Lam, Su Shiung, Howard A. Chase. 2012. A Review on Waste to Energy Processes Using Microwave Pyrolysis. *Energies.* doi:10.3390/en5104209.

Li, Liang, Ryan Diederick, Joseph R.V. Flora, Nicole D. Berge. 2013. Hydrothermal Carbonization of Food Waste and Associated Packaging Materials for Energy Source Generation. *Waste Management.* doi:10.1016/j.wasman.2013.05.025.

Liang, Shaobo, Yinglei Han, Liqing Wei, Armando G. McDonald. 2015. Production and Characterization of Bio-Oil and Bio-Char from Pyrolysis of Potato Peel Wastes. *Biomass Conversion and Biorefinery.* doi:10.1007/s13399-014-0130-x.

Liu, Haili, Xiaoqian Ma, Longjun Li, Zhi Feng Hu, Pingsheng Guo, Yuhui Jiang. 2014. The Catalytic Pyrolysis of Food Waste by Microwave Heating. *Bioresource Technology.* doi:10.1016/j.biortech.2014.05.020.

Lombardi, Lidia, Ennio Carnevale, Andrea Corti. 2015. A Review of Technologies and Performances of Thermal Treatment Systems for Energy Recovery from Waste. *Waste Management* 37: 26–44. doi:10.1016/j.wasman.2014.11.010.

McGaughy, Kyle, M. Toufiq Reza. 2018. Hydrothermal Carbonization of Food Waste: Simplified Process Simulation Model Based on Experimental Results. *Biomass Conversion and Biorefinery.* doi:10.1007/s13399-017-0276-4.

Miura, Masakatsu, Harumi Kaga, Akihiko Sakurai, Toyoji Kakuchi, Kenji Takahashi. 2004. Rapid Pyrolysis of Wood Block by Microwave Heating. *Journal of Analytical and Applied Pyrolysis* 71 (1): 187–99. doi:10.1016/S0165-2370(03)00087-1.

Nandan, Abhishek, Bikarama Prasad Yadav, Soumyadeep Baksi, Debajyoti Bose. 2017. *Recent Scenario of Solid Waste Management in India.* WSN World Scientific News.

Nomanbhay, Saifuddin, Bello Salman, Refal Hussain, Mei Yin Ong. 2017. Microwave Pyrolysis of Lignocellulosic Biomass––A Contribution to Power Africa. *Energy, Sustainability and Society.* doi:10.1186/s13705-017-0126-z.

Oh, Shinyoung, Hang Seok Choi, In Gyu Choi, Joon Weon Choi. 2017. Evaluation of Hydrodeoxygenation Reactivity of Pyrolysis Bio-Oil with Various Ni-Based Catalysts for Improvement of Fuel Properties. *RSC Advances.* doi:10.1039/c7ra01166k.

Okumura, Yukihiko, Toshiaki Hanaoka, Kinya Sakanishi. 2009. Effect of Pyrolysis Conditions on Gasification Reactivity of Woody Biomass-Derived Char. Proceedings of the Combustion Institute. doi:10.1016/j.proci.2008.06.024.

Opatokun, Suraj Adebayo, Tao Kan, Ahmed Al Shoaibi, C. Srinivasakannan, Vladimir Strezov. 2016. Characterization of Food Waste and Its Digestate as Feedstock for Thermochemical Processing. *Energy and Fuels.* doi:10.1021/acs.energyfuels.5b02183.

Papargyropoulou, Effie, Rodrigo Lozano, Julia K. Steinberger, Nigel Wright, Zaini Bin Ujang. 2014. The Food Waste Hierarchy as a Framework for the Management of Food Surplus and Food Waste. *Journal of Cleaner Production.* doi:10.1016/j.jclepro.2014.04.020.

Park, Chanyeong, Nahyeon Lee, Jisu Kim, Jechan Lee. 2020. Co-Pyrolysis of Food Waste and Wood Bark to Produce Hydrogen with Minimizing Pollutant Emissions. *Environmental Pollution.* doi:10.1016/j.envpol.2020.116045.

Parth, Rajput, Tripathi Akash, Neelancherry Remya. 2020. Microwave Co-Pyrolysis of Biomass and Plastic – Effect of Microwave Susceptors on the Yield and Property of Biochar. 2nd International Conference on Bioprocess for Sustainable Environment and Energy.

Pavlovič, Irena, Željko Knez, Mojca Škerget. 2013. Hydrothermal Reactions of Agricultural and Food Processing Wastes in Sub- and Supercritical Water: A Review of Fundamentals, Mechanisms, and State of Research. *Journal of Agricultural and Food Chemistry.* doi:10.1021/jf401008a.

Pham, Thi Phuong Thuy, Rajni Kaushik, Ganesh K. Parshetti, Russell Mahmood, Rajasekhar Balasubramanian. 2015. Food Waste-to-Energy Conversion Technologies: Current Status and Future Directions. *Waste Management.* doi:10.1016/j.wasman.2014.12.004.

Praveen Kumar, Kanduri, Seethamraju Srinivas. 2020. Catalytic Co-Pyrolysis of Biomass and Plastics (Polypropylene and Polystyrene) Using Spent FCC Catalyst. *Energy and Fuels.* doi:10.1021/acs.energyfuels.9b03135.

Qiao, Wei, Xiuyi Yan, Junhui Ye, Yifei Sun, Wei Wang, Zhongzhi Zhang. 2011. Evaluation of Biogas Production from Different Biomass Wastes with/without Hydrothermal Pretreatment. *Renewable Energy.* doi:10.1016/j.renene.2011.05.002.

Rajasekhar, M., N. Venkat Rao, G. Chinna Rao, G. Priyadarshini, N. Jeevan Kumar. 2015. Energy Generation from Municipal Solid Waste by Innovative Technologies – Plasma Gasification. *Procedia Materials Science* 10 (Cnt 2014): 513–18. doi:10.1016/j.mspro.2015.06.094.

Rehrah, D., M. R. Reddy, J. M. Novak, R. R. Bansode, K. A. Schimmel, J. Yu, D. W. Watts, M. Ahmedna. 2014. Production and Characterization of Biochars from Agricultural By-Products for Use in Soil Quality Enhancement. *Journal of Analytical and Applied Pyrolysis.* doi:10.1016/j.jaap.2014.03.008.

Remya, Neelancherry, Jih Gaw Lin. 2011. Current Status of Microwave Application in Wastewater Treatment-A Review. *Chemical Engineering Journal.* doi:10.1016/j.cej.2010.11.100.

Sahoo, Diptiprakash, Neelancherry Remya. 2020. Influence of Operating Parameters on the Microwave Pyrolysis of Rice Husk: Biochar Yield, Energy Yield, and Property of Biochar. *Biomass Conversion and Biorefinery.* doi:10.1007/s13399-020-00914-8.

Salman, Bello, Saifuddin Nomanbhay, Arshad Adam Salema. 2019. Microwave-Synthesised Hydrothermal Co-Pyrolysis of Oil Palm Empty Fruit Bunch with Plastic Wastes from Nigeria. *Biofuels.* doi:10.1080/17597269.2019.1626000.

Sankaranarayanan, Thangaraju M., Antonio Berenguer, Cristina Ochoa-Hernández, Inés Moreno, Prabhas Jana, Juan M. Coronado, David P. Serrano, Patricia Pizarro. 2015. Hydrodeoxygenation of Anisole as Bio-Oil Model Compound over Supported Ni Co Catalysts: Effect of Metal and Support Properties. *Catalysis Today.* doi:10.1016/j.cattod.2014.09.004.

Saqib, Najam Ul, Saeid Baroutian, Ajit K. Sarmah. 2018. Physicochemical, Structural and Combustion Characterization of Food Waste Hydrochar Obtained by Hydrothermal Carbonization. *Bioresource Technology.* doi:10.1016/j.biortech.2018.06.112.

Schanes, Karin, Karin Dobernig, Burcu Gözet. 2018. Food Waste Matters – A Systematic Review of Household Food Waste Practices and Their Policy Implications. *Journal of Cleaner Production.* doi:10.1016/j.jclepro.2018.02.030.

Shukla, Neha, Diptiprakash Sahoo, Neelancherry Remya. 2019. Biochar from Microwave Pyrolysis of Rice Husk for Tertiary Wastewater Treatment and Soil Nourishment. *Journal of Cleaner Production.* doi:10.1016/j.jclepro.2019.07.042.

Soongprasit, Kanit, Viboon Sricharoenchaikul, Duangduen Atong. 2019. Pyrolysis of Millettia (Pongamia) Pinnata Waste for Bio-Oil Production Using a Fly Ash Derived ZSM-5 Catalyst. *Journal of Analytical and Applied Pyrolysis.* doi:10.1016/j.jaap.2019.02.012.

Suriapparao, Dadi V., R. Vinu. 2015a. Bio-Oil Production via Catalytic Microwave Pyrolysis of Model Municipal Solid Waste Component Mixtures. *RSC Advances.* doi:10.1039/c5ra08666c.

———. 2015b. Bio-Oil Production via Catalytic Microwave Pyrolysis of Model Municipal Solid Waste Component Mixtures. *RSC Advances.* 5 (71): 57619–31. doi:10.1039/C5RA08666C.

Tang, Yijing, Qunxing Huang, Kai Sun, Yong Chi, Jianhua Yan. 2018. Co-Pyrolysis Characteristics Kinetic Analysis of Organic Food Waste and Plastic. *Bioresource Technology.* doi:10.1016/j.biortech.2017.09.210.

Waluyo, Joko, I. G.B.N. Makertihartha, Herri Susanto. 2018. Pyrolysis with Intermediate Heating Rate of Palm Kernel Shells: Effect Temperature and Catalyst on Product Distribution. In AIP Conference Proceedings. doi:10.1063/1.5042882.

Wan, Yiqin, Paul Chen, Bo Zhang, Changyang Yang, Yuhuan Liu, Xiangyang Lin, Roger Ruan. 2009. Microwave-Assisted Pyrolysis of Biomass: Catalysts to Improve Product Selectivity. *Journal of Analytical and Applied Pyrolysis* 86 (1): 161–67. doi:10.1016/j.jaap.2009.05.006.

Wang, Shurong, Gongxin Dai, Haiping Yang, Zhongyang Luo. 2017. Lignocellulosic Biomass Pyrolysis Mechanism: A State-of-the-Art Review. *Progress in Energy and Combustion Science.* doi:10.1016/j.pecs.2017.05.004.

Wang, Xianhua, Hanping Chen, Kai Luo, Jingai Shao, Haiping Yang. 2008. The Influence of Microwave Drying on Biomass Pyrolysis. *Energy and Fuels*, 22:67–74. doi:10.1021/ef700300m.

Xiong, Xinni, Iris K.M. Yu, Leichang Cao, Daniel C.W. Tsang, Shicheng Zhang, Yong Sik Ok. 2017. A Review of Biochar-Based Catalysts for Chemical Synthesis, Biofuel Production, and Pollution Control. *Bioresource Technology.* doi:10.1016/j.biortech.2017.06.163.

Xiu, Shuangning, Abolghasem Shahbazi. 2012. Bio-Oil Production and Upgrading Research: A Review. *Renewable and Sustainable Energy Reviews.* doi:10.1016/j.rser.2012.04.028.

Yadav, Krishna, Megha Tyagi, Soni Kumari, Sheeja Jagadevan. 2019. Influence of Process Parameters on Optimization of Biochar Fuel Characteristics Derived from Rice Husk: A Promising Alternative Solid Fuel. *Bioenergy Research*. doi:10.1007/s12155-019-10027-4.

Zhang, Huiyan, Jian Zheng, Rui Xiao, Dekui Shen, Baosheng Jin, Guomin Xiao, Ran Chen, et al. 2013. A Review on Pyrolysis of Plastic Wastes. *Energy Conversion and Management* 3 (17): 5769. doi:10.1016/j.enconman.2016.02.037.

4 Biochar Production and Its Characterization to Assess Viable Energy Options and Environmental Co-Benefits from Wood-Based Wastes

Rajat Gaur,[1] Sailesh N. Behera,[1,2]
Vishnu Kumar,[1] and Jagabandhu Dixit[3]
[1]Environmental Engineering Laboratory, Department
of Civil Engineering, Shiv Nadar University, Greater
Noida, Gautam Buddha Nagar, Uttar Pradesh, India
[2]Centre for Environmental Sciences and
Engineering, Shiv Nadar University, Greater Noida,
Gautam Buddha Nagar, Uttar Pradesh, India
[3]Disaster Management Laboratory, Department of
Civil Engineering, Shiv Nadar University, Greater
Noida, Gautam Buddha Nagar, Uttar Pradesh, India

CONTENTS

DOI: 10.1201/9781003201076-4

4.1   INTRODUCTION

India is a developing country with a high rate of growth in the industrial sector as well as the population specifically in urban regions. Although increasing urbanization and industrialization have helped in India's growth as an economic state, it has brought into the picture a lot of challenges with it such as an increase in waste generation, environmental pollution issues, and higher energy demand (Goswami et al., 2019; Singh and Behera, 2019; Yadav et al., 2020). As a consequence of population outbursts coupled with rapid industrialization and urbanization, waste generation from various sources has increased significantly over the years. The requirement of energy sources and management of huge solid waste generation have urged the research community to explore viable options that can fulfill the energy demand and reduce the burden of wastes on the land surface (Dhar et al., 2017; Lohri et al., 2017). In the same line of research, the production of waste-derived biochar and its application in multiple purposes are working well, and researchers are finding its potential usages in laboratory studies (Das et al., 2016; Goswami et al., 2019). Biochar is generally referred to as charred organic matter which is prepared with the help of various processing techniques to use it further in various applications such as improving soil properties or sequestration of carbon or co-combustion process (Parshetti et al., 2014; Yargicoglu et al., 2015). The International Biochar Initiative (IBI) states "Biochar is a solid material obtained from the carbonization of biomass" (Manyà, 2012; Madej et al., 2016).

When the terminology carbonization comes up, we first need to understand various processing techniques which are developed to convert biomass materials into biochar. With the help of such techniques, many categories of biochar can be produced which are further utilized as fuel or adsorbents, or absorbents on the field. The potential applications of biochar are being practiced in many areas including water holding capacity, nutrient availability, microbial activity, soil organic matter, water retention, crop yields in soil, reduction in fertilizer needs, nutrient leaching and erosion, and reduction in greenhouse gas emissions (Sohi et al., 2010; Woolf et al., 2010; Agegnehu et al., 2017).

There have been two major types of technologies including biological and thermal methods, which are practiced to produce biochar from biomass raw materials or feedstock, or precursor materials (Foong et al., 2020; Qin et al., 2020). The biological methods include anaerobic digestion, hydrolysis, and fermentation, whereas thermal methods include combustion techniques, pyrolysis process, liquefaction techniques, torrefaction method, and gasification process. This is to be noted that biochar materials recovered or produced from thermal methods or techniques have higher energy values and densities than material recovered from biological method (Mohan et al., 2014; Liu et al., 2015; Anupam et al., 2016).

Among various thermal methods for the production of biochar, pyrolysis is a process in which thermal decomposition of biomass material takes place in the absence of oxygen or comparatively with less oxygen than the normal combustion process. The three main products derived through pyrolysis of biomass materials include a liquid part called bio-oil, a solid product termed as biochar, and gaseous products such as carbon dioxide (CO_2), carbon monoxide (CO), and methane (CH_4) (Zhang et al., 2010; Sullivan and Ball, 2012). Considering the conditions of operation which include temperature, heating rate, pressure, and vapor residence time, the pyrolysis process has been further conducted in two major ways, i.e., slow pyrolysis and fast pyrolysis. The selection of the method depends on several factors including varying parameters during the process, availability of different categories of raw materials and their physicochemical properties, and ultimately on the performance of recovered biochar for designated purposes such as soil remediation, land fertility enhancement, and its potential as energy fuel in combustion (Antal and Gronli, 2003; Zhang et al., 2010). As a result, several challenges are being faced in biochar science and technology for any specific raw material using any pyrolysis technology and process conditions to predict and assure the quality of recovered biochar, their agronomic benefits, environmental effects, and co-benefits (Mašek et al., 2013; Ronsee et al., 2013; Hasan et al., 2021). Under slow pyrolysis process, the biomass raw material is heated in an oxygen-free or oxygen-limited environment with slower heating rates (1–30°C/min) and with relatively long vapor residence time and usually a lower temperature than fast pyrolysis, typically 400°C (Mohan et al., 2006; Durak, 2015). Under this slow pyrolysis process, the desired product is the solid fraction called char (biochar) and also, the accompanying product liquid and gaseous materials (Uddin et al., 2014; Rajamohan and Kasimani, 2018). Whereas, during the fast pyrolysis process, the liquid fraction production called bio-oil is favored more and the process is characterized by high heating rates and short vapor residence time (Czernik and Bridgwater, 2004; Montoya et al., 2015).

To produce high-quality biochar for the objective of use in agricultural purposes and use for a co-combustion process for obtaining energy with reliable and consistent product qualities, the process of slow pyrolysis is normally considered as the most feasible production technique (Song and Guo, 2012; Wang et al., 2020). Looking into Indian perspectives in growing concerns for energy demand, the waste-derived biochar can be a viable option in using as a co-mixture with fuel coal in many combustion processes. On one more note, the reason why there is a necessity to shift of energy supply from fossil to renewable sources is because of the excessive use of fossil fuel energy that may be exhausted in the future. Such problematic future prediction is a matter of concern for the country. One of the advantages of the utilization of biochar lies in the fact that the application of biochar as a co-mixture with coal can be used during the combustion process for heat and power generation. As a result, a substantial reduction in greenhouse gas (CO_2) emissions can be achieved and the heating value of low-grade fuel coal will be enhanced through co-mixture during the process of combustion (Woolf et al., 2010; Yousaf et al., 2017; Wang et al., 2019).

Although research on various aspects of biochar production from solid waste and its application in the energy sector is increasing in developed and emerging countries, such viable option studies are poorly understood in the Indian context. To fill the gaps in the literature, this study is the first of its kind that provides results from experimental investigations to assess the quality of waste-derived biochar recovered from two wood-based waste samples. These waste samples include woodchips and sawdust available in the local market. The specific objectives of this book chapter are provided as follows: (1) presentation of procedures of the method to produce wood-based waste-derived biochar using slow pyrolysis process at six different temperatures of 200°C, 250°C, 300°C, 350°C, 400°C, and 450°C, (2) method of assessment for physicochemical characterization of both raw materials and recovered biochar to arrive at an optimum condition of operating temperature, and selection of the better raw material out of two feedstock (woodchip and sawdust) for further analysis, and (3) presentation of preliminary examination to look at the feasibility of the use of biochar as a co-mixture with coal for firing in combustion process through the determination of the energy values and other properties at different proportions.

4.2 MATERIALS AND METHODS

4.2.1 Sample Collection and Preparation

The first step for the preparation of biochar was the selection of precursor or raw material. We selected two raw materials, i.e., woodchip and sawdust. These two raw materials were the solid wastes generated from the woodwork and industry in the local market. As explained in the aforementioned section, the primary objective of this study was to assess the potential of the production of biochar from possible wastes generated from wood-based manufacturing units. The comparative examinations were performed to check on which feedstock would be more effective in terms of the production of biochar. The two types of solid waste-based raw materials (woodchip and sawdust) were collected from the local market situated at Dadri, Greater Noida, Uttar Pradesh, India. The woodchip material was collected from the waste generated from a furniture shop that manufactures plywood sheets, and this manufacturing unit uses Marandi and Poplar plant wood (local names) for producing the same. The sawdust material was collected from the wastes of a shop that produces wood sticks which are used for the cremation of human bodies after death, and these wastes are generated from chipping of the Neem and Jamun plant wood (local names). The crucible meant for the preparation of samples was thoroughly cleaned with soap solution followed by acetone solution and was dried properly using a blower in the laboratory. The acetone cleaning was necessary to remove the background water content present in the crucible (Janković et al., 2019; Miliotti et al., 2020). The collected raw materials of woodchip and sawdust were cleaned to make them fit for usable purposes for conducting planned experiments meant for this study. The raw materials were chopped into smaller sizes using a cutting mill and passed over a 2 mm screen (Yargicoglu et al., 2015; Gabhi et al., 2020). These materials were prepared in batches of 25 g each followed by air drying for a substantial period. After oven drying, the samples were kept in humidity-controlled silica gel-containing desiccators for cooling and conditioning purposes, and the samples were then sent for further production of biochar (Yousaf et al., 2017; Wang et al., 2019).

4.2.2 CHARACTERISTICS OF WOOD WASTES

The physical characteristics of raw materials from both woodchip and sawdust were undergone through proximate analysis following methods and procedures recommended by American Society for Testing and Materials (ASTM) standards, S1762-84 (ASTM, 2007; Ronsse et al., 2013; Aller et al., 2017). Under the scope of this analytical category, moisture content, ash content, volatile content, and fixed carbon were determined. These parameters were determined for both raw materials and recovered biochar after their production. The procedures for the determination of these physical properties are summarized as follows. A substantial quantity (40 g) of air-dried sample was heated in a porcelain crucible and the weight difference before and after heating was considered for calculation of the particular parameter for each of the samples. A specific sample was dried at 110°C for 2 hours in an oven for the moisture content test, and the readings were taken accordingly. For the determination of the volatile matter, the sample was heated to 950°C for 10 min with a covered crucible in a muffle furnace and respective readings were recorded. To determine ash content, the sample was heated to 750°C for 3–4 hours with uncovered crucible in a muffle furnace and respective results were observed for further calculation.

The fixed carbon is the measure of the amount of non-volatile carbon remaining in a material sample. This is the carbon residue that remains after heating the specimen in a prescribed manner to decompose thermally unstable components and distill volatiles. The weight of the original sample considered for proximate analysis, subtracted by its volatile matter content on a dry-ash basis (considering the ash content) corresponds to fixed carbon (Liu et al., 2012; Aller et al., 2017). For proximate analysis, all properties of woodchip and sawdust samples and their recovered biochar were determined three times and the average of every run is reported in this book chapter.

In this study, we expressed the fixed carbon based on the dry-ash basis, as described in the following equation (Ronsse et al., 2013):

$$M_{fc}\,(\%) = \frac{M_{dry} - M_{vm} - M_{ash}}{M_{dry} - M_{ash}} \times 100 \qquad (4.1)$$

where M_{fc} represents a fraction of fixed carbon as wt.%, M_{dry} represents the oven-dry weight of the sample (g), M_{vm} represents the weight of volatile matter in the sample (g), and M_{ash} weight of the ash residue of the sample (g).

4.2.3   METHOD FOR PREPARATION OF BIOCHAR PRODUCTION

The instrument used for the preparation of biochar was a fixed bed reactor. The schematic diagram of the experimental setup has been shown in Figure 4.1 and the slow pyrolysis process was conducted in a vertical, tubular, and high temperature-resistant reactor which was heated by an electric furnace in a controlled manner. The biochar was produced using the process of slow pyrolysis which involved heating the feedstock in an inert atmosphere at relatively low temperature (250–450°C) keeping a residence time (10 min) for producing biochar (Parke et al., 2014; Rajamohan et al., 2018; Wang et al., 2019). Initially, the required operating temperature was set with the help of a temperature controller and then nitrogen gas was allowed to pass through preheater and reactor to maintain an inert atmosphere. The instrument was calibrated to maintain a heating rate of 10°C/min. As soon as the required operating temperature for running the experiment was attained, the feedstock of either of the raw materials (woodchip or sawdust) weighing 20 g was put into the hooper and with the help of an iron stick it was placed into the reactor, and the feeding time lasted for 60 s. After that, the residence time of 10 min was taken into account, and with the help of a stopwatch, the experiment was carefully conducted accordingly. The oil produced was collected in respective flasks and later stored in vials after recording the weights. After the residence time was maintained, the reactor was opened and the removal time was approximately 60 s. The waste-derived recovered biochar material obtained from the reactor was allowed to cool in a desiccator and the weight of the biochar was taken.

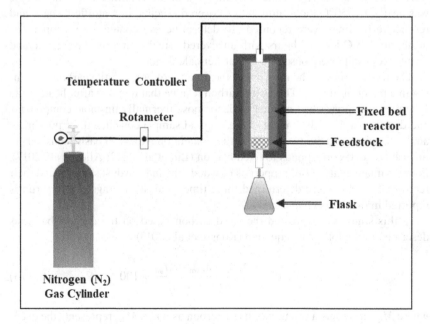

FIGURE 4.1   Schematic diagram of steps of the experimental setup used for the production of waste-derived biochar.

The above procedures were repeated for each of the feedstock samples (wood-chip and sawdust) at six different temperatures of 200°C, 250°C, 300°C, 350°C, 400°C, and 450°C. Every experimental test was conducted three times to obtain better reproducibility of the results and to collect a sufficient amount of waste-derived biochar product for further post-pyrolysis analyses and interpretation of results. The mass yield of recovered biochar was estimated as a percentage by comparing the obtained weight of biochar with the original dry weight of the spe-cific sample. In other words, the yield was expressed as a weight percentage of dry ash-free biochar recovered to dry ash-free initial biomass material. In this study, a dry ash-free (d_{af}) basis for estimation of yield was considered to avoid positive bias in yield in case of using biomass samples with a high mineral (ash) content (Ronsse et al., 2013; Menya et al., 2018; Martinez et al., 2021). The biochar yield (wt.%) was estimated using the following expression:

$$\eta(\%) = \frac{M_c - M_{ash,c}}{M_{dry,b} - M_{ash,b}} \times 100 \tag{4.2}$$

where η represents the yield of recovered char (biochar) from pyrolysis process in %, M_c represents the weight of char (biochar) recovered from the pyrolysis reactor (g) after which the pyrolysis process was considered to be oven-dry, $M_{dry,b}$ represents oven-dry weight of raw biomass (raw) material (g), and $M_{ash,c}$ and $M_{ash,b}$ represent the respective ash contents (i.e., weight in g) in char (biochar) and bio-mass (raw) material.

The yield of biochar and bio-oil were calculated by measuring the weight using the weighing balance. However, to determine the yield of the non-condensable gases (NCG), a mass balance equation was used, expressed as follows (Naqvi et al., 2014; Mazlan et al., 2015):

$$\text{Biochar}(\text{wt.\%}) + \text{Bio-oil}(\text{wt.\%}) + \text{NCG}(\text{wt.\%}) = 100(\text{wt.\%}) \tag{4.3}$$

4.2.4 Analysis of Pyrolyzed Biochar

The higher heating value (HHV) is an important parameter that defines the energy content and efficiency of a fuel required for the combustion process. To perform the ultimate analysis of the raw material and recovered biochar, a mathemati-cal correlation has been used from the previously reported studies (Parikh et al., 2005; Sheth and Babu, 2009). The reason for choosing this correlation expression was to have a minimum error in the computation of HHV which is 3.74% and its error is less than the errors given by other correlations. The following equation was used for the computation of HHV in the unit of MJ/kg:

$$\text{HHV} = 0.3536 \times \text{FC} + 0.1559 \times \text{VM} - 0.0078 \times \text{Ash} \tag{4.4}$$

where HHV represents the HHV of a sample in the unit of MJ/kg, FC represents the fixed carbon in %, VM represents volatile matter in %, and Ash represents ash content in %.

For the recovered biochar from the pyrolysis process at six different temperature conditions, the proximate analysis was conducted again for each of the five set of samples along with HHV calculation. In the end, biochar with the highest calorific value obtained at 400°C was selected for further characterization and analysis along with the comparative study with selected raw materials of wood-chip waste and sawdust waste.

To obtain insight into the surface morphology or topographic characteristics, the samples were characterized through SEM (scanning electron microscopy) and the instrument which was used was SEMEVO18 by ZEISS. For studying the thermal stability of the raw material and recovered biochar obtained at 400°C, thermal gravimetric analysis was conducted. The instrument used for the analysis was Shimadzu DTG60. The experiment was conducted in an inert condition of nitrogen atmosphere with a flow rate of 50 ml/min maintaining a temperature rate of 10°C/min. The CHNOS analysis was performed to get an idea about the elemental composition of the biochar samples. These related analytical parameters are the crucial factors in affecting the mechanism of the combustion process and influence further for patterns of emission of pollutants to the atmosphere. This ultimate analysis was carried out in two phases, while the CHNOS analysis was conducted with the help of an instrument, Thermo Scientific, FLASH EA 1112 Series, and the oxygen estimation was carried using the instrument using the mass balance expression.

After waste-derived biochar was recovered and characterized, the samples of both woodchip-derived biochar and sawdust-derived biochar obtained at 400°C were undergone for further analysis to assess the suitability of biochar in the process of co-mixture with conventional fuel coal. Under this scenario, the blends of biochar at 30%, 50%, and 70% were co-mixed with thermal combustible fuel coal and proximate analysis was conducted with an estimation of HHVs. The examinations were done with coal and three co-mixtures of (i) 30% coal and 70% biochar, (ii) 50% coal and 50% biochar, and (iii) 70% coal and 30% biochar. The coal taken for analysis was obtained from the local market at Dadri, Greater Noida, Uttar Pradesh, India. According to the grades of coal as specified in the Coal Directory of India, 2011–12 Coal Statistics, the coal obtained had the grade of G11–G12. The grade of the coal was determined with the help of proximate analysis parameters and the conversion formulae as mentioned in the Directory.

4.3 RESULTS AND DISCUSSION

4.3.1 CHARACTERISTICS OF WASTE-BASED RAW MATERIALS

Following the procedures explained in the section of materials and methods, proximate analytical parameters (moisture content, volatile matter content, ash content, and fixed carbon), heating value in terms of HHV with a unit of MJ/kg, and ultimate analytical parameters (carbon, hydrogen, nitrogen, sulfur, and oxygen) were determined for both raw materials of woodchip and sawdust wastes. Figure 4.2 shows the results of proximate and ultimate analyses with HHVs of woodchip

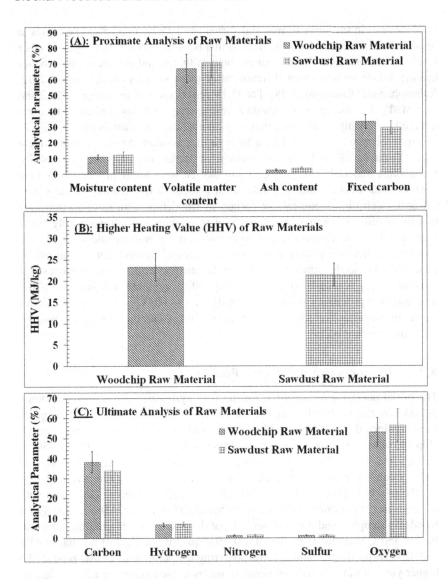

FIGURE 4.2 Results of woodchip and sawdust raw materials from (A) proximate analysis, (B) higher heating values, and (C) ultimate analysis.

and sawdust raw materials. The woodchip material had lower moisture content (varied from 9.0% to 12.3%) than sawdust materials (varied from 10.3% to 14.1%). These results showed a significant difference between these two samples in terms of existing moisture content. The volatile matter content in woodchip and sawdust materials was observed in the range of 57.4% to 76.1%, and 62.3% to 79.8%, respectively. The ash content in the raw material of the woodchip was less than that of sawdust. The fixed carbon in woodchip raw material varied from 28.7%

to 37.9%, and in sawdust raw material varied from 25.2% to 33.5%. The trends in results of the proximate analysis showed that both these raw materials had higher levels of moisture and volatile matter contents that would obstruct to have better heating values without thermal treatment of these samples (Enders et al., 2012; Rajamohan and Kasimani, 2018). The HHV results showed an average of 23.2 and 21.4 MJ/kg for woodchip and sawdust, respectively, and this confirmed that one material (woodchip) is advantageous over another material (sawdust).

From the results of the ultimate analysis, it was observed that carbon content varied from 32.9% to 43.7% for woodchip material and from 28.3% to 38.7% for sawdust material. The hydrogen and oxygen were lower in woodchips than sawdust, and similarly nitrogen and sulfur present in woodchips were less than sawdust. Such higher contents of carbon in woodchips than sawdust indicated that the quality of combusting property of woodchips was comparatively better than sawdust. And, the presence of more nitrogen and sulfur contents in sawdust can generate relatively more quantities of oxides of nitrogen (NO_x) and oxides of sulfur (SO_x) to the atmosphere during the burning or combustion process for possible energy generation (Guo and Zhong, 2018; Nyashina and Strizhak, 2018). The results of proximate and ultimate analyses, and HHV values went in favor of woodchip showing it as a better potential fuel fitting into the combustion process compared to sawdust.

4.3.2 PRODUCTS OF THE PYROLYSIS PROCESS

The initial operating temperature for the slow pyrolysis process was considered as 200°C in this study. The three product streams, i.e., biochar, bio-oil, and gases were obtained during this production process for both feedstock samples (woodchip and sawdust) at six different temperatures of 200°C, 250°C, 300°C, 350°C, 400°C, and 450°C. Figure 4.3 shows the percentages of biochar, bio-oil, and gases produced from these two raw materials at six different temperatures during the pyrolysis process. The production percentage of biochar decreased with increase in temperature, specifically 50.5% at 200°C, and 34.0% at 450°C for woodchip samples, and it was observed that the process at an operating temperature of 350°C and 400°C produced a similar quantity of yield of biochar at 43% and 42%, respectively. Although the operating temperature of 200°C produced a higher yield of biochar, this temperature was not able to produce a better quality of carbonized material which would provide higher calorific value (Song and Guo, 2012; Liu et al., 2013). In other words, the generated biochar should have more HHV value at any specific operating temperature in order to obtain more carbonized biomass. The percentage of production of bio-oil for woodchip material increased with the rise in temperature with 22.0% at 200°C, and 31.3% at 450°C. Similarly, the production of gases increased with the rise in temperature with 27.5% at 200°C, and 34.7% at 450°C. Considering the composition of all these products, the characteristics of the slow pyrolysis process were more defined at 350°C and 400°C with less bio-oil and gas production at 25.8% and 26.7%, and 31.2% and 31.5%, respectively, compared to an operating temperature of 450°C.

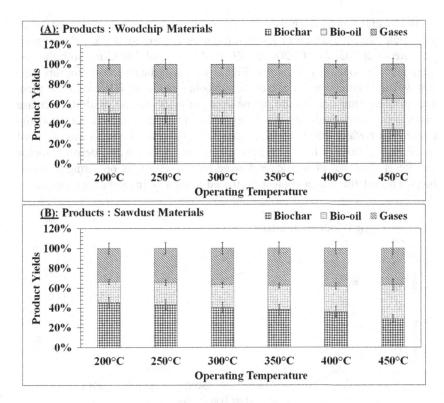

FIGURE 4.3 Products at different operating temperatures during the generation of biochar from (A) woodchip-derived material, and (B) sawdust-derived raw material.

It should be noted that a sharp increase in the generation of bio-oil and gases with a production yield of 31.3% and 34.7%, respectively, was observed at an operating temperature of 450°C (Manyà, 2012; Foong et al., 2020). Therefore, the operating temperature of 400°C for biochar production from woodchip waste had more advantages over other operating temperatures in the perspective of optimizing the best product quality of biochar (this assessment parameter was checked more at later interpretation).

The trends in production yields of biochar, bio-oil, and gases were observed for sawdust material, which was similar to woodchip material. Specifically, a decrease in biochar yield with the rise in operating temperature and increase in yields of bio-oil and gases with the temperature rise was observed during the production of sawdust-derived biochar. Comparing two waste-derived biochar products, the slow pyrolysis process generated 6% more woodchip-derived biochar than that of sawdust-derived biochar at an optimized operating temperature of 400°C. Additionally, this optimized condition of the production process generated a higher percentage of gases (38.3%) for sawdust material than that of woodchip (31.3%). Therefore, woodchip-derived biochar was advantageous than sawdust-derived biochar under the optimized operating temperature of 400°C.

To affirm the optimizing condition for operating temperature, proximate analysis, yield, and estimation of HHV value were conducted at six different temperature of 200°C, 250°C, 300°C, 350°C, 400°C, and 450°C for both the raw materials of woodchip and sawdust. Figure 4.4 shows the results of moisture content, volatile matter content, ash content, biochar yield, and HHV values at these six operating temperatures during the slow pyrolysis process. With the increase in operating temperature, the moisture content of biochar decreased for woodchip material by 6.8% at 200°C and 2.4% at 450°C, and a similar trend was observed in the case of sawdust biochar. The moisture content of sawdust-derived biochar was lower than that of woodchip-derived biochar at all six operating temperatures. The volatile matter content decreased in the rise in operating temperature

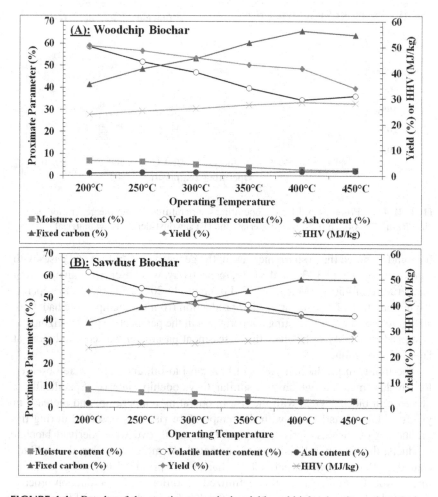

FIGURE 4.4 Results of the proximate analysis, yield, and higher heating values during the production of biochar at a different operating temperature from (A) woodchip raw material, and (B) sawdust raw material.

during the production of woodchip-derived biochar with 58.6% at 200°C and 36.2% at 450°C, and a similar trend was observed in the case of sawdust-derived biochar. The volatile matter content in sawdust-derived biochar at 400°C operating temperature was much higher (more than 12%) than that of woodchip-derived biochar indicating lower fixed carbon content (less carbonized biomass) in sawdust-derived biochar (Liu et al., 2012; Ronsse et al., 2013; Aller et al., 2017).

The ash content increased with a rise in temperature for woodchip-derived biochar with 1.2% at 200°C and 2.1% at 450°C, and a similar trend was observed in the case of sawdust-derived biochar. However, the pyrolysis process generated a significantly more percentage of ash content in sawdust-derived biochar than that in woodchip-derived biochar. Looking at the trend of fixed carbon present in woodchip-derived biochar at different temperatures, it was found that there was a constant increase of fixed carbon from 41.4% at 200°C to 65.6% at 400°C followed by an insignificant drop at 450°C. The sawdust-derived biochar showed a similar pattern on the variation of fixed carbon with operating temperature having an increasing trend up to 400°C and a slight drop at 450°C. Overall, the fixed carbon profile of woodchip-derived biochar showed better results than that of sawdust-derived recovered biochar, indicating that woodchip would fulfill better prospectus in terms of energy recovery (Demirbas, 2006; Park et al., 2014).

As explained in the aforementioned paragraph, the production yield of biochar has shown decreasing trend with a rise in operating temperature for both categories of biochar, however, the yield is significantly more for woodchip-derived biochar than sawdust-derived biochar at 400°C. The HHV value increased with a rise in operating temperature during the production of woodchip-derived biochar with 23.8 MJ/kg at 200°C and 28.5 MJ/kg at 400°C, and a similar trend was observed in the case of sawdust-derived biochar. There was a slight insignificant drop in HHV at 450°C compared to 400°C for woodchip-derived biochar. With a more cognizant view of three operating conditions of temperature at 350°C, 400°C, and 450°C, the quality of biochar produced at an operating temperature of 400°C was better in terms of higher carbon content (better-carbonized fuel), less volatile matter, less moisture content, higher yield, and higher HHV. Therefore, based on these results to achieve better carbonization of biomass in recovered biochar, we considered maintaining an operating temperature of 400°C for further tests, analysis, and interpretation of test results (Song and Guo, 2012; Yargicoglu et al., 2015; Guo and Zhong, 2018).

4.3.3　Proximate Analytical Results of Recovered Biochar

Figure 4.5 shows the results of proximate analysis, yield, and HHVs of waste-derived recovered biochar at an operating temperature of 400°C from both the raw materials (woodchip and sawdust). Comparing the results of waste-derived recovered biochar with its respective raw materials, it was observed that moisture content was reduced significantly from 10.8% to 2.7% in the case of woodchip material, and 12.2% to 3.8% for sawdust material. The contents of volatile materials were reduced substantially for both these biochar materials compared to their

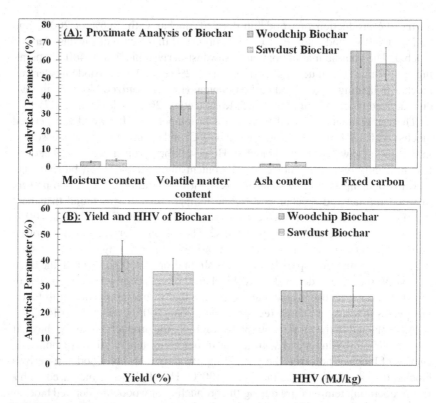

FIGURE 4.5   Results of (A) the proximate analysis, and (B) yield and higher heating values of waste-derived recovered biochar.

respective raw materials. As a result, fixed carbon content increased in recovered biochar that helped in achieving HHVs through carbonization of biomass during the slow pyrolysis process (Park et al., 2014). The substantial increase in values of HHV in biochar (22.1% to 28.5% for woodchip material and 21.4% to 26.5% for sawdust material) inferred that recovered biochar would be worth providing higher energy compared to their respective raw materials. The reduction in contents of volatile matter and ash contents in the process of slow pyrolysis supported our inference to obtain a better quality of recovered biochar material from these two wood-based raw materials.

Comparing proximate analytical parameters and heating values of woodchip-derived recovered biochar and sawdust-derived recovered biochar at an operating temperature of 400°C, woodchip-derived biochar was more advantageous in terms of better fuel quality with HHVs than sawdust-derived biochar. Fewer quantities of volatile matter (average of 34.4%) and ash content (average of 1.8%) were present in woodchip-derived recovered biochar than sawdust-derived biochar (42.6% and 2.9%, respectively), indicating that emissions of pollutants and generation of ash during possible combustion process would be comparatively lower in woodchip-derived recovered biochar (Woolf et al., 2010; Liu et at al., 2012; Ronsse et al.,

2013). More detailed explanations with reasons for such trends in both these waste-derived recovered biochar are provided in the previous section.

4.3.4 Ultimate Analytical Results of Recovered Biochar

Figure 4.6 shows results of ultimate analysis (contents of carbon, hydrogen, nitrogen, sulfur, and oxygen) of waste-derived recovered biochar at an operating temperature of 400°C from woodchip and sawdust materials. Comparing the results of raw materials with their respective recovered biochar, it was observed that carbon content increased from 38.2% (raw) to 69.6% (char) for woodchip material, and from 33.8% (raw) to 63.1% (char) for sawdust material. The contents of nitrogen and sulfur were substantially reduced in recovered biochar compared to their respective waste raw materials. The contents of oxygen and hydrogen were also reduced in recovered biochar compared to their respective waste raw materials. The presence of higher elements of pollutant origin (nitrogen and sulfur) in both the raw materials indicated that direct burning of these raw materials (without any thermal treatment) could cause larger emissions of pollutants like NO_x and SO_x to the atmosphere (Guo and Zhong, 2018; Nyashina and Strizhak, 2018). Comparing the results of ultimate analytical parameters of woodchip-derived biochar with sawdust-derived biochar, carbon content varied from 59.7% to 80.2% for woodchip-derived biochar, and from 53.5% to 72.4% for sawdust-derived biochar. Such higher contents of carbon in woodchip-derived biochar than sawdust-derived biochar indicated that the quality of combusting property of recovered biochar from woodchip was comparatively better than that of sawdust. Hydrogen content varied from 3.4% to 4.4% for woodchip-derived biochar, and from 3.8% to 5.2% for sawdust-derived biochar. The contents of nitrogen and sulfur were observed to be more in sawdust-derived biochar than the woodchip-derived biochar.

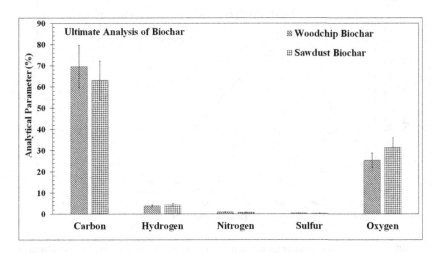

FIGURE 4.6   Results of ultimate analysis of waste-derived recovered biochar.

4.3.5   RESULTS OF TGA OF PREFERRED MATERIAL AND ITS BIOCHAR

As described and confirmed in the previous section that woodchip waste-derived recovered biochar produced at an operating temperature of 400°C had the best quality among all operating temperatures, we selected samples of raw material and recovered woodchip-derived biochar at 400°C to conduct TGA analysis to assess the thermal stability profiles of these two samples. Figure 4.7 shows TGA results of woodchip raw material and recovered biochar at 400°C in the N_2 atmosphere. The trends in the weight loss curve from the TGA profile of raw material revealed that the material was stable up to 250°C and could sustain this temperature with less amounts of mass loss (4%). After this temperature of 250°C, the weight loss started decreasing faster with 80% loss occurring at 400°C. The loss

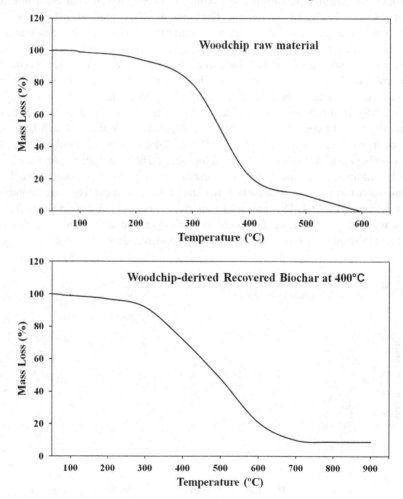

FIGURE 4.7   Thermal gravimetric analysis results of woodchip raw material and its recovered biochar at 400°C in N_2 atmosphere.

of weight of raw material could be attributed to the degradation and decomposition of the carbonaceous biomass to pollutant gases like CO_2, CO, and CH_4 (Sun et al., 2014; Nanda et al., 2016). Thus this trend showed that if the raw material is subjected to combustion without any carbonization of biomass using any thermal agitating technique, the material would emit higher quantities of gaseous pollutants to the atmosphere. In addition to the above trend, it was also assessed that decomposition of all materials finished around 550–600°C. Therefore, it could be inferred that the raw material can't sustain higher temperature and it became unstable after 400°C. Such similar trends of results of TGA profiles from wood-based raw materials have been reported in previous studies (Poletto et al., 2012; Chen et al., 2019; Ren et al., 2020).

Unlike trends of TGA profile of raw material, the woodchip-derived recovered biochar was more thermally stable compared to its raw material. Specifically, around 3% mass loss occurred until the temperature reached 300°C, and then after a gradual decrease in mass loss occurred till 600°C. After 600°C, the mass loss dropped from 80% to 90% at 800°C showing a steady trend. Thus trends in thermal profile showed that biochar can sustain more temperature compared to its raw material. Also, if the mass loss of material occurred at a higher temperature, the quantities of carbon-based gaseous pollutants emitted to the atmosphere would be less. This could be due to the reason that the carbonization process of fuel material would generate more energy rather than emission to the atmosphere (Sun et al., 2014; Zhao et al. 2017). To obtain better profiles and affirmative conclusions, more in-depth studies on thermal profile analysis are required to be conducted. However, these preliminary investigations revealed that waste-derived biochar recovered through a slow pyrolysis process at an operating temperature of 400°C was thermally more stable than its raw material.

4.3.6 Results of SEM of Preferred Material and Its Biochar

Figure 4.8 shows the images of SEM of woodchip raw material and its recovered biochar at 400°C to assess the morphological patterns of these materials. The shapes of images of raw material were observed as spacious aligned with honeycomb-like groups showing larger vacuum space available in this material. Such vacuum spaces were representative of the carbonaceous skeleton from the biological capillary structure of the raw material. Moreover, cross-linking patterns were observed, confirming the raw material type as hardwood material through SEM results (Kumar et al., 2006). The images of recovered biochar at both diameters showed that the structure was tightly packed with the presence of evenly distributed pores compared to that of the raw material. This could be possible because of the carbonization process that was supportive in filling the spacious vacuum with evenly distributed micro-pores, which were composed of condensed volatile matter and other decomposition products causing a reduction in porosity of recovered biochar (Bourke et al., 2007; Ghani et al., 2013). Overall, the SEM micrographs detected the development of well-defined pores of recovered biochar compared to raw material. In a comparative way

FIGURE 4.8 Images of scanning electron microscopy of woodchip raw material and its recovered biochar at 400°C.

of assessment, it was observed that the surface of raw material was dull and uneven, while the biochar surface had a smoother surface with ridges and wood texture. Such an assessment inferred that the recovered biochar material would be composed of less volatile matter compared to its raw material. Thus, from this preliminary assessment, it could be inferred that the quality of recovered biochar in terms of its suitability for the combustion process would be a better option than its raw material.

4.3.7 Characteristic Results of Co-Mixtures of Preferred Biochar and Coal

As explained in the section on materials and methods, co-mixture of fuel coal and woodchip-derived biochar (produced at an operating temperature of 400°C) were made at different ratios on a mass basis as 30:70, 50:50, and 70:30. Figure 4.9 presents the results of proximate and ultimate analyses of fuel coal and its mixtures with biochar. Under proximate analysis, it was observed that fuel coal had significantly higher volatile matter content (in the range of 45.8% to 60.3%) compared to woodchip-derived biochar. The fixed carbon content in woodchip-derived biochar was significantly higher than that of fuel coal. The heating value, HHV result showed woodchip-derived biochar had significantly better calorific values (~4%) than that of fuel coal. The ultimate analysis results showed that higher carbon contents in woodchip-derived biochar than that of fuel coal. Whereas, fuel coal had higher nitrogen and sulfur contents compared to woodchip-derived biochar. This indicated that during the combustion process fuel coal can emit more gaseous pollutants of NO_x and SO_x than that of woodchip-derived biochar. Such trends of results inferred that biochar could be a viable option for its use in

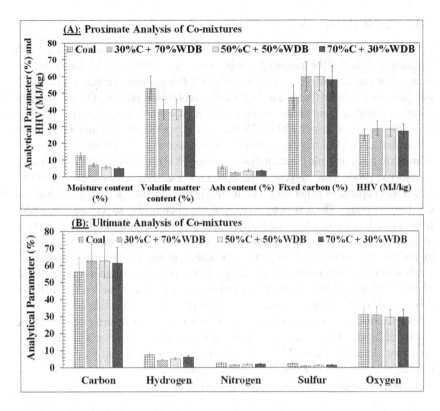

FIGURE 4.9 Physicochemical properties of coal and co-mixture of biochar at different ratios with results from (A) proximate analysis, and (B) ultimate analysis. (Abbreviations: C = coal; WDB = woodchip-derived biochar.)

the combustion process along with fuel coal to improve the calorific value and to reduce the rate of emission of pollutants to the atmosphere.

The analytical results showed that co-mixture fuel of 50% fuel coal and 50% woodchip-derived biochar gave better performance compared to co-mixture of 30% fuel coal and 70% woodchip-derived biochar, and of 70% fuel coal and 30% woodchip-derived biochar. Although the 30:70 mixture had provided slightly higher values of HHV, this was insignificant. Additionally, opting for such a ratio may not be possible due to the high demand in supply of biochar. Hence, it is better to keep the viable option for co-mixture fuel of 50% fuel coal and 50% woodchip-derived biochar in the combustion of fuel in related industrial utilities. From this preliminary investigation, it was inferred that the co-mixture concept of fuel coal and recovered biochar could work well in combustion processes to obtain a better quality of fuel and to reduce pollutant emissions to the atmosphere. Such similar trends or assessments have been reported by previous studies (Liu et al., 2012; Wang et al., 2019). To arrive at more confirmative inferences, more in-depth experimental tests and property analyses are required on a long-term basis.

4.4   CONCLUSIONS

This unique preliminary experimental laboratory-scale study was planned and conducted to assess the potential use of wood waste-derived biochar recovered from two waste samples (woodchip and sawdust raw materials) under a slow pyrolysis process for its further use as fuel in the combustion process. During the production of biochar from waste samples under slow pyrolysis process, five different operating temperatures of 200°C, 250°C, 300°C, 350°C, 400°C, and 450°C were considered to conduct the carbonization of biomass in waste samples. The results from proximate and ultimate analyses, values of HHV, and quantities of yields for biochar production suggested the optimum operating temperature as 400°C, which was capable of generating a better quality of biochar for both woodchip and sawdust materials. The assessing analytical parameters also suggested that woodchip-derived recovered biochar had advantages over sawdust-derived recovered biochar in terms of providing good quality of fuel with more yield of production, higher calorific values, and fewer pollutant components. The thermal profile study confirmed that woodchip-derived biochar had better stability than its raw material. The recovered biochar from the woodchip was co-mixed with fuel coal at different mass ratios to check the combustion potential of biochar using responsible assessing parameters. It was observed that the co-mixing process favored improving the quality of fuel coal in terms of generating more calorific values and fewer emission scenarios. Therefore, it could be concluded from this study that waste recovered biochar is having all favorable properties to make it fit for use as a co-mixture with fuel coal in the combustion process of industrial utilities.

REFERENCES

Agegnehu, G., Srivastava, A.K., Bird, M.I. 2017. The role of biochar and biochar-compost in improving soil quality and crop performance: A review. *Applied Soil Ecology*, 119:156–170.

Aller, D., Bakshi, S., Laird, D.A. 2017. Modified method for proximate analysis of biochars. *Journal of Analytical and Applied Pyrolysis*, 124: 335–342.

Antal, M.J., Grønli, M. 2003. The art, science, and technology of charcoal production. *Industrial & Engineering Chemistry Research*, 42(8): 1619–1640.

Anupam, K., Sharma, A.K., Lal, P.S., Dutta, S., Maity, S. 2016. Preparation, characterization and optimization for upgrading *Leucaena leucocephala* bark to biochar fuel with high energy yielding. *Energy*, 106: 743–756.

ASTM 2007. *D1762-84: Standard Method for Chemical Analysis of Wood Charcoal*. American Society for Testing and Materials International, West Conshohocken, PA.

Bourke, J., Manley-Harris, M., Fushimi, C., Dowaki, K., Nunoura, T., Antal, M.J. 2007. Do all carbonized charcoals have the same chemical structure? 2. A model of the chemical structure of carbonized charcoal. *Industrial & Engineering Chemistry Research*, 46(18): 5954–5967.

Chen, H., Forbes, E.G.A., Archer, J., De Priall, O., Allen, M., Johnston, C., Rooney, D. 2019. Production and characterization of granules from agricultural wastes and comparison of combustion and emission results with wood based fuels. *Fuel*, 256: 115897.

Czernik, S., Bridgwater, A.V. 2004. Overview of applications of biomass fast pyrolysis oil. *Energy & Fuels*, 18(2): 590–598.

Das, O., Bhattacharyya, D., Sarmah, A.K. 2016. Sustainable eco-composites obtained from waste derived biochar: A consideration in performance properties, production costs, and environmental impact. *Journal of Cleaner Production*, 129: 159–168.

Demirbas, A. 2006. Effect of temperature on pyrolysis products from four nut shells. *Journal of Analytical and Applied Pyrolysis*, 76(1–2): 285–289.

Dhar, H., Kumar, S., Kumar, R. 2017. A review on organic waste to energy systems in India. *Bioresource Technology*, 245: 1229–1237.

Durak, H. 2015. Thermochemical conversion of *Phellinus pomaceus* via supercritical fluid extraction and pyrolysis processes. *Energy Conversion and Management*, 99: 282–298.

Enders, A., Hanley, K., Whitman, T., Joseph, S., Lehmann, J. 2012. Characterization of biochars to evaluate recalcitrance and agronomic performance. *Bioresource Technology*, 114: 644–653.

Foong, S.Y., Liew, R.K., Yang, Y., Cheng, Y.W., Yek, P.N.Y., Mahari, W.A.W., Lee, X.Y., Han, C. S., Vo, D-V. N., Le, Q.V., Aghbashlo, M., Tabatabaei, M., Sonne, C., Peng, W., Lam, S.S. 2020. Valorization of biomass waste to engineered activated biochar by microwave pyrolysis: Progress, challenges, and future directions. *Chemical Engineering Journal*, 389: 124401.

Gabhi, R., Basile, L., Kirk, D.W., Giorcelli, M., Tagliaferro, A., Jia, C.Q. 2020. Electrical conductivity of wood biochar monoliths and its dependence on pyrolysis temperature. *Biochar*, 2(3): 369–378.

Ghani, W.A. W.A.K., Mohd, A., da Silva, G., Bachmann, R.T., Taufiq-Yap, Y.H., Rashid, U., Ala'a, H. 2013. Biochar production from waste rubber-wood-sawdust and its potential use in C sequestration: Chemical and physical characterization. *Industrial Crops and Products*, 44: 18–24.

Goswami, L., Manikandan, N.A., Taube, J.C.R., Pakshirajan, K., Pugazhenthi, G. 2019. Novel waste-derived biochar from biomass gasification effluent: Preparation, characterization, cost estimation, and application in polycyclic aromatic hydrocarbon biodegradation and lipid accumulation by *Rhodococcus opacus. Environmental Science and Pollution Research*, 26(24): 25154–25166.

Guo, F., Zhong, Z. 2018. Co-combustion of anthracite coal and wood pellets: Thermodynamic analysis, combustion efficiency, pollutant emissions and ash slagging. *Environmental Pollution*, 239: 21–29.

Hasan, M.M., Rasul, M.G., Khan, M.M.K., Ashwath, N., Jahirul, M.I. 2021. Energy recovery from municipal solid waste using pyrolysis technology: A review on current status and developments. *Renewable and Sustainable Energy Reviews*, 145: 111073.

Janković, B., Manić, N., Dodevski, V., Popović, J., Rusmirović, J.D., Tošić, M. 2019. Characterization analysis of Poplar fluff pyrolysis products. Multi-component kinetic study. *Fuel*, 238: 111–128.

Kumar, B. P., Shivakamy, K., Miranda, L.R., Velan, M. 2006. Preparation of steam activated carbon from rubberwood sawdust (*Hevea brasiliensis*) and its adsorption kinetics. *Journal of Hazardous Materials*, 136(3): 922–929.

Liu, W. J., Jiang, H., Yu, H. Q. 2015. Development of biochar-based functional materials: Toward a sustainable platform carbon material. *Chemical Reviews*, 115(22): 12251–12285.

Liu, Z., Quek, A., Hoekman, S.K., Balasubramanian, R. 2013. Production of solid biochar fuel from waste biomass by hydrothermal carbonization. *Fuel*, 103: 943–949.

Liu, Z., Quek, A., Hoekman, S.K., Srinivasan, M. P., Balasubramanian, R. 2012. Thermogravimetric investigation of hydrochar-lignite co-combustion. *Bioresource Technology*, 123: 646–652.

Lohri, C.R., Diener, S., Zabaleta, I., Mertenat, A., Zurbrügg, C. 2017. Treatment technologies for urban solid biowaste to create value products: A review with focus on low-and middle-income settings. *Reviews in Environmental Science and Bio/Technology*, 16(1): 81–130.

Madej, J., Hilber, I., Bucheli, T.D., Oleszczuk, P. 2016. Biochars with low polycyclic aromatic hydrocarbon concentrations achievable by pyrolysis under high carrier gas flows irrespective of oxygen content or feedstock. *Journal of Analytical and Applied Pyrolysis*, 122: 365–369.

Manyà, J.J. 2012. Pyrolysis for biochar purposes: A review to establish current knowledge gaps and research needs. *Environmental Science & Technology*, 46(15): 7939–7954.

Martinez, C.L.M., Sermyagina, E., Saari, J., de Jesus, M.S., Cardoso, M., de Almeida, G.M., Vakkilainen, E. 2021. Hydrothermal carbonization of lignocellulosic agroforest based biomass residues. *Biomass and Bioenergy*, 147: 106004.

Mašek, O., Brownsort, P., Cross, A., Sohi, S. 2013. Influence of production conditions on the yield and environmental stability of biochar. *Fuel*, 103: 151–155.

Mazlan, M.A.F., Uemura, Y., Osman, N.B., Yusup, S. 2015. Fast pyrolysis of hardwood residues using a fixed bed drop-type pyrolyzer. *Energy Conversion and Management*, 98: 208–214.

Menya, E., Olupot, P.W., Storz, H., Lubwama, M., Kiros, Y. 2018. Characterization and alkaline pretreatment of rice husk varieties in Uganda for potential utilization as precursors in the production of activated carbon and other value-added products. *Waste Management*, 81: 104–116.

Miliotti, E., Casini, D., Rosi, L., Lotti, G., Rizzo, A.M., Chiaramonti, D. 2020. Lab-scale pyrolysis and hydrothermal carbonization of biomass digestate: Characterization of solid products and compliance with biochar standards. *Biomass and Bioenergy*, 139: 105593.

Mohan, D., Pittman Jr, C.U., Steele, P.H. 2006. Pyrolysis of wood/biomass for bio-oil: A critical review. *Energy & Fuels*, 20(3): 848–889.

Mohan, D., Sarswat, A., Ok, Y.S., Pittman Jr, C.U. 2014. Organic and inorganic contaminants removal from water with biochar, a renewable, low cost and sustainable adsorbent – A critical review. *Bioresource Technology*, 160: 191–202.

Montoya, J.I., Valdés, C., Chejne, F., Gómez, C. A., Blanco, A., Marrugo, G., Osorio, J., Castillo, E., Aristóbulo, J., Acero, J. 2015. Bio-oil production from *Colombian bagasse* by fast pyrolysis in a fluidized bed: An experimental study. *Journal of Analytical and Applied Pyrolysis*, 112: 379–387.

Nanda, S., Dalai, A.K., Berruti, F., Kozinski, J.A. 2016. Biochar as an exceptional bioresource for energy, agronomy, carbon sequestration, activated carbon and specialty materials. *Waste and Biomass Valorization*, 7(2): 201–235.

Naqvi, S.R., Uemura, Y., Yusup, S.B. 2014. Catalytic pyrolysis of paddy husk in a drop type pyrolyzer for bio-oil production: The role of temperature and catalyst. *Journal of Analytical and Applied Pyrolysis*, 106: 57–62.

Nyashina, G., Strizhak, P. 2018. Impact of forest fuels on gas emissions in coal slurry fuel combustion. *Energies*, 11(9): 2491.

Parikh, J., Channiwala, S.A., Ghosal, G.K. 2005. A correlation for calculating HHV from proximate analysis of solid fuels. *Fuel*, 84(5): 487–494.

Park, J., Lee, Y., Ryu, C., Park, Y.K. 2014. Slow pyrolysis of rice straw: Analysis of products properties, carbon and energy yields. *Bioresource Technology*, 155: 63–70.

Parshetti, G. K., Quek, A., Betha, R., Balasubramanian, R. 2014. TGA–FTIR investigation of co-combustion characteristics of blends of hydrothermally carbonized oil palm biomass (EFB) and coal. *Fuel Processing Technology*, 118: 228–234.

Poletto, M., Zattera, A.J., Santana, R.M. 2012. Structural differences between wood species: Evidence from chemical composition, FTIR spectroscopy, and thermogravimetric analysis. *Journal of Applied Polymer Science*, 126(S1): E337–E344.

Qin, C., Wang, H., Yuan, X., Xiong, T., Zhang, J., Zhang, J. 2020. Understanding structure-performance correlation of biochar materials in environmental remediation and electrochemical devices. *Chemical Engineering Journal*, 382: 122977.

Rajamohan, S., Kasimani, R. 2018. Analytical characterization of products obtained from slow pyrolysis of *Calophyllum inophyllum* seed cake: Study on performance and emission characteristics of direct injection diesel engine fuelled with bio-oil blends. *Environmental Science and Pollution Research*, 25(10): 9523–9538.

Ren, X., Guo, J., Li, S., Chang, J. 2020. Thermogravimetric analysis–Fourier transform infrared spectroscopy study on the effect of extraction pretreatment on the pyrolysis properties of eucalyptus wood waste. *ACS Omega*, 5(36): 23364–23371.

Ronsse, F., Van Hecke, S., Dickinson, D., Prins, W. 2013. Production and characterization of slow pyrolysis biochar: Influence of feedstock type and pyrolysis conditions. *GCB Bioenergy*, 5(2): 104–115.

Sheth, P.N., Babu, B.V. 2009. Experimental studies on producer gas generation from wood waste in a downdraft biomass gasifier. *Bioresource Technology*, 100(12): 3127–3133.

Singh, S., Behera, S.N. 2019. Development of GIS-based optimization method for selection of transportation routes in municipal solid waste management. In *Advances in Waste Management* (pp. 319–331). Springer, Singapore.

Sohi, S.P., Krull, E., Lopez-Capel, E., Bol, R. 2010. A review of biochar and its use and function in soil. *Advances in Agronomy*, 105: 47–82.

Song, W., Guo, M. 2012. Quality variations of poultry litter biochar generated at different pyrolysis temperatures. *Journal of Analytical and Applied Pyrolysis*, 94: 138–145.

Sullivan, A.L., Ball, R. 2012. Thermal decomposition and combustion chemistry of cellulosic biomass. *Atmospheric Environment*, 47: 133–141.

Sun, Y., Gao, B., Yao, Y., Fang, J., Zhang, M., Zhou, Y., Chen, H., Yang, L. 2014. Effects of feedstock type, production method, and pyrolysis temperature on biochar and hydrochar properties. *Chemical Engineering Journal*, 240: 574–578.

Uddin, M.N., Daud, W.W., Abbas, H.F. 2014. Effects of pyrolysis parameters on hydrogen formations from biomass: A review. *RSC Advances*, 4(21): 10467–10490.

Wang, Q., Lai, Z., Mu, J., Chu, D., Zang, X. 2020. Converting industrial waste cork to biochar as Cu (II) adsorbent via slow pyrolysis. *Waste Management*, 105: 102–109.

Wang, W., Wen, C., Li, C., Wang, M., Li, X., Zhou, Y., Gong, X. 2019. Emission reduction of particulate matter from the combustion of biochar via thermal pre-treatment of torrefaction, slow pyrolysis or hydrothermal carbonisation and its co-combustion with pulverized coal. *Fuel*, 240: 278–288.

Woolf, D., Amonette, J.E., Street-Perrott, F.A., Lehmann, J., Joseph, S. 2010. Sustainable biochar to mitigate global climate change. *Nature Communications*, 1(1): 1–9.

Yadav, A., Behera, S.N., Nagar, P.K., Sharma, M. 2020. Spatio-seasonal concentrations, source apportionment and assessment of associated human health risks of PM2.5-bound polycyclic aromatic hydrocarbons in Delhi, India. *Aerosol and Air Quality Research*, 20: 2805–2825.

Yargicoglu, E.N., Sadasivam, B.Y., Reddy, K.R., Spokas, K. 2015. Physical and chemical characterization of waste wood derived biochars. *Waste Management*, 36: 256–268.

Yousaf, B., Liu, G., Abbas, Q., Wang, R., Ali, M.U., Ullah, H., Liu, R., Zhou, C. 2017. Systematic investigation on combustion characteristics and emission-reduction mechanism of potentially toxic elements in biomass-and biochar-coal co-combustion systems. *Applied Energy*, 208: 142–157.

Zhang, L., Xu, C.C., Champagne, P. 2010. Overview of recent advances in thermo-chemical conversion of biomass. *Energy Conversion and Management*, 51(5): 969–982.

Zhao, S.X., Ta, N., Wang, X.D. 2017. Effect of temperature on the structural and physicochemical properties of biochar with apple tree branches as feedstock material. *Energies*, 10(9): 1293.

5 Bio-Medical Waste Management

Need, Handling Rules, and Current Treatment Technologies

Geetanjali Rajhans,[1] Adyasa Barik,[1]
Sudip Kumar Sen,[2] and Sangeeta Raut[1]
[1]Center for Biotechnology, School of Pharmaceutical
Sciences, Siksha O Anusandhan (Deemed to be
University), Bhubaneswar, Odisha, India
[2]Biostadt India Limited, Waluj,
Aurangabad, Maharashtra, India

CONTENTS

DOI: 10.1201/9781003201076-5

5.1   INTRODUCTION

In the unprecedented crisis of COVID-19, the security of life and livelihoods has become the central decision and policy of governments at all levels. The tremendous challenge for the healthcare community lies in its need to cope with existing facilities and insufficient protective resources for patients in need of emergency care. The approach to address this public health crisis has so far been to update medical standards, major testing projects, and recalibration of public policy (WHO 2020). Although the planet has seen the optimistic environmental effects of COVID-19 during national lockdowns, like cleaner waterways and clearer skies (Gardiner 2020), the management of bio-medical waste (BMW) has become a menace. The dynamics of waste generation has changed, triggering woe to policy makers and staff engaged in disinfection and sanitization (Mallapur 2020). During an epidemic, several kinds of bio-medical and harmful waste are created, which include contaminated gloves, masks as well as other protective equipment and a large amount of non-infected products of similar type (UNEP 2020). Inappropriate collection activities may result in infection with the virus in general municipal waste, which could create a transmission risk.

Contagious wastes are not restricted to hospitals and health centers, since people with mild or asymptomatic symptoms often produce virus-loaded wastes, because a virus can survive on cardboard, plastic, and metal for hours or even weeks (Kampf et al. 2020; van Doremalen et al. 2020), it can threaten the life of waste managing staff when tossed or dumped indiscriminately. The situation can deteriorate even further in developing countries where BMW management staff lack proper personal protective equipment (PPE).

Ensuring minimum health and safety threats in the collection, transport, and recycling of waste has also been a demanding endeavor in many developed countries. Other limiting factors are the lack of technological expertise and other scientific and economic resources to manage BMW in a developing nation. Besides the interpretation of existing global practices in the management of BMW, this chapter also aims to identify new approaches to the emerging threats of the current crisis, while at the same time proposing viable improvements to conventional procedures in order to prevent and solve similar problems during future pandemics.

5.2 SOURCES, IDENTIFICATION, AND CLASSIFICATION OF BMW

Though urban solid waste has drawn interest of urban policy makers, environmental advocates along with civil servants, there is still a need of concern about certain unique waste sources and their management. The wastes from hospitals, clinics, general practitioners, dispensaries, nursing homes, animal houses, veterinary hospitals, blood banks and research institutes come under the category of BMW. Households, factories, educational center, research facilities, blood banks, pharmaceutical labs and healthcare services are other sources of BMW (Anonymous 2000; Chitnis et al. 2000). The various sources of BMW have been listed in Figure 5.1.

BMW identification is the first step toward proper segregation of waste. In the early 1990s, laws were established concerning the detection and processes involved in the segregation, processing/treatment and disposal of such waste. Although 10% are infectious and 5% are non-infectious, yet they have been considered as toxic waste. WHO reported that 85% of hospital wastes are not dangerous and roughly 15–35% of hospital waste is identified as contagious. Approximately, 2 kg of BMW is generated per bed per day in India, which includes anatomical waste, cytotoxic waste, and sharps which could trigger different forms of lethal diseases if not carefully handled, and additionally cause earth disruptions and has antagonistic effects on biological parity. Approximately 600 g of untreated/infective waste would be generated at a rate of 5–10 kg/day per bed in a general hospital (Nath et al. 2010). The net production of BMW in India is 405,702 kg/day, of which only 291,983 kg/day is disposed of, as indicated by the Ministry of Environment and Forest (MoEF). As a result, roughly 28% of

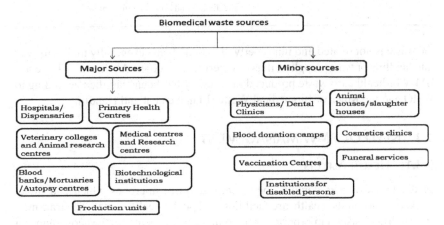

FIGURE 5.1 Biomedical waste sources.

TABLE 5.1

Bio-Medical Waste Categories and Components

Waste Category	Waste Type	Components
Category No. 1	Human anatomical waste	Human organs, tissues, body parts
Category No. 2	Animal waste	Animal organs, tissues body parts, bleeding parts, carcasses, blood, fluid, and experimental animals used in research, waste generated from animal houses, veterinary hospitals, etc.
Category No. 3	Biotechnology and Microbiology wastes	Waste generated in laboratory cultures, stocks or specimen of microorganisms, live or attenuated vaccines, human and animal cell cultures used in research and infectious agents and industrial labs, wastes from production of biological, from research toxins, dishes and devices used for transfer of culture
Category No. 4	Sharp wastes	Needles, scalpels, syringes, glass, blades, etc. which might cause cuts and puncture. Sharps, both used and new, are included
Category No. 5	Discarded medicines and cytotoxic drugs	Wastes comprising of discarded, contaminated and outdated medical drugs
Category No. 6	Solid wastes	Cotton, dressings, soiled plaster casts, lines, beddings, and other items polluted with body fluids and blood
Category No. 7	Solid wastes	Disposable products other than waste sharps, such as tubings, catheters, intravenous sets, etc.
Category No. 8	Liquid wastes	Waste produced in the laboratories, as well as activities such as washing, cleaning, housekeeping, and disinfection
Category No. 9	Incineration ash	Any bio-medical waste incineration ash
Category No. 10	Chemical wastes	Chemicals used in biological processing, as well as insecticides and other disinfectants

the waste is not treated and improperly disposed of, and eventually makes its way into landfills or water supplies, thus re-entering our bodies. As shown in Table 5.1, BMW or healthcare waste produced in a facility has been classified according to WHO and the Rules of Biomedical Waste (Management and Handling).

5.3 NEED OF BMW MANAGEMENT

5.3.1 Health Aspects

Medical care wastes contain competently harmful microbes which can infect workers at hospitals, healthcare facilities, and public health center. The transmission of drug-resistant microorganisms from the health center to the ecosystem can also include other possible infectious threats.

5.3.2 Ethical Aspects

Healthcare professionals have a moral responsibility to keep hospitals from being centers of sickness rather than centers of cure (Prüss et al. 1999). Public health awareness and environmental risks linked to improper segregation, waste assortment, storage, transportation, handling, treatment and disposal is necessary. This is necessary for all healthcare sectors to have a routine training program, with a particular focus on waste handlers (Kumar and Duggal 2015).

5.3.3 Environmental Impacts

The improper management and disposal of waste from healthcare units can root for indirect health threats by releasing pathogens and toxic pollutants into the atmosphere.

5.3.4 Schedules under BMW (Handling Rules)

The 2016 BMW management rules include 4 schedules, 5 forms, and 18 rules that provide information on applications; definitions; operator and authority responsibilities; waste segregation, packaging, transport, and storage; treatment and disposal standards; record and responsibilities of the approved officials; authorization process; advisory commission; and tracking of the enforcement of the rules in the BMW Management Rules (CPCB 2016).

The schedules include:

- Bio-medical waste management categories, segregation, color coding, collection, treatment and disposal – Schedule I (Table 5.2)
  Table 5.2 represents the Schedule I of BMW management which includes the categories of BMW, its segregation methods, color coding, collection, treatment and disposal.
- Standards for treatment and disposal – Schedule II (Table 5.3)
  Table 5.3 represents the Schedule II of BMW management which illustrates the standards for treatment and disposal of BMW.
- Prescribed authorities and responsibilities – Schedule III (Table 5.4)
  Table 5.4 represents the Schedule III of BMW management which lists the concerned authorities
- Labels for BMW containers and bags – Schedule IV (Figure 5.2)

The forms include:

- Accidental reporting – Form I
- Application and authorization documents – Forms II and III
- Annual reporting – Form IV (Figure 5.2)
- Appeal – Form V

5.3.4.1 Yearly Report

Send a report on the categories and the amount of BMW generated to the designated authority, annually, in Form IV prior to 30 June every year.

TABLE 5.2

Schedule I – BMW Management Categories, Segregation, Color Coding, Collection, Treatment, and Disposal

Category	Waste Type	Type of Bag or Container to Be Used	Disposal or Treatment Options
Yellow	• Human anatomical waste • Animal anatomical waste • Soiled waste • Expired or discarded medicines • Chemical wastes • Discarded linen, mattresses, beddings, contaminated with blood or body fluid • Microbiology, biotechnology and other clinical laboratory waste	Yellow-colored non-chlorinated plastic bags or containers	Incineration or plasma pyrolysis or deep burial
	Chemical liquid waste	Separate collecting unit for the disposal of wastewater	Following resource recovery, chemical liquid waste must be pretreated before being mixed with wastewater, and the combined discharge must meet discharging requirements
Red	Contaminated waste(recyclable) tubing, bottles, intravenous tubes and sets, catheters, urine bags, syringes (without needles) and vacutainers (with their needles cut) and gloves	Red-colored non-chlorinated plastic bags or containers	Autoclaving, microwaving, and shredding, or sterilization and shredding; plastic waste must not be dumped in landfills
White (translucent)	Sharp wastes including metals, needles, needles from needle-tip cutter or burner, scalpels, blades	Puncture-proof, leak-proof, tamper-proof containers	Following autoclaving or dry heat sterilization, shredding, mutilation or entrapment in a metal tank or cement concrete is used
Blue	• Glassware • Metallic body implants	Cardboard boxes with blue-colored marking	Disinfection by autoclaving, microwaving or hydroclaving, followed by recycling

TABLE 5.3
Schedule II – Standards for Treatment and Disposal

- Standards for incineration
- Operating and emission standards for disposal by plasma pyrolysis or gasification
- Standards for autoclaving of bio-medical wastes
- Standards of microclaving
- Standards for deep burial
- Standards for efficacy of chemical disinfection
- Standards for dry heat sterilization
- Standards for liquid wastes

TABLE 5.4
Schedule III – List of Prescribed Authorities

- Ministry of Environment, Forest and Climate change, Government of India
- Central or State Ministry of Health and Family Welfare, Central Ministry for Animal Husbandry and Veterinary or State Department of Animal Husbandry and Veterinary
- Ministry of Defense
- Central Pollution Control Board
- State Government of Health
- State Pollution Control Boards
- Municipalities or Corporations

(a)

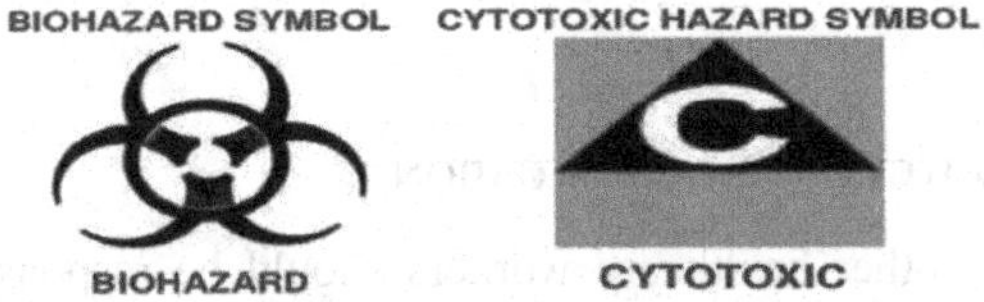

BIOHAZARD SYMBOL   CYTOTOXIC HAZARD SYMBOL

BIOHAZARD   CYTOTOXIC

(b)   Label shall be non-washable and prominently visible

Day.......Month.........Year........   Date of generation............

Waste Category   In case of emergency please
No...................   contact
Waste   Name and Address
Quantity.............   Phone No.

Sender's Name and Address   Receiver's Name and Address
Phone no..............   Phone no....................
Telex no.................   Telex no. ....
Contact Person....................   Fax no.........................
                                            Contact Person....................

FIGURE 5.2   (a,b): Schedule IV- Label for Biomedical waste containers.

5.3.4.2 Accident Report

In any accident, an approved person must promptly convey the remedial action in Form I to the specified authority and shall send the statement in writing within 24 hours.

5.3.4.3 Authorization

A 3-year license with an initial 1-year trial period beginning from the issue date, shall be issued.

5.3.4.4 Penalty

Penalties are levied on people, owners or occupants who are found to be causing pollution. Under the Indian Penal Code, a person may be sentenced to five years in prison or a fine of up to one lakh rupees, or both under Section 15, 16 and 17. If not executed, an additional fine of Rs. 5,000 a day and if not executed for longer than 1 year, then prison sentence will be continued for up to 7 years.

5.4 HOW TO MANAGE BMW

5.4.1 SURVEY AND MINIMIZATION

Medical operations produce waste that should often be discarded by the person who used the object and disposed of at the point of use. It is often necessary to minimize the quantity of hospital waste produced as well as precautions should be taken for its handling. For the sake of mitigating attempts in succeeding operations of handling, treatment and disposal, investigations should be done prior to waste generation.

5.4.2 COLLECTION AND SEGREGATION

Nursing and other healthcare workers should be responsible for ensuring the waste-carrying bags should be closed or sealed until roughly three quarters are full. Light-gauge bags may be closed by tying the neck, but heavier-gauge bags would almost certainly need a self-locking plastic sealing tag. Bags should not be stapled. Before removing it from the hospital ward or department, sealed sharp containers must be stored in a labeled yellow BMW bin. Wastes at the point of output must not be allowed to accumulate. As part of a BMW management strategy, a regular program for their collection should be developed. One of the most critical measures in effectively handling BMW is segregation. Because only about 10–25% of BMW is harmful, treatment and storage costs will be significantly decreased if adequate segregation is carried out. The risk of infecting workers managing BMW is also significantly minimized by the isolation of hazardous waste from non-hazardous waste. Currently, if the dangerous component is automatically isolated from the other waste, the portion of the BMW that is harmful and needs special handling can be decreased to about 2–5%. Segregation consists of identifying the various sources of waste on the grounds of their toxic property, the treatment method and the disposal procedures used.

5.4.3 TRANSPORTATION AND STORAGE

In addition, it was also found that waste was not disposed of in a timely manner, and that in certain hospitals, waste was kept in the same waste containers for days. Although in rulebook section 19, it is mentioned that waste should not be kept exceeding 20 hours. According to earlier studies (Pruss and Townsend 1998), the removal of waste bags in time is an integral part of waste management. Waste dispersion was also observed in several clinics near waste containers where waste has been stored for a long time.

Transportation should always be accurately recorded and all automobiles from the collection point to the treatment center, should bear a consignment label. In addition, the vehicles used to gather dangerous/contagious BMW should not be used for any other purpose. They should be free of sharp edges, simple to load and unload, clean/disinfected and completely protected to avoid hospital or road spillage when transported. Cytotoxic waste should be processed in a specified protected area apart from all BMW. Radioactive waste should be contained behind lead insulation in tanks to escape dispersion. Waste to be deposited after radioactive decay ought to be labeled with the radionuclide form, date and particulars of the storing conditions required (IAEA 1996).

5.4.4 TREATMENT AND DISPOSAL

5.4.4.1 On-Site Treatment

In rural healthcare facilities, on-site treatment of BMW may also be carried out. In places where healthcare facilities are located at far distances from each other and where the condition of roads is bad, on-site care facilities are especially suitable. The benefits of providing an on-site treatment center for each healthcare facility include simplicity and mitigation of threat to community well-being as well as the environment by storing harmful/contagious BMW in hospital facilities.

5.4.4.2 Off-Site Treatment

When centralized regional facilities exist, the BMW produced in a hospital could be handled off-site.

5.5 CURRENT TREATMENT AND DISPOSAL TECHNIQUES

5.5.1 CHEMICAL PROCESSES

5.5.1.1 Simple Chemical Decontamination Processes

Chemical methods are currently being utilized for hospital waste disposal, as it is routinely used in healthcare facilities for the destruction of microorganisms on surgical instruments, floors and walls. To destroy or inactivate the pathogens, chemical compounds are added to the waste; this treatment typically leads to disinfection, not sterilization. For handling liquid waste like blood, vomit, feces or sewage from hospitals, chemical disinfection is most acceptable. Though, solid as well as exceptionally toxic BMW, comprising micro-biological cultures, sharps, etc., can too be chemically sterilized, with the following restrictions:

- Until disinfection, waste shredding and/or milling is typically required; the shredder is always the weaker spot in the treatment system, result of continuous mechanical failure
- Strong disinfectants, which are often harmful on their own, are needed and can only be used by properly qualified and sufficiently trained staff

However, if alternative recycling services are not readily available, they can be shredded and then chemically disinfected. The criteria for potential disposal of residues should be regarded carefully in the proposed use of chemical disinfection; inappropriate disposal could cause significant environmental problems. Microbial tolerance has been examined and the classes of most resistant to the least resistant microbes are as follows: bacterial spores – mycobacteria – hydrophilic viruses – lipophilic viruses – fungal spores – vegetative bacteria. A disinfectant which is proven to have antagonistic action toward a specific class of microbes would also function with other less immune groups. Most of the parasites are extremely resistant to disinfection, such as *Giardia* and *Cryptosporidium* spp., and are typically rated between mycobacteria and viruses. In normal microbiological studies, disinfection efficacy is derived from the survival rates of indicator species. Currently, in developed nations, there are restricted chemical disinfections of healthcare waste. This is a mandate in developing countries to treat extremely contagious body fluids; for instance, stools from cholera patients. Chemical disinfection is normally carried out in hospital facilities. In rural primary healthcare facilities, this alternative is always special, but BMW generated at main healthcare facilities can also be treated on-site. In the case of hospitals far from each other and bad roads, the care facilities on-site are especially suitable.

Prior to disinfection, solid BMW must be shredded for the following factors:

- To enhance interaction between waste and disinfectant by augmenting the surface area and avoiding any confined areas
- To prevent any undesirable visual impacts on disposal, all items must be unidentifiable
- To reduce the volume of waste

5.5.1.2 Chemical Disinfectant Types

The goal of disinfection is to kill microorganisms to a "satisfactory" level. Some disinfectants work to destroy or inactivate some species of microorganisms and others work against some sort of organism. The identification of target microorganisms to be killed is therefore important. The preference of disinfectants, however, relies not just on their potency but also on their corrosiveness and other dangers involved with their handling. The use of ethylene oxide for waste disposal is no longer recommended due to the major risks involved with its handling. Utilization of ozone (O_3) for waste sterilization is under study at the moment. This is a solid and reasonably healthy disinfectant. The major disinfectants that have been in use are stable for notably 5 years and stay effectual for 6–12 months after opening the bottle, with the exception of sodium hypochlorite. Sometimes, effective disinfectants are volatile and toxic; many are poisonous to mucous membranes and skin. Therefore, staff need to dress in protective gear, including gloves and glasses or goggles for protective vision. In some cases, disinfectants are also aggressive and should be treated and processed accordingly. Without pretreatment, minor quantities of disinfectants can be directed into sewers, as long as an appropriate method for treatment of sewage is available; at no cost huge volumes of disinfectants be released to the sewers. There should be no disinfectants discharged into water resources.

5.5.2 Thermal Processes

5.5.2.1 Wet Thermal Treatment

The wet thermal disinfections or steam disinfections rely on higher temperatures, high-pressure steam exposure to shredded bacterial waste and are similar to the autoclave sterilization method. If adequate contact time and temperatures are available, several groups of microbes are inactivated; a minimum temperature of 121°C is required for sporulated bacteria. An inactivation of microorganisms of around 99.99% is required, compared to an autoclave sterilization of 99.999%. The wet heat method allows waste to be shredded before treatment and to improve disinfection productivity where sharply, molding or crushing is recommended. The procedure is unsuitable for the handling of animal and anatomical waste and does not handle chemical or medicinal waste effectively.

The comparatively low investment and maintenance costs and low environmental effects of the wet thermal process, however, are undoubted benefits that should be taken into consideration when incineration is not feasible. When disinfected, waste can be gathered and disposed of by municipality.

5.5.2.1.1 Operation and Technology

A horizontal steel cylinder attached to a steam engine, with 6 bar (600 kPa) and temperature 160°C, can withstand the reaction tank for the wet thermal phase. A vacuum pump and electrical supply are also needed for the system. During the process, pressures and temperatures are managed and tracked and the machine operation can be automated. Wet heat methods normally are batch methods; however, they can be continuous as well. The waste is shredded at the outset of the process and the splinter is split or milled until it is inserted in the tank. The conditions of vacuum in the tank are established; this raises the partial pressure of steam and hence the effect of interaction between steam and waste. During total contact time of 1–4 hours, a minimum temperature of 121°C and normally pressure between 2 and 5 bar (200–500 kPa) is required. After the waste is completely shredded and fills no more than half the tank, optimal working conditions can be obtained. The reaction tank is cooled down at the end of the touch period and drained and washed afterwards. Contact time needed to achieve disinfection is less than the required theoretical times – 20 min above 121°C and 2 bar (200 kPa) and 5 min above 134°C and 3.1 bar (310 kPa). Owing to the fact that steam may require more time to get through certain waste elements, like microbiological cultures or hypodermic needles. The successful use of *Bacillus subtilis* or *Bacillus stearothermophilus* tests should be regularly monitored for the effect of a water-subject thermal disinfection. The equipment should be run by qualified technicians and maintained; maintenance of the shredder is particularly crucial.

5.5.2.1.2 Autoclaving

Autoclaves typically are used in reusable surgical devices sterilization hospitals. They allow only small amounts of waste to be handled and hence widely utilized only for extremely contagious waste, such as microbial cultures or sharps. All general hospitals should be fitted with automobile systems, including those which have minimal funding. In this segment, autoclaving waste has the same advantages and drawbacks as other wet thermal methods. Research has shown that the successful inactivation of limited amounts (about 5–8 kg) of all vegetative micro organisms and most bacterial spores involves a period of 60 min at 121°C and 1 bar (100 kPa); therefore, maximum steam penetration should be assured.

5.5.2.2 Dry Heat Treatment Technologies

Glasswares and similar reusable devices and contagious health waste have been sterilized using hot air ovens. Waste is heated at elevated temperatures (up to 185°C) as well as prolonged contact periods (90–150 min) in comparison to steam-dependent methods by conducting, natural or forced convection or heat radiation. The biological predictor *Geobacillus stearothermophilus* should be absolutely and reliably removed (Widmer and Frei 2011; Emmanuel and Stringer 2007).

The screw-feed technology is basically a non-burning method, dry thermal decontamination where waste is shredded in a spinning auger and heated. Continuously operating systems are commercially viable and currently in service in many hospitals, often called continuous feed augers. The major steps involve:

- Shredding of waste into particles (approx. 25 mm diameter).
- Heating the shredded waste into an auger at 110–140°C (heat is provided by the oil circulating throughout the central shaft).

- Auger rotates the waste for about 20 min, and the debris is then compacted. The volume and weight of the waste was decreased by 80%
and 20–35%. It is an effective procedure for the management of infectious waste and sharps, but not for the treatment of pathological, cytotoxic or radioactive waste. Filtering of the drainage air is necessary
and pre-discharge treatment of condensing water produced during the
process.

5.5.3 MECHANICAL PROCESSES (WHO MODULE)

Mechanical treatments to rip BMWs apart go by verbs: pulverizes, granulates,
grinds, shreds, agitates, crushes and mixes. The treatment will decrease the
bulk volume of waste by at least 60%. Wastes can be transferred by augers, belt-
conveyors and other material-handling devices via the processing facility.

For further storage or recycling, mechanical treatment may not destroy bacteria or clean devices, but it can minimize the amount of waste. Mechanical smash-
ing or breaking of waste can affect its appearance, that may help in reducing the
psychological effect of waste on workers using crusher and milling machines to
manage waste. Mechanical systems may also minimize the amount of bulk waste
and improve the surface area of the solid parts before chemical or heat treatment.
Frequently, the waste is combined with other materials (cemetery or polymers),
but thus the waste mass will expand, and the engineer involved should analyze
very closely if it is worth mixing waste with an external agent. Yet mechanical
devices may also be an operational hassle – more things could go wrong and more
things would need to be decontaminated. In addition, solid waste may be dust-
producing. It can be an occupational threat and environmental danger if this dust
becomes airborne. This is why machines are always kept in a closed space with
marginally lower air pressure or under a hood.

5.5.4 IRRADIATION PROCESSES

Through the operation of microwaves with around 2,450 MHz frequency and
12.24 cm of wavelength, most microorganisms are killed. These microwaves
quickly heat water stored within the waste, destroying the contagious components
by heat conduction.

A loading system moves the waste to a shredder in a microwave treatment
facility, reducing the waste to tiny bits. Then it is humidified, sent to an irradiation chamber fitted with a collection of microwave generators and irradiated
for roughly 20 min. The waste is put into a container after irradiation and joins
the stream of municipality waste. Bacteriological and virological tests can regularly validate the efficacy of microwave disinfection. A regular bacteriological test utilizing *Bacillus subtilis* is suggested in the United States to prove a
decline in viable spores by 99.99%. The research method is comparable to the
protocol described for wet thermal disinfection. In many countries, the microwave method is widely used and is becoming progressively more mainstream.

Fairly higher prices and possible issues with installation and servicing indicate its unsuitability in developing countries.

5.5.5 INERTIZATION

In order to reduce the possibility of hazardous compounds found in the waste from being released into water bodies, this process called 'inertization' includes combining BMW with cement and other materials before disposing off. In case of pharmaceutical and incineration ash with higher metal content, inertization is particularly suitable (here the procedure is often known as 'stabilization'). The covering should be separated, the pharmaceutical grounds and a combination of acid, lime, and cement added for the inertization of waste from pharmaceutical sites. A homogeneous mass is formed and on-site cubes (for instance, 1 m^3) or pellets are manufactured on-site and can then be shipped to an appropriate storage location. The homogeneous mixture may alternatively be carried to a landfill in a liquid state and dumped into industrial waste. Typical proportions for the mixture are as follows: 65% pharmaceutical waste, 15% cement, 15% lime and 5% water. The method is fairly economical and could be operated with comparatively simple equipment. The major merits as well as demerits associated with different treatment and disposal methods are outlined in Table 5.5.

5.5.6 BIOLOGICAL PROCESSES

An evolving technology named "Bio-converter" 9 (Bio-medical Disposal, Inc.) involves biological approaches for the BMW disposal. To decontaminate BMW, it uses an enzyme solution and the resultant sludge is pushed into an extruder used to eliminate water to drain the sewage whereas the solid waste is transferred to the landfills. The use of biodegradable plastics is another form of environmental BMW management. Many bio-medical implants made of biodegradable plastics are biologically degraded by extracellular microbial enzymes. These bacteria use such biodegradable polymers as a substrate in the event of starvation and in the absence of an adequate substrate. For the large-scale economic production of biodegradable plastics, more research needs to be performed (Emmanuel 2001; Nema and Ganeshprasad 2002).

Moreover, membrane bioreactors (MBRs) incorporate biologically activated sludge with a sludge water separation membrane filtration step. MBRs come in many different types, including aerobic, anaerobic, and organic pollutant MBRs (Alrhmoun 2014; Judd 2011). Other emerging technologies for destruction of BMWs include gas-phase chemical reduction, base-catalyzed decomposition, supercritical water oxidation, sodium reduction, verification, superheated steam reforming, Fe-TAML/peroxide treatment (pharmaceutical waste), biodegradation (using mealworm or bacteria to eat plastics), mechanochemical treatment, sonic technology, electrochemical technologies, solvated electron technology and phytotechnology. These technological advances aren't yet suitable for widespread use in the BMW treatment (EPA 2005).

TABLE 5.5
Bio-Medical Waste Treatment/Disposal Methods

Treatment/ Disposal Method	Advantages	Disadvantages
Rotary kiln	Suitable for all infectious waste, most chemical and pharmaceutical waste	High expenditure and cost of operation
Pyrolytic incineration	High efficiency of disinfection Adequate for all infectious waste and most pharmaceutical and chemical waste	Degradation of cytotoxics is incomplete Relatively high investment and operating costs
Single-chamber incineration	Good disinfection efficiency Substantial reductions in weight and volume of waste The residues may be disposed of in landfills Relatively low investment and operating costs	Significant pollutant emissions into the atmosphere Cleaning of slag and soot on a regular basis is needed Inadequacy in the destruction of chemical products (thermal resistance) and cytotoxics
Drum or brick incineration	Drastic reduction of weight and volume of the waste Very low investment and operating costs	Destroys only 99% of microorganisms No destruction of various chemicals and drugs Massive emission of black smoke, fly ash, toxic flue gas and odors
Chemical disinfection	Highly efficient disinfection under good operating conditions Some chemical disinfectants are relatively inexpensive Drastic reduction in waste volume	Requires highly trained technicians Employs harmful substances requiring rigorous security measures Inadequate for pharmaceutical, chemical and some types of infectious waste
Wet thermal treatment	Environmentally sound Drastic reduction in waste volume Relatively low investment and operating costs	Shredders are prone to malfunctioning and breakdowns on a regular basis Trained technicians are required for operation Unsuitable for anatomical, pharmaceutical, and chemical waste, as well as wastes which are not easily steam-permeable
Microwave irradiation	Good disinfection efficiency under appropriate operating conditions Drastic reduction in waste volume Environmentally sound	Relatively high investment and operating costs Challenges associated with service and maintenance
Encapsulation	Simple, low-cost and safe May also be applied to pharmaceuticals	Not recommended for non-sharp infectious waste
Safe burying	Low costs Relatively safe if access to site is restricted and where natural infiltration is limited	Safe only if access to site is limited and certain precautions are taken
Inertization	Relatively inexpensive	Not suitable to infectious waste

5.6 CONCLUSIONS

The national engaged facilities and the cradle-to-grave legislation, proficient regulatory authorities and well-trained workers should, preferably, concentrate on the disposing of BMW. The progress in the handling of biological waste continues to minimize waste. In the interests of protecting staff, patients, the public and ecosystem, in order to meet the strong commitment to the fundamental right to live in a safe and secure environment, these rules, norms and regulations covering BMW management in each country are intended to modulate the disposal of several forms of BMW. A novel waste created but not reported in the rules should be entitled to a buyback program for an agency or should be treated in line with the guidelines of the Centers for Disease Control and Prevention or the WHO. Moreover, development of models for the regulation of BMW treatment procedures and studies of environmental, recycling, and PVC-free non-burn-friendly solutions will go a long way toward a sustainable climate. In order to recognize the very critical field of public health, more analysis on BMW is needed globally. Moreover, the need of the hour is to emphasize on advance green technology for successful management of BMW.

REFERENCES

Anonymous. 2000. Guidelines for common hazardous waste incineration Central Pollution Control Board Ministry of Environment & Forests Hazardous Waste Management Series HAZWAMS/30/2005-06.

Chitnis, V., Patil, S., and Chitnis, D. S. 2000. Ravikant Hospital Effluent: A Source of Multi drug Resistant Bacteria. *Current Sciences* 79: 535–40.

CPCB. 2016. *Bio-Medical Waste (Management and Handling) Rules, 2016.* New Delhi: Ministry of Environment and Forests, Government of India. http://cpcb.nic.in/Bio_medical.php.

Emmanuel, J. 2001. *Non-Incineration Medical Waste Treatment Technologies.* Washington, DC: Health Care without Harm.

Emmanuel, J., and Stringer, R. 2007. *For Proper Disposal: A Global Inventory of Alternative Medical Waste Treatment Technologies.* Arlington: Health Care without Harm.

Gardiner, B. 2020. Pollution Made COVID-19 Worse. Now, Lockdowns Are Clearing the Air. https://www.nationalgeographic.com/science/2020/04/%0Apollution-made-the-pandemic-worse-but-lockdowns-clean-the-sky/.

Himabindu, P. A., Madhukar, M., and Udayashankara, T. H. 2015. A Critical Review on Bio-medical Waste and Effect of Mismanagement. *International Journal of Engineering Research and Technology* 4, Nr. 03 (10. März). doi:10.17577/IJERTV4IS030179, http://www.ijert.org/view-pdf/12546/a-critical-review-on-bio-medical-waste-and-effect-of-mismanagement.

IAEA. 1996. *Regulations for the Safe Transport of Radioactive Material.* Vienna: International Atomic Energy Agency (IAEA Safety Standards Series, No. ST-1).

Kampf, G., Todt, D., Pfaender, S., and Steinmann, E. 2020. Persistence of Coronaviruses on Inanimate Surfaces and Their Inactivation with Biocidal Agents. *Journal of Hospital Infect* 104, Nr. 3: 246–51.

Kumar, A., and Duggal, S. 2015. Safe Transportation of Bio-Medical Waste in a Health Care Institution. *Indian Journal of Medical Microbiology* 33, Nr. 3: 383–6.

Mallapur, C. 2020. Sanitation Workers at Risk from Discarded Medical Waste Related to COVID-19. https://www.indiaspend.com/sanitation-workersat-%0Arisk-from-discarded-medical-waste-related-tocovid-19/.

Nath, A. P., Prashanthini, V., and Visvanathan, C. 2010. Healthcare Waste Management in Asia. *Waste Management* 30: 154–61.

Nema, S. K., and Ganeshprasad, K. S. 2002. Plasma Pyrolysis for Medical Waste. *Current Sciences* 83: 271–8.

Prüss, A., Giroult, E., and Rushbrook, P. 1999. *Safe Management of Wastes from Health-Care Activities.* Geneva, Switzerland.

Pruss, A., and Townsend, W. K. 1998. Teacher's Guide. *Management of Wastes from Health Care Activities.* Geneva: World Health Organization.

Singh, A., Joshi, H. S., Katyal, R., Singh, R., and Singh, H. 2017. Bio-medical Waste Management Rules, 2016: A Brief Review. *International Journal of Advanced and Integrated Medical Sciences* 2, Nr. 4 (December): 201–4. doi:10.5005/jp-journals-10050-10107. https://www.ijaims.com/doi/10.5005/jp-journals-10050-10107.

UNEP. 2020. BASEL: Waste Management an Essential Public Service in the Fight to Beat COVID-19. http://www.basel.int/Implementation/PublicAwareness/%0APress Releases/WastemanagementandCOVID19/tabid/8376/Default.aspx.

van Doremalen, N., Bushmaker, T., Morris, D. H., Holbrook, M. G., Gamble, A., Williamson, J. O., and Lloyd-Smith, B. N. 2020. Aerosol and Surface Stability of SARS-CoV-2 as Compared with SARS-CoV-1. *New England Journal of Medicine* 382, Nr. 16: 1564–7.

WHO. 2020. COVID-19 Strategy Up Date. The World Health Organization. https://www.who.int/emergencies/diseases/novel-coronavirus-2019.

Widmer, A. F., and Frei, R. 2011. *Decontamination, Disinfection, Sterilization.* Hg. von Landry ML Versalovic J, Carroll KC, Funke G, Jorgensen JH and Warnock DW. 10th Aufl. Washington, DC: ASM Press.

6 Advances in the Recycling of Polymer-Based Plastic Materials

Muhammad Kashif Shahid,[1] Ayesha Kashif,[2] and Younggyun Choi[3]
[1]Research Institute of Environment & Biosystem, Chungnam National University, Daejeon, Republic of Korea
[2]Department of Senior Health Care, Eulji University, Daejeon, Republic of Korea
[3]Department of Environmental & IT Engineering, Chungnam National University, Daejeon, Republic of Korea

CONTENTS

6.1 INTRODUCTION

Today, human life is dependent on polymer-based materials ranging from food packaging to the advanced heavy industries. The promising physical and chemical properties of plastics including durability, low density, cost effectiveness, and molding features make them applicable in multiple everyday activities such as farming, packaging material, electronics, automobile, aerospace, construction material, membrane water treatment, etc. (Wei and Zimmermann 2017; Shahid, Kashif, et al. 2020). Generally, the plastic materials are made up of polypropylene (PP), polyethylene (PE), polystyrene (PS), polyethylene terephthalate (PET),

DOI: 10.1201/9781003201076-6

polyvinyl chloride (PVC), polyurethane (PUR), and high-density polyethylene (HDPE). A variety of plastic products are being used globally, and an excessive demand and production of plastic materials is a continuous threat to the life and ecosystem (Geyer, Jambeck, and Law 2017; Gigault et al. 2018).

The non-biodegradable nature of plastic materials significantly contributes to environmental toxicity, as landfilling of plastic debris promotes the channels for the fate of pollutants. As dye agents carry several toxic components, they can infiltrate the environment through plastic (Gewert, Plassmann, and MacLeod 2015). Metals and glass can be used as an alternative to plastics, especially in packaging; however, they have several unavoidable adverse effects on the environment including heavy weight and excessive transportation leading to higher CO_2 emissions. As per an estimate, about 6.30 billion tons of plastic waste was globally produced from 1950 to 2015 (Brooks, Wang, and Jambeck 2018). Around 80% of this waste is accumulated in natural environment or landfills, and merely 9% is recycled.

Generally, plastic recycling is categorized as primary, secondary, tertiary, and quaternary recycling. Primary recycling includes the extrusion of pure polymers, whereas secondary recycling involves sorting and size reduction prior to extrusion of plastic waste. Under optimized operational conditions, polymers can pass through many rounds of primary and secondary recycling without the concerns regarding loss in efficiency. Tertiary recycling is a chemical process that is applied when primary and secondary recycling remain ineffective. The conversion of polymers into monomers is also a kind of tertiary recycling. Quaternary recycling is used for the plastic materials, inappropriate for all other recycling processes and have potential for energy recovery. The major limitation of this process is the production of greenhouse gas (GHG) emissions.

Plastic materials are synthesized from polymers and complex mixture of additives (plasticizers, pigments, and stabilizers) and may comprise unintended contaminants and impurities. For instance, bisphenol A is an emerging contaminant that is used in the synthesis of polycarbonate plastic (Shahid et al. 2021).

This chapter discusses the types and application of different plastic materials used in daily life and the available technologies for their recycling. The mechanical and chemical approaches for recycling of plastic wastes are discussed briefly. Mechanisms involved in the recycling processes are presented. A special attention is given to the energy recovery from the plastic wastes.

6.2 PLASTICS AND THEIR APPLICATIONS

Generally, plastics are referred to as polymeric synthetic or semi-synthetic materials (Cai et al. 2019). The majority of polymer plastics are made up of long chains of carbon atoms, with or without attached nitrogen, oxygen, or sulfur atoms. The long carbon chain consists of several repeating small units of monomers and a single polymer chain contains several thousand repeating units. Plastics are generally categorized based on the chemical structure of the backbone chain and side chains of the polymer (Chamas et al. 2020). They can also be categorized based on the synthesis process (e.g., cross-linking, polyaddition, and condensation) and

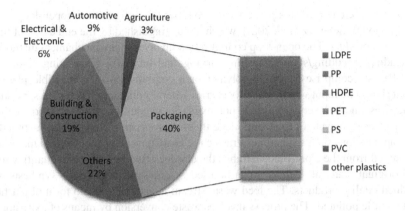

FIGURE 6.1 Practical application of polymers and plastics. The particular fraction of polymers is based on data from polymeric constituents of plastic packaging in Europe (Kaiser, Schmid, and Schlummer 2018). (Image reused under the Creative Commons Attribution License.)

the physical characteristics (e.g., density, hardness, tensile strength, glass transition temperature, and thermal resistance). Other categorizations are based on design configuration or manufacturing objectives such as engineering plastics, conductive polymers, thermoplastics, biodegradable plastics, elastomers, and thermosets.

Thermoplastics are the kind of plastics that turn soft when heated and can be molded under pressure and solidify on cooling. PET is the most common example of thermoplastics that is widely used in packaging, beverage bottles, polyester clothes, etc. due to the specific properties (e.g., clear, hard, high strength and stiffness, heat and chemical resistance) (Chu et al. 2021). Another type of thermoplastic is HDPE that has remarkable process aptitude, tremendous balance of impact strength and rigidity, higher chemical resistance, and remarkable water vapor barrier characteristics. HDPEs are applied in blow molded products, injection molded products, pipes, films, etc. Other examples of thermoplastics are PVC, low-density polyethylene (LDPE), polypropylene (PP), polystyrene (PP), acrylonitrile butadiene styrene (ABS), nylon, and polycarbonate. Thermoset is a type of plastic that cannot be remolded or softened under the effect of heat (Schwarz et al. 2021). Phenol, unsaturated polyester, urea formaldehyde, polyurethanes, and epoxy are the common examples of thermosets. Figure 6.1 highlights the share of several polymers in the polymer market of Europe.

6.3 RECYCLING OF PLASTIC WASTE

6.3.1 MECHANICAL RECYCLING

Nowadays, mechanical recycling is considered a foremost option for managing the plastic waste. It involves grinding, washing, and re-melting of plastic waste into secondary plastic without changing the chemical structure of primary plastic waste. Mechanical recycling is a widely applied process due to simple and low-cost processing. Generally, this process has two types of configurations (open-loop and closed-loop) (Schwarz et al. 2021), which are distinguished by the end usage of the recycled product. The open-loop configuration involves many mechanical methods including shredding/slicing, washing, drying, and finally re-granulating. Usually, feed waste comprises a single polymer or a combination of compatible plastic materials. The input waste generally carries several pollutants such as inks, plastic additives, and leftovers of unsuited polymers. These contaminations deteriorate the properties of recycled plastic and make them appropriate merely for low-demanding applications such as pipelines, trash bags, etc. Conversely, the recycled material obtained from the closed-loop exhibits the closeness in properties and quality with an original material. Hence, it can be used as a raw material in the synthesis of valued quality products. The feed waste stream is generally a solo form of plastic and a little polluted. The process involves waste conversion by means of extrusion.

Extrusion is a primary process applied in mechanical recycling that converts the plastic waste into re-granulated material. It is a low-cost and solvent-free method that can be applied to several polymers on a large scale (Maris et al. 2018). Generally, the extruder applies heat energy and rotating screws to persuade plasticization or thermal softening that is later passed from temperature-controlled barrel units. Process efficiency and the quality of the end product significantly depend on the screw speed and temperature. Extruding under extreme conditions (high temperature and screw speed) can expedite the chain scission and produce the polymers, which remain unable to be processed (Oblak et al. 2015). The contaminated recycled material causes a loss in quality and a rise in discrepancy in the regenerated polymer (Eriksen et al. 2019).

It is noteworthy that domestic plastic waste, especially household waste generally includes a range of polymers, which are mostly contaminated. As the recyclability significantly depends on the quality of the reprocessed matter, possible quality losses signify a restriction in the progress associated with circular economy. Therefore, to enhance the recyclability, it is obligatory to improve the quality of recycled plastic to a level where the sustainable properties of plastic material can be achieved. Dyes applied to color plastic materials can speed up the degradation reactions in extruders. The printing toners and papers or plastic labels can induce volatile ink constituents in the end product (recyclate pellet). Some plastic lubricants contain fatty acids that can be oxidized to generate undesirable smell in the recyclate (Horodytska, Valdés, and Fullana 2018).

6.3.2 CHEMICAL RECYCLING

This process involves the breakdown of polymers into small subunits (monomers) or other valued chemicals, and the recycled material can be utilized as raw material to produce new polymer materials (Hamad, Kaseem, and Deri 2013). In general, the chemical recycling methods include depolymerization (glycolysis, methanolysis, and hydrolysis), cracking (pyrolysis), and partial oxidation (gasification). Several studies are reported on the thermal catalytic depolymerization of plastic waste to form hydrocarbon products including pyrolysis gas, pyrolysis oil, and carbon residue. Pyrolysis involves thermal decomposition of materials containing organic carbon in the absence of oxygen (Shahid, Kim, et al. 2020). The plastic material is heated to 400–500°C to remove the volatile compounds. When the feed material is heated above the thermal stability of its organic matter, the macromolecular matrix breaks down into stable low-molecular-weight gaseous and liquid products along with solid residue.

Several types of catalysts are applied to increase the process efficiency of pyrolysis of plastic waste. Catalysts play a very important role in improving process proficiency, pointing to the definite reaction and lowering the duration and temperature of the process (Ratnasari, Nahil, and Williams 2017). During pyrolysis of plastic materials, the catalytic reactions may include oligomerization, cracking, aromatization, cyclization, and isomerization (Serrano, Aguado, and Escola 2012). Many studies described the application of mesoporous and microporous catalysts for the transformation of plastic wastes into char and oil. The application of synthetic catalysts was found to be effective in improving the efficiency of the pyrolysis process; however, their application is conditioned with higher operational cost. Figure 6.2 presents the reaction schemes for the depolymerization of PET using different catalysts.

The plasma pyrolysis technology (PPT) is another advanced method that integrates the pyrolysis process with the thermo-chemical characteristics of plasma (Solis and Silveira 2020). The extreme and versatile heat production capability of PPT assists it in disposing of all kinds of plastic waste. Initially, the plastic waste is placed in the primary compartment at 850°C, where it dissociates into higher hydrocarbons, hydrogen, carbon monoxide, methane, etc. Later, combustion of pyrolysis gases occurs in the secondary compartment, in the presence of oxygen.

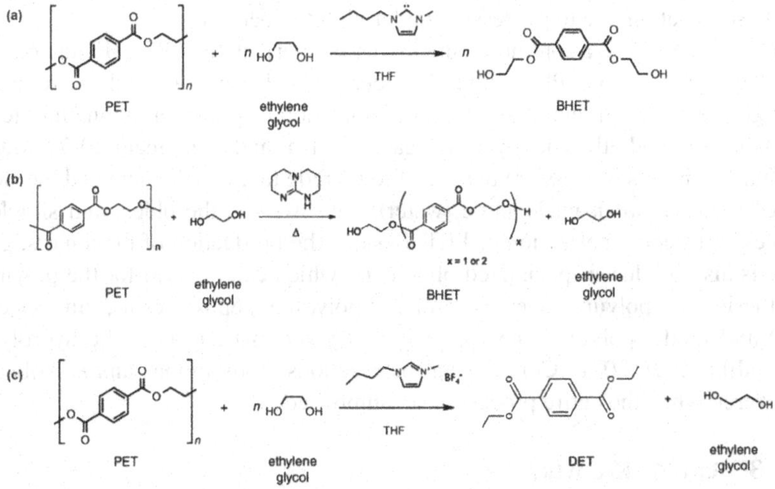

FIGURE 6.2 Reaction outline of the depolymerization of polyethylene terephthalate via (a) N-heterocyclic carbenes, (b) guanidine 1,5,7-Triazabicyclo[4.4.0]dec-5-ene, and (c) 1-Butyl-3-methylimidazolium tetrafluoroborate (Miao, von Jouanne, and Yokochi 2021). (Image reused under the Creative Commons Attribution License.)

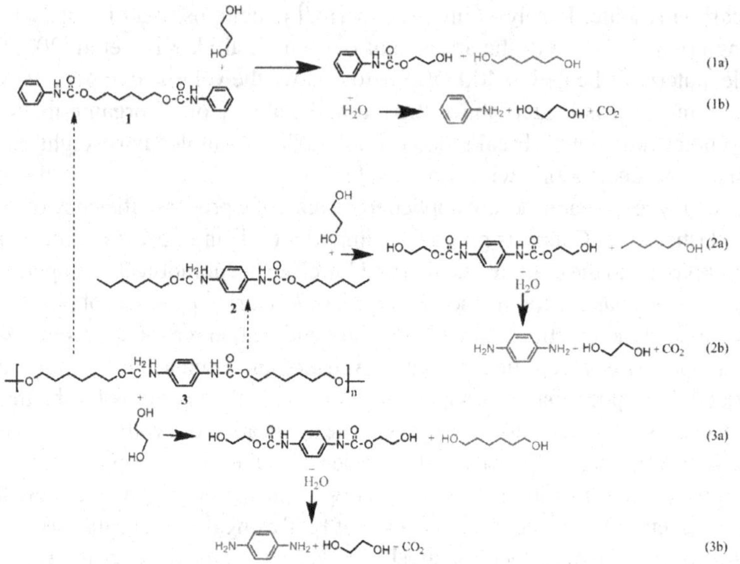

FIGURE 6.3 A schematic representation of multistage chemical recycling of model dicarbamates. (Zahedifar et al. 2021). (Image reused under the Creative Commons Attribution License.)

Due to the high voltage spark, the inflammable gases burn out. Temperature remains at 1050°C in the secondary compartment. Hydrocarbons, carbon monoxide, and hydrogen are combusted into H_2O and safe CO_2. PPT does not require waste segregation due to processing under high temperatures.

Hydrolysis is also a common route for depolymerization of PET (Figure 6.3). It involves the reaction of PET with H_2O under acidic, basic, or neutral environment at high pressure and temperature, and dissociation of polyester chains into terephthalic acid and ethylene glycol (Ragaert, Delva, and Van Geem 2017). Major limitations in this process are the use of low-grade terephthalic acid and the slow process due to weak nucleophile (water). Glycolysis is the oldest and simplest process for depolymerization of PET. Besides the production of monomers, glycolysis also produces specialized oligomers, which can be used for the polymer synthesis (e.g., polyurethanes, unsaturated polyesters, epoxy resins, vinyl esters, etc.) and model polyurethane applying first glycolysis and later the hydrolysis (Zahedifar et al. 2021). Considering the reactions, both amines and polyols are recovered when the entire process is accomplished.

6.3.3 ENERGY RECOVERY

Incineration is a known process for the recovery of energy from plastic waste. Due to the high value of calorific, plastic materials are regarded as an enriched source of energy. Followed by incineration, 90–99% reduction can be achieved in the waste volume that is very helpful in meeting the challenges of insufficient space and restricted landfilling. Several methods are known for the incineration of plastic waste. Co-incineration of domestic solid waste containing major share of plastic waste is achieved via direct single-stage, double-stage, and fluidized bed combustion operation. Plastic waste is commonly applied as a fuel in cement industry to minimize the energy costs (Aldrian et al. 2016). Refused derived fuel (RDF) is a newly coined term for the fuel or energy generated from the incineration of solid wastes such as plastic materials. As plastic waste mainly contains carbon-based organic substance, it discharges energy on reaction with oxygen (combustion) (Schwarzböck et al. 2016). Primary constraints that direct the selection of the waste material as RDF are the greater heating value (input energy), the components of ash residue, and the toxicity of the discharged gases. Studies have reported gas emissions and pollution caused by combustion of waste in cement kilns (Zemba et al. 2011).

Plastic solid waste has also been applied in blast furnaces (Al-Salem, Lettieri, and Baeyens 2010). Initially, plastic wastes go through pretreatment to remove the contaminates and later, crushed plastic waste enters into the blast furnace for gasification in the high-temperature zone. Plastic waste processed in blast furnaces results 50% loss in weight at about 390°C, similar to the performance of thermogravimetric chambers. The imperfect combustion of plastic waste is the foremost challenge in the operation of a blast furnace, which happens due to inadequate O_2 supply (sub-stoichiometric combustion environments) or O_2 restricted access due to diffusion-controlled methods (mass transfer). The combustion efficiency also depends on the particle size of plastic and the O_2 content. Figure 6.4

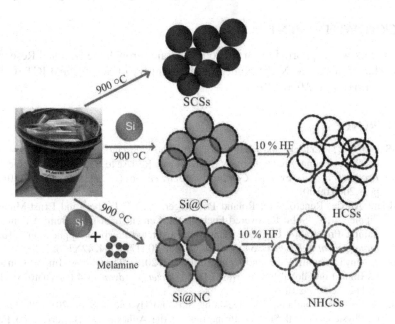

FIGURE 6.4 A schematic illustration of solid carbon spheres, hollow carbon spheres, and N-doped hollow carbon spheres derived from waste polypropylene (Tripathi, Durbach, and Coville 2019). (Image reused under the Creative Commons Attribution License.)

represents an outcome of the study focused on the improving value of plastic waste by applying it as a source of hydrocarbon in synthesis of several spherical carbon nanomaterials.

6.4 CONCLUSIONS

Plastic waste results in a serious environmental impression on the planet and it is imperious to control this. Subsequently, recycling of plastic waste has been considered as an imperative method to address this issue. The efficiency of recycling depends on the type of plastic waste and the targeted product. Recycling can be attained by mechanical or chemical processes. The pretreatment of plastic waste (e.g., sorting, segregation) might be required in some cases, prior to chemical or mechanical recycling. Various polymer materials exhibit diverse physiochemical characteristics, and therefore choice of recycling approach varies based on particular plastic waste. This chapter emphasizes the latest research and progresses in the recycling of plastic wastes.

DECLARATION OF COMPETING INTERESTS

The authors declare that they have no known competing financial interests or personal relationships that could have appeared to influence the work reported in this article.

ACKNOWLEDGMENT

This work was supported by Brain Pool Program through the National Research Foundation of Korea (NRF) funded by the Ministry of Science and ICT (Grant No.: 2019H1D3A1A02071191).

REFERENCES

Al-Salem, S M, P Lettieri, and J Baeyens. 2010. "The Valorization of Plastic Solid Waste (PSW) by Primary to Quaternary Routes: From Re-Use to Energy and Chemicals." *Progress in Energy and Combustion Science* 36 (1): 103–29. doi: 10.1016/j. pecs.2009.09.001.

Aldrian, Alexia, Renato Sarc, Roland Pomberger, Karl E Lorber, and Ernst-Michael Sipple. 2016. "Solid Recovered Fuels in the Cement Industry – Semi-Automated Sample Preparation Unit as a Means for Facilitated Practical Application." *Waste Management & Research* 34 (3): 254–64. doi: 10.1177/0734242X15622816.

Brooks, Amy L, Shunli Wang, and Jenna R Jambeck. 2018. "The Chinese Import Ban and Its Impact on Global Plastic Waste Trade." *Science Advances* 4 (6): eaat0131. doi: 10.1126/sciadv.aat0131.

Cai, Li, Dan Wu, Jianhong Xia, Huahong Shi, and Hyunjung Kim. 2019. "Influence of Physicochemical Surface Properties on the Adhesion of Bacteria onto Four Types of Plastics." *Science of the Total Environment* 671: 1101–7. doi: 10.1016/j. scitotenv.2019.03.434.

Chamas, Ali, Hyunjin Moon, Jiajia Zheng, Yang Qiu, Tarnuma Tabassum, Jun Hee Jang, Mahdi Abu-Omar, Susannah L Scott, and Sangwon Suh. 2020. "Degradation Rates of Plastics in the Environment." *ACS Sustainable Chemistry & Engineering* 8 (9): 3494–511. doi: 10.1021/acssuschemeng.9b06635.

Chu, Jianwen, Yanpeng Cai, Chunhui Li, Xuan Wang, Qiang Liu, and Mengchang He. 2021. "Dynamic Flows of Polyethylene Terephthalate (PET) Plastic in China." *Waste Management* 124: 273–82. doi: 10.1016/j.wasman.2021.01.035.

Eriksen, M K, J D Christiansen, A E Daugaard, and T F Astrup. 2019. "Closing the Loop for PET, PE and PP Waste from Households: Influence of Material Properties and Product Design for Plastic Recycling." *Waste Management* 96: 75–85. doi: 10.1016/j.wasman.2019.07.005.

Gewert, Berit, Merle M Plassmann, and Matthew MacLeod. 2015. "Pathways for Degradation of Plastic Polymers Floating in the Marine Environment." *Environmental Science: Processes & Impacts* 17 (9): 1513–21. doi: 10.1039/C5EM00207A.

Geyer, Roland, Jenna R Jambeck, and Kara Lavender Law. 2017. "Production, Use, and Fate of All Plastics Ever Made." *Science Advances* 3 (7): e1700782. doi: 10.1126/sciadv.1700782.

Gigault, Julien, Alexandra ter Halle, Magalie Baudrimont, Pierre-Yves Pascal, Fabienne Gauffre, Thuy-Linh Phi, Hind El Hadri, Bruno Grassl, and Stéphanie Reynaud. 2018. "Current Opinion: What Is a Nanoplastic?" *Environmental Pollution* 235: 1030–34. doi: 10.1016/j.envpol.2018.01.024.

Hamad, Kotiba, Mosab Kaseem, and Fawaz Deri. 2013. "Recycling of Waste from Polymer Materials: An Overview of the Recent Works." *Polymer Degradation and Stability* 98 (12): 2801–12. doi: 10.1016/j.polymdegradstab.2013.09.025.

Horodytska, O, F J Valdés, and A Fullana. 2018. "Plastic Flexible Films Waste Management – A State of Art Review." *Waste Management* 77: 413–25. doi: 10.1016/j.wasman.2018.04.023.

Kaiser, Katharina, Markus Schmid, and Martin Schlummer. 2018. "Recycling of Polymer-Based Multilayer Packaging: A Review." *Recycling.* doi: 10.3390/recycling3010001.

Maris, Joachim, Sylvie Bourdon, Jean-Michel Brossard, Laurent Cauret, Laurent Fontaine, and Véronique Montembault. 2018. "Mechanical Recycling: Compatibilization of Mixed Thermoplastic Wastes." *Polymer Degradation and Stability* 147: 245–66. doi: 10.1016/j.polymdegradstab.2017.11.001.

Miao, Yu, Annette von Jouanne, and Alexandre Yokochi. 2021. "Current Technologies in Depolymerization Process and the Road Ahead." *Polymers.* doi: 10.3390/polym13030449.

Oblak, Pavel, Joamin Gonzalez-Gutierrez, Barbara Zupančič, Alexandra Aulova, and Igor Emri. 2015. "Processability and Mechanical Properties of Extensively Recycled High Density Polyethylene." *Polymer Degradation and Stability* 114: 133–45. doi: 10.1016/j.polymdegradstab.2015.01.012.

Ragaert, Kim, Laurens Delva, and Kevin Van Geem. 2017. "Mechanical and Chemical Recycling of Solid Plastic Waste." *Waste Management* 69: 24–58. doi: 10.1016/j.wasman.2017.07.044.

Ratnasari, Devy K, Mohamad A Nahil, and Paul T Williams. 2017. "Catalytic Pyrolysis of Waste Plastics Using Staged Catalysis for Production of Gasoline Range Hydrocarbon Oils." *Journal of Analytical and Applied Pyrolysis* 124: 631–37. doi: 10.1016/j.jaap.2016.12.027.

Schwarz, A E, T N Ligthart, D Godoi Bizarro, P De Wild, B Vreugdenhil, and T van Harmelen. 2021. "Plastic Recycling in a Circular Economy; Determining Environmental Performance through an LCA Matrix Model Approach." *Waste Management* 121: 331–42. doi: 10.1016/j.wasman.2020.12.020.

Schwarzböck, Therese, Philipp Aschenbrenner, Helmut Rechberger, Christian Brandstätter, and Johann Fellner. 2016. "Effects of Sample Preparation on the Accuracy of Biomass Content Determination for Refuse-Derived Fuels." *Fuel Processing Technology* 153: 101–10. doi: 10.1016/j.fuproc.2016.07.001.

Serrano, D P, J Aguado, and J M Escola. 2012. "Developing Advanced Catalysts for the Conversion of Polyolefinic Waste Plastics into Fuels and Chemicals." *ACS Catalysis* 2 (9): 1924–41. doi: 10.1021/cs3003403.

Shahid, Muhammad Kashif, Ayesha Kashif, Ahmed Fuwad, and Younggyun Choi. 2021. "Current Advances in Treatment Technologies for Removal of Emerging Contaminants from Water – A Critical Review." *Coordination Chemistry Reviews* 442: 213993. doi: 10.1016/j.ccr.2021.213993.

Shahid, Muhammad Kashif, Ayesha Kashif, Prangya Ranjan Rout, Muhammad Aslam, Ahmed Fuwad, Younggyun Choi, Rajesh Banu J, Jeong Hoon Park, and Gopalakrishnan Kumar. 2020. "A Brief Review of Anaerobic Membrane Bioreactors Emphasizing Recent Advancements, Fouling Issues and Future Perspectives." *Journal of Environmental Management* 270 (June): 110909. doi: 10.1016/j.jenvman.2020.110909.

Shahid, Muhammad Kashif, Jun Young Kim, Gwyam Shin, and Younggyun Choi. 2020. "Effect of Pyrolysis Conditions on Characteristics and Fluoride Adsorptive Performance of Bone Char Derived from Bone Residue." *Journal of Water Process Engineering* 37: 101499. doi: 10.1016/j.jwpe.2020.101499.

Solis, Martyna, and Semida Silveira. 2020. "Technologies for Chemical Recycling of Household Plastics – A Technical Review and TRL Assessment." *Waste Management* 105: 128–38. doi: 10.1016/j.wasman.2020.01.038.

Tripathi, Pranav K, Shane Durbach, and Neil J Coville. 2019. "CVD Synthesis of Solid, Hollow, and Nitrogen-Doped Hollow Carbon Spheres from Polypropylene Waste Materials." *Applied Sciences.* doi: 10.3390/app9122451.

Wei, Ren, and Wolfgang Zimmermann. 2017. "Microbial Enzymes for the Recycling of Recalcitrant Petroleum-Based Plastics: How Far Are We?" *Microbial Biotechnology* 10 (6): 1308–22. doi: 10.1111/1751-7915.12710.

Zahedifar, Pegah, Lukasz Pazdur, Christophe M L Vande Velde, and Pieter Billen. 2021. "Multistage Chemical Recycling of Polyurethanes and Dicarbamates: A Glycolysis–Hydrolysis Demonstration." *Sustainability.* doi: 10.3390/su13063583.

Zemba, Stephen, Michael Ames, Laura Green, Maria João Botelho, David Gossman, Igor Linkov, and José Palma-Oliveira. 2011. "Emissions of Metals and Polychlorinated Dibenzo(p)Dioxins and Furans (PCDD/Fs) from Portland Cement Manufacturing Plants: Inter-Kiln Variability and Dependence on Fuel-Types." *Science of the Total Environment* 409 (20): 4198–4205. doi: 10.1016/j.scitotenv.2011.06.047.

7 Utilization of Reclaimed Asphalt Pavement (RAP) Material as a Part of Bituminous Mixtures

Ramya Sri Mullapudi,[1] *Gottumukkala Bharath,*[2] *and Narala Gangadhara Reddy*[3]

[1]Department of Civil Engineering, IIT Hyderabad, India
[2]Scientist, CSIR – Central Road Research Institute Delhi, New Delhi, India
[3]Civil Engineering Department, Kakatiya Institute of Technology and Science, Warangal, India

CONTENTS

DOI: 10.1201/9781003201076-7

7.1 INTRODUCTION – IMPORTANCE OF RAP USAGE IN PAVEMENTS

A good road network is an important criterion for the rapid development of a country. Huge money is being spent for the expansion of roads especially by the evolving countries of the world. India has a road network of about 6 lakh kilometers. The traditional methods for preparing the top surface layer is through bituminous material, which needs higher energy in the form of extraction of binder from the crude oil, preparation of good aggregate materials, and at the later stage preparation of bituminous mix at a hot or cold mix plant.

Of late, there has been huge increase in the costs of flexible pavement materials and also due to excessive mining of quarries lead to depletion of natural quality materials prompted to search for alternative and elective materials (Reddy et al., 2021). Rehabilitation of existing roads produces large amounts of waste materials and disposal of these materials is not only an economic burden but also possesses environmental and ecological issues.

The term reclaimed asphalt pavement (RAP) refers to the extracted materials of the existing flexible pavement materials and then these are used for reconstruction purpose as a valuable partial replacement of aggregate and material with added rejuvenators (Gottumukkala et al., 2018). There are reports that existed from 1920s which speak about recycling of asphalt concrete pavement for constructing base or surface courses and some reports which discuss about the design procedures adopted and about the laboratory performance of RAP mixes (Kennedy and Perez, 1978; Taylor, 1978). RAP can be used as a subbase, and will be mixed with virgin pavement materials and also as a fill material for various geotechnical application. The first usage of RAP in asphalt mixes was in 1973 with 3% of it allowed to replace the new materials in an asphalt mix, from that time due to the escalation in the prices of the binders and aggregates and also due to their availability, higher percentages are being allowed, which has been further enhanced now to 20–30% and even 50% (Abdo, 2016). The materials from existing pavement are recycled to produce fresh pavement layers resulting in significant saving of material, money, and energy and also helps to find a better alternative for the disposal of old materials. Recycling also offers considerable savings in cost of material and haulage.

The binder undergoes aging (evaporation of volatile fractions and oxidative reactions) during the design life of the asphalt pavement layer resulting in reduction in the penetration and increase in the viscosity and stiffness of the binder. The final stiffness of the reclaimed RAP binder depends on the initial properties of the binder and the intensity of aging undergone by the binder (Ma et al., 2010; Khosla et al., 2012; Zaumanis et al., 2013). While the proportion of RAP material used in the RAP mixes was considerably small (less than 15%) for the initial few decades (West, 2010), the use of RAP in pavements, the scarcity of road quality aggregates has been driving the use of increased proportions of RAP in hot mixes (Izaks et al., 2015). The grade of virgin binder and the type of rejuvenator to be added depends on the properties of the target binder (for the type of pavement and climatic conditions) and the RAP binder properties. The grade and proportion of virgin binders are usually selected with the help of appropriate blending charts or binder/mix evaluation methods (MS-2, 2015).

7.2   UTILIZATION OF RAP IN DIFFERENT COUNTRIES

Various countries have explored the use of RAP in different road layers. Most state departments of transportation specify the use of RAP material in HMA mixes, the national average percentage of RAP usage was estimated to be around 12% (Copeland, 2011). There are only a few state agencies in USA and Canada that don't allow usage of RAP in bituminous layers, rest of the states allow usage of RAP in bituminous as well as base layers. The restriction on usage of RAP is due to its performance when it is used in surface layers (Copeland, 2011). Florida DOT developed guidelines which suggest that 80% RAP can be used for the base and subbase as well as filler material without compromising the engineering properties and showed that RAP does not have any environmental concerns. Various leaching test methods over a period of one year show that the measured heavy metals were within the limits of USEPA guidelines. Table 7.1 shows the RAP usage by various guidelines/countries.

TABLE 7.1

Recommendation of RAP Usage by Various Transportation Agencies

Major Countries	Permitted RAP (%)	Surface Layer	Base Layer	Remarks	Reference
USA	0–10	Yes	Yes	Almost all states	(Copeland, 2011)
	11–19	Yes	Yes	Almost all states	
	20–29	Yes	Yes	Few states	
	> 30	Yes	Yes	Very few states	
India	10	Yes	Yes	Widely used in bituminous macadam road	(MoRTH, 2013)
UK	10	–	–	Used in stone mastic asphalt (SMA) mixes	
New Zealand and Australia	15	–	–	Used in dense graded bituminous mixes	(Austroads, 2015)

7.3 METHODS OF RECYCLING

The recycling of RAP is very important for various infrastructure application such as pavement, embankment, building blocks, etc. The recycling of RAP saves million dollars. It is believed that the RAP matches the specifications of new materials or even better quality than locally available natural materials. The Asphalt Recycling and Reclaiming Association (1992) and IRC 120-2015 have listed most used methods for recycling of asphalt: (a) hot in-place recycling, (b) hot in-plant recycling, (c) cold in-place recycling, and (d) full depth reclamation.

7.3.1 HOT IN-PLACE RECYCLING

In the method of hot in-place recycling, pavement surface is milled to a certain depth by heating it with the help of the heating plates. Once the bituminous material from the pavement is milled, required amounts of the fresh aggregates and binder are added to this milled RAP material. The mixture of the RAP material, fresh aggregates, and the binder are mixed thoroughly, laid and compacted to reach the target thickness. The advantages of hot in-place recycling consume less time and a very low level of interruption is caused to the traffic. Additional savings are in terms of the transportation cost that is incurred to transport the material to and fro to the plant and construction site in case of in-plant methods. In this method, surface cracks, ruts, and bumps can be eliminated easily and accurately. Generally, the depths of recycling ranging from 20 to 50 mm are the depths for which the hot in-place recycling method is adopted.

7.3.2 HOT IN-PLANT RECYCLING

In the hot in-plant recycling method, recycled material is transported to the mixing plant where it is mixed with the virgin aggregate and binder to prepare recycled mixture. Mixing is carried out at appropriate mixing temperatures. This method requires minimal alteration in the mixing plants used for normal hot mix asphalt mixtures. RAP mixes using hot mix asphalt technique can be produced by using both batch and drum type mixing plants. In this method typically 10–30% RAP is used for hot mix recycling. The advantage of this method is that the RAP has similar durability performance as that of fresh HMA.

7.3.3 COLD IN-PLACE RECYCLING

The bituminous material removed to certain depth from the existing bituminous layers is reused along with the fresh aggregates and binders to prepare cold recycled mixes in cold in-place recycling method. To remove the material from the bituminous layers, the material is scarified and crushed. To the crushed RAP material appropriate amounts of fresh aggregates and binder are added. General depth that can be used for cold in-place recycling ranges between 70 and 100 mm. As the process is an *in situ* process, transportation can be avoided partially except for any recycling agent hence material transportation cost is less. Other advantages of this method is an important treatment for distress, better quality of ride, and reduced air quality problems.

7.3.4 Full-Depth Reclamation

In this method, complete set of bituminous layers and a prefixed depth from the base layer are used to lay a stabilized base layer. This method is basically a cold recycling method, in which various kinds of additives such as foam, cement, emulsion, lime, and fly ash are used to achieve a better performing base layer. There are four major steps to produce the base layer mixture, those are: crushing of the existing layers, incorporation of additive, laying and compaction, and laying of a wearing course. If it so happens that the material obtained from the existing pavement layers is not sufficient to match the thickness of the layers that are to be laid, new aggregates are added to match the thickness requirement. This type of recycling method is generally used to lay a thickness of layers ranging between 100 and 300 mm. Few advantages of this method are hauling costs, improved base, reduced waste disposal, and easy to treat pavement distress.

7.4 MIX DESIGN GUIDELINES FOR RAP MIXES

With incorporation of higher RAP content in the hot asphalt mixtures, mix design becomes a significant issue to be examined. Bituminous mixtures containing RAP material are usually designed using the same mix design methods as those used for conventional bituminous mixes. The first step of the hot recycled asphalt mixes mix design is to determine the blending proportion of the RAP material and the required amount of fresh aggregate fractions in order to meet the target gradation. Next step is to select the type of the virgin binder and the quantity. Blending charts based upon viscosity or Superpave binder rutting parameter are given from which the virgin binder grade can be selected (IRC 120, 2015; MS-2, 2015). After selecting the virgin binder grade, trial mixes are prepared at various binder contents using Marshall or Superpave gyratory compactors and optimum binder content is selected based on the appropriate criteria. Flow chart recommended for the design of RAP mixes (MS-20, 1986) is given in Figure 7.1.

Ministry of Road Transport and Highways (MoRTH, 2013) specifies that the minimum penetration of the recovered binder has to be 15 dmm to use a RAP content greater than 10%.

Kandhal and Foo (1997) indicated for use of RAP content below 20%, virgin binder grade can remain constant as the target binder; RAP content between 20 and 30% requires one grade softer virgin binder and RAP content above 30% requires a virgin grade binder that can produce target binder properties while usage. According to EN 13108-1 (2006), the same virgin binder grade can be used for a mixture containing up to 10% RAP for surface courses and 20% for base courses. For mixtures containing higher RAP percentages, penetration and softening points have to be estimated using the blending equations for the selection

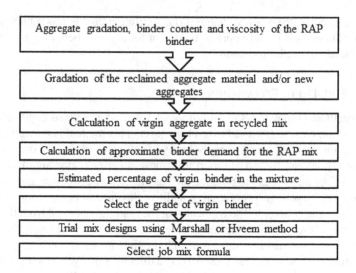

FIGURE 7.1   Flow chart of mix design procedure for RAP mixes.

of virgin binder grade. AASHTO M323 (2013) recommends same grade of virgin binder for RAP content below 15%, one grade softer binder for 15–25% RAP and use of blending charts for mixes containing more than 25% RAP. The minimum values of penetration lie between 5 and 15 dmm and maximum softening point values lie between 70 and 77°C for different European countries in order to select RAP material suitable for recycling.

NCHRP 452 report (McDaniel and Anderson, 2001) provides two types of methods for virgin binder identification: (1) choose virgin binder grade based upon the RAP content and the RAP binder quality and (2) virgin binder grade is initially identified through which the RAP binder percentage is determined for a given target binder grade. Blending charts and equations are provided for PG upper, intermediate, and lower temperatures to calculate the virgin binder grade or RAP binder content.

MS-2 (2015) gives the guidelines for the mixing temperatures of RAP material and virgin aggregates and binders. RAP material is heated to 110°C, virgin binder at its mixing temperature and the virgin aggregates are superheated to 0.5°C for every 1% of RAP usage. For example, if one uses 30% RAP to prepare the bituminous mixture, virgin aggregates are superheated by 15°C. Determination of theoretical maximum specific gravity (Gmm) of RAP mixtures has to be done by remixing the sample with 1–3% binder in order to remove the uncoated faces created during the processing of RAP material (MS-2, 2015). The Gmm value is determined through similar process done for regular mix designs but by correcting the volume and mass for the binder added.

Applicability of rule of mixes concept is a simple concept where the compound materials property will be equal to the weighted mean of the individual materials used to prepare the compound material. Applicability of rule of mixes concept for RAP binder blend properties has been examined by Mullapudi et al. (2019), have reported that the rule is observed to be valid except for the PG (performance grade) upper temperatures. Usually applicability of the rule of mixes concept is assumed to be valid by MS-2 (2015), to select the virgin binder grade for RAP mixes.

7.5 CHALLENGES FACED WHEN USING RAP MATERIAL IN BITUMINOUS MIXES

7.5.1 Aggregate Characteristics

Target gradation for the RAP mixtures is similar to that of any of the conventional gradations used in a particular country or a state. The aggregate gradation of the RAP material is characterized after extracting the binder from the RAP material. Usually the target gradations are prepared by adding required fractions of virgin aggregates and a selected percentage of RAP material. During the milling process if excessive fines are generated, the fractionation technique can be used to separate aggregates of different sizes according to the requirement and then mixed at different proportions to keep the RAP aggregate gradation within the limits (McDaniel et al., 2002; Copeland, 2011; Shannon, 2012; Sabahfer and Hossain, 2015; Gottumukkala, 2021). Use of higher amounts of coarser fraction is generally encouraged to reduce the finer-sized fractions which tend to appear more after milling of RAP. But, using higher amounts of finer RAP content can reduce the amount of virgin binder content to achieve the design criteria (Brock and Richmond, 2007; Bhrath, 2016; Bharath et al., 2021).

7.5.2 Virgin and RAP Binder Blending Characteristics

Blending of the virgin and RAP binder is going to decide the properties as well as the performance of the RAP mixes. Researchers conceptualize three different scenarios of blending in the RAP mixes: (1) Complete blending (100%), (2) Black rock method (0% blending), and (3) Partial blending. Complete blending is easy, calculating the combined binder's property for the identification of virgin binder and other purposes, which was not happening in the RAP mixtures being prepared (McDaniel et al., 2000; Shirodkar et al., 2013). Shirodkar et al. (2011) reported that with 25% RAP content and PG 70-28 as virgin binder the degree of blending as 70% while for 35% RAP content and PG 58-28 virgin binder was used the degree of blending was observed to be 96%. Also, with increase in the difference in virgin binder and RAP binder properties the degree of blending reduces (Shirodkar et al., 2011). Shirodkar et al. (2013) reported that for higher percentages of RAP (such as 50%) when the degree of blending is assumed to be 100% but in reality it is as less than 50% degrees of blending, there was a change in the critical temperature grade of the resultant binder. This difference in the critical temperature grade of the binder will have its effect on the performance characteristics of the RAP mixes. Hence degree of blending between the RAP and virgin binder is one of the critical issues that has to be taken care before one starts to work on the selection of the virgin binders for different percentages of RAP.

7.5.3 PROPERTIES OF THE RAP BINDER

In the course of usage of the bituminous mixtures in the pavements, binder undergoes aging which will increase the amount of asphaltene content resulting in increase of viscosity, stiffness, and decrease in ductility. This increase in stiffness of the RAP binder will increase the susceptibility to cracking, if proper care is not taken while selecting virgin binder grade. Also, the stiffness of the RAP binder is not constant throughout the width of the film thickness. Noureldin and Wood (1987) adopted a method called staged extraction by using TCE as solvent, in order to determine the stiffness of the binder across its film thickness. They observed that the outermost layer is 7 times stiffer than the innermost layer and had 2 times lower penetration value. Use of rejuvenators is also another technique to soften the aged RAP binders, so that the resultant binder's stiffness can be within the limits of the target binder.

7.6 PERFORMANCE CHARACTERISTICS OF RAP MIXES

7.6.1 RUTTING OF RAP MIXES

With increase in the RAP percentage in the mix, the stiffness as well as the rutting resistance of the mix generally increase until unless there is no rejuvenator used to unstiffen the aged RAP binder (Kim et al., 2007; Xiao et al., 2007; Zhou et al., 2011; Bernier et al., 2012; Boriack, 2014).

7.6.2 FATIGUE CRACKING CHARACTERISTICS OF RAP MIXES

Shu et al. (2008) conducted indirect tensile strength, resilient modulus, and beam fatigue tests on mixes with 0%, 10%, 20%, and 30% RAP content and observed that indirect tensile strength and resilient modulus increased but failure strain, toughness index, and dissipated creep strain energy decreased with increase in RAP percentage. Longer fatigue life was obtained for mixes with 30% RAP content compared to mixes with 0%, 10%, and 20% RAP.

Widyatmoko (2008) evaluated the fatigue resistance of asphalt mixtures containing different percentages of RAP (0%, 10%, 30%, and 50%) using indirect tensile fatigue test (ITFT). Virgin binder of 80/100 penetration grade and a rejuvenating oil were used. The fatigue resistance of the RAP mixtures was observed to be equal to or greater than that of the mixtures prepared without RAP. Increase in the proportion of the rejuvenated binder seemed to improve fatigue resistance of the mixes.

Hajj et al. (2009) evaluated the fatigue characteristics of the mixtures containing different percentages (0%, 15%, and 30%) of RAP material obtained from three different sources prepared using two different grades of virgin binders (PG64-22 neat binder and PG64-28 polymer modified bitumen) using beam fatigue test in control strain mode at a temperature of 22°C. The fatigue lives of the mixes prepared with RAP and PG64-22 were longer than those of the mixes without RAP material. For the mixes prepared using PG64-28 binder, the fatigue lives obtained were smaller than those of the virgin mixes. To explain the fatigue performance of the RAP mixes more realistically, a typical pavement structure was analyzed considering different RAP mixes. The initial strain values obtained from the analysis were used in appropriate fatigue models developed for different RAP mixes. In the case of PG64-22 virgin binder, use of 15% RAP improved the fatigue life and 30% RAP reduced the fatigue life for two sources of RAP and for the third RAP source, the fatigue life increased. For mixes with PG64-28 virgin binder, irrespective of the RAP percentage and RAP source, the fatigue lives reduced significantly.

Tabaković et al. (2010) conducted cyclic wheel tracking (CWT) test and controlled stress ITFT on mixes containing various percentages of RAP and reported that the fatigue resistance of mixes improved with increase in the RAP content. From the crack length and crack area obtained from the CWT test, it was observed that the fatigue resistance was more for the mixtures containing RAP compared to the virgin mixes. When compared at equal number of load passes, the mixtures prepared with 30% and 20% RAP contents had smaller crack area followed by the mixes prepared with 10% RAP. The crack area was the maximum for the section with virgin mixes.

Huang et al. (2011) evaluated the cracking resistance of mixtures containing RAP material (0%, 10%, 20%, and 30%) with two types of virgin binders (neat and polymer modified) through beam fatigue test conducted in control strain mode. With the inclusion of RAP material, the fatigue resistance of the mixtures, evaluated in terms of the plateau value of the dissipated energy vs. cycles plot, reduced for mixes with neat as well as modified virgin binders. The fatigue life of the mixtures determined using the criterion of 50% stiffness reduction was higher for mixes with higher percentages of RAP. The field sections constructed using different percentages of RAP were monitored at the end of four years. The section having 30% RAP content exhibited marginally more cracks than those of the section with 20% or less RAP.

Ajideh et al. (2013) conducted uni-axial fatigue test in control stress mode on the specimens prepared using 0 and 50% RAP. The fatigue lives of the mixtures determined using 50% stiffness reduction criterion was higher for the mixes prepared with 50% RAP at all the tested stress levels. Willis et al. (2013) evaluated the fatigue resistance of mixes prepared with 0%, 25%, and 50% RAP contents using overlay tester. The mixtures were prepared using PG67-22 as virgin binder. The inclusion of RAP content reduced the number of cycles to failure (defined as reduction in stiffness by 93%) showing that increase in RAP content drastically decreased the fatigue resistance in this extremely high strain test. Willis et al. (2013) proposed two ways of improving the fatigue cracking resistance of RAP mixes; using softer grade binder or increasing the amount of virgin binder content. Both the approaches improved the fatigue resistance of the RAP mixes. The authors recommended that increasing the virgin binder content can be considered for RAP contents up to 30% and soft grade virgin binder can be used beyond 30% RAP content.

Boriack (2014) conducted beam fatigue test in control strain mode on the beam specimens prepared with 0%, 20%, 40%, and 100% RAP. The fatigue lives of the mixes prepared with RAP material were lower than the fatigue life of the mix prepared without RAP material. Additional specimens were prepared and tested for 100% RAP mixes at binder contents which were 0.5%, 1.0%, and 1.5% higher than the design binder content. The fatigue life of the 100% RAP mix prepared with 1.5% additional binder was comparable to that of the 20% RAP mixture prepared at the design binder content.

Mannan et al. (2015) examined the fatigue life of asphalt mixtures (beam fatigue) as well as binders (linear amplitude sweep) in control strain mode containing RAP material at different percentages of RAP (0% and 35%). The fatigue lives of the mixtures containing RAP material reduced when compared to the mixtures without RAP. In contrast, the fatigue resistance of the binders containing RAP binder increased when compared in terms of the fatigue lives obtained from linear amplitude sweep as well as time sweep tests. The lower fatigue resistance of the RAP mixtures was attributed to the poor interaction between virgin binder and RAP aggregates whereas the binders used for the fatigue testing were obtained by extracting and recovering the binder from the prepared mixtures which facilitated proper blending of the virgin and RAP binders.

Pasetto and Baldo (2017) tested mixtures containing different contents (0%, 20%, and 40%) of RAP material using four-point bending test in control strain mode. Compared to the mixes without any RAP material, the mixes containing RAP material exhibited higher resistance to fatigue. The improved resistance to fatigue was attributed to the greater affinity of virgin binder toward the RAP particles. In the mixture prepared with RAP material, the aggregates were coated with stiffer RAP binder which reduces the stress concentrations which in turn improves the fatigue resistance of RAP mixes. Mullapudi et al. (2020a) examined the fatigue characteristics of the RAP mixes using ITFT in constant stress mode and reported that the fatigue lives increased with incorporation of RAP into the mixes.

As can be seen from the afore-mentioned discussion, contradictory results were reported about the fatigue characteristics of the asphalt mixtures containing RAP material. Due to this, there is a need to examine the fatigue characteristics of the mixtures containing RAP materials.

7.7   USE OF WARM MIX ADDITIVES (WMAs) FOR THE PREPARATION OF RAP MIXES

WMA usage is one of the new age technologies used to reduce the production temperature (mixing and compaction), which reduces the aging caused during the production stage (short-term aging). WMA can be classified as foam-based technologies, rheological modifiers, and chemical modifiers based on the principle through which they reduce the viscosity of bitumen in the range of construction temperatures (Shiva Kumar and Suresha, 2019). Mixes containing RAP and WMA showed better rutting resistance than those of the mixes containing only WMA (Doyle & Howard, 2013; Hill, 2011; Mogawer et al., 2013, Zhao et al., 2013). Oliveira et al. (2012) reported that fatigue resistance of the open graded mixes with WMA and RAP material was better than that of the mixes not having RAP material. Moisture damage resistance of the WMA mixes can be a bit tricky as the production temperatures are lower compared to the hot mix asphalt due to which complete removal of moisture from the aggregates has to be a mandatory check to enhance the resistance to moisture damage.

7.8 PREPARATION OF HIGH MODULUS MIXES USING RAP MATERIAL

The requirements of high modulus bituminous mixes are to have a minimum dynamic modulus of 14,000 MPa at testing conditions of 15°C temperature and 10 Hz frequency. To produce mixes having high dynamic modulus, binder having penetration value of 30 or lower is generally used. In order to improve the flexibility of the mixture, fine gradation and higher binder contents (using lower design air voids) are used. Due to the good rutting resistance and moisture damage resistance, high modulus mixes are popularly used in European countries. RAP usage to produce high modulus mixes has been examined by few researchers. Ma et al. (2016) reported that the rutting resistance improved, moisture resistance reduced but still was within the limiting criteria and anti-fatigue along with low temperature properties reduced with incorporation of RAP for the preparation of high modulus asphalt mixtures.

7.9 RELATION BETWEEN CHEMICAL MAKEUP AND MECHANICAL PROPERTIES OF RAP MIXES

Chemical makeup of the binder has an effect on the mechanical and rheological properties of the binders as well as the mixtures. For the normal bituminous mixtures, the mechanical and rheological properties were studied in the past with respect to SARA (saturates, aromatics, resins, and asphaltenes) fractions, size of molecules and quantity of functional groups. The studies conducted to understand the dependency of binder and mixture properties on the chemical makeup of the RAP binder blends are very limited. Mullapudi and Sudhakar Reddy (2020) examined the relations between the cohesive surface free energy and indices obtained for aging through Fourier transform infrared spectroscopy. Mullapudi et al. (2019) examined the relationship between the aging indices determined from Fourier transform infrared spectroscopy and mechanical properties of the binders such as softening point, viscosity, and complex modulus. The oxides and aromatics were observed to have direct proportionality with the softening point, viscosity, and complex modulus. Mullapudi et al. (2020b) have observed that there is an inverse relationship between fatigue and healing ability of RAP binder blends with oxides and aromatics; and a direct relation with surface free energy of the binder blends. Relationship between rheological properties and cohesive surface free energy values was examined and reported by Mullapudi et al. (2020c). Matolia et al. (2020) examined the relationship between surface free energy of the binder blends with different components of the binders (SARA). Mullapudi et al. (2020a) have reported that there is an inverse relation between the oxides and aromatics with the healing ability of the RAP mixtures. Dependence of fatigue and healing ability of the RAP mixes on the chemical characteristics such as functional groups and surface free energy values have been reported by Mullapudi et al. (2020a, d).

7.10 DISCUSSION ON RAP MATERIAL AND MEASURES BEING FOLLOWED TO REDUCE VARIABILITY

Variability of the RAP mixes is usually examined by collecting samples at different intervals and determining the asphalt content and gradation of the aggregates. Estakhri et al. (1999) has examined 13 stockpiles and reported that the variability in the RAP material is less than that of the virgin aggregates and incorporating higher RAP content also didn't increase the variability. NAPA (1996) RAP material from a single source is typically consistent. Zaumanis et al. (2018) reported that there was an improvement in the limiting value of RAP percentage by milling and further improvement was observed by stockpile mixing in terms of variability. When more than one source of RAP material is usually used to produce RAP mixes, the RAP material has to be homogenized by mixing and characterize the mixed material for the aggregate gradation and properties of recovered bitumen. Newcomb et al. (2007) recommends to have the coefficient of variability less than 20%. West (2015) in their research report specifies to have the binder content in the sampled RAP material to have standard deviation less than 0.5%, variation in aggregates passing medium size sieve should be less than 5% and passing through sieve. No. 200 should be less than 1.5%. Valdés et al. (2011) and Khosla et al. (2009) also emphasized the importance of homogenization of the RAP stockpile when more than one source of RAP is being used. For larges-cale projects handling RAP mixes may not be able to obtain RAP material from a single source when they use larger percentages of RAP. For this purpose, when more than one source of RAP is required to be used, one source of the RAP material can be used after another to produce the mixtures (Montanez et al., 2020). While doing so the design of the mixtures has to be adjusted whenever change in RAP source is made. Failing to do so will take a toll on the mixtures performance and durability.

Fractionation of the RAP material is one of the techniques used to reduce the variability of the RAP mixes apart from the stockpile mixing. In the fractionation technique the stockpiled RAP is sieved and kept in separate sizes. This will facilitate the usage of required proportions of each fraction to achieve the target gradation. By controlling different fractions of the RAP material, the percentage of RAP that can be added to the mix is enhanced. Several researchers have demonstrated the advantages of fractionation technique to enhance the percentage of RAP usage (Shannon, 2012; Bhrath, 2016).

7.11 USE OF DIFFERENT TECHNOLOGIES FOR PRODUCING RAP MIXES IN THE FIELD

For producing RAP mixes using hot mix asphalt technology, there are different techniques used to maintain the RAP materials mixing temperature at a lower temperature than that of the virgin aggregates. This is to further retain the RAP binder properties by not damaging it during the production stage and certain minimum temperature is used to activate the binder present in RAP material. Double barrel drum technology is used in some production plants where in the inner drum contains and heats virgin aggregates and the outer drum will contain RAP material. The heating of RAP material is done through the heat transferred from the inner drum. The heat energy transferred from the inner drum is captivated with the help of insulation material used to produce outer drum and this heat energy is utilized to heat the RAP material. Introduction of RAP material at the center of the drum (central entry method) is used to avoid the blue smoke from the RAP material. In this method RAP enters approximately at mid-point of the drum away from the burner. Heat from the superheated virgin aggregates is utilized to heat the RAP to mixing temperatures. With this technique direct contact of RAP material with the burner flame is avoided to safeguard RAP material from damage that can be caused during production stage by improper handling.

7.12 CONCLUSIONS

The chapter summarizes the characteristics of RAP material generated from rehabilitation of flexible pavement. Different types of recycling techniques used to reclaim the RAP material into bituminous mixes have been discussed. The variance between the virgin binders and the binder present in the RAP material has also be summarized. The performance characteristics of the RAP mixes with increasing RAP contents have been discussed briefly. The dependence of different properties of RAP binder blends and mixes upon the chemical makeup of the binders has been reported. Different technologies like warm mix asphalt additives and usage of RAP for high modulus asphalt mixes preparation has also been discussed along with the benefits offered by each of the techniques. Various techniques followed to incorporate RAP into the mixing process in the field and also fractionation techniques used to reduce the variability in the produced RAP mixes have been discussed.

REFERENCES

AASHTO M323. 2013. *Specification for Superpave Volumetric Mix Design*. Washington, DC: American Association of State Highway and Transportation Officials.

Abdo, A. A. 2016. Utilizing reclaimed asphalt pavement (RAP) materials in new pavements – A review. *International Journal of Thermal & Environmental Engineering*, 12(1): 61–66.

Ajideh, H., Bahia, H., Carnalla, S., Earthman, J. 2013. Evaluation of fatigue life of asphalt mixture with high RAP content utilizing innovative scanning method. In *Airfield and Highway Pavement, Sustainable and Efficient Pavements*, pp. 1112–1121.

Austroads. 2015. *Maximising the Re-Use of Reclaimed Asphalt Pavement Outcomes of Year Two: RAP Mix Design, AP-T286-15*. Sydney, NSW: Austroads.

Bernier, A., Zofka, A., Yut, I. 2012. Laboratory evaluation of rutting susceptibility of polymer-modified asphalt mixtures containing recycled pavements. *Construction and Building Materials*, 31: 58–66.

Bharath, G., Reddy, K. S., Tandon, V., Reddy, M. A. 2021. Aggregate gradation effect on the fatigue performance of recycled asphalt mixtures. *Road Materials and Pavement Design*, 22(1): 165–184.

Bhrath, G. 2016. *Performance characteristics of bituminous mixtures containing reclaimed asphalt pavement material*. Ph.D. thesis submitted to Indian Institute of Technology Kharagpur, India.

Brock, J. D., Richmond, J. L. 2007. *Milling and Recycling*. Chattanooga, TN: ASTEC, Technical Paper T-127.

Boriack, P. C. 2014. *A laboratory study on the effect of high RAP and high asphalt binder content on the performance of asphalt concrete*. Ph.D. thesis, Virginia Tech, USA.

Copeland, A. 2011. Reclaimed asphalt pavement in asphalt mixtures: State of the practice (No. FHWA-HRT-11-021).

Doyle, J. D., Howard, I. L. 2013. Rutting and moisture damage resistance of high reclaimed asphalt pavement warm mixed asphalt: Loaded wheel tracking vs. conventional methods. *Road Materials and Pavement Design*, 14(2): 148–172.

EN 13108-1. 2006. *Bituminous Mixtures: Material Specifications: Part 1: Asphalt Concrete*. European Committee for Standardization.

Estakhri, C., Spiegelman, C., Gajewski, B., Yang, G., Little, D. 1999. *Recycled Hot-Mix Asphalt Concrete in Florida: A Variability Study*. Austin, TX: International Center for Aggregates Research.

Gottumukkala, B., Kusam, S. R., Tandon, V., Muppireddy, A. R. 2018. Estimation of blending of rap binder in a recycled asphalt pavement mix. *Journal of Materials in Civil Engineering*, 30(8): 04018181.

Hajj, E. Y., Sebaaly, P. E., Shrestha, R. 2009. Laboratory evaluation of mixes containing recycled asphalt pavement (RAP). *Road Materials and Pavement Design*, 10(3), 495–517.

Hill, B. 2011. *Performance evaluation of warm mix asphalt mixtures incorporating reclaimed asphalt pavement* (Master of Science in Civil Engineering thesis). University of Illinois, Illinois.

Huang, B., Shu, X., Vukosavljevic, D. 2011. Laboratory investigation of cracking resistance of hot-mix asphalt field mixtures containing screened reclaimed asphalt pavement. *Journal of Materials in Civil Engineering*, 23(11): 1535–1543.

IRC 120. 2015. *Recommended Practice for Recycling of Bituminous Pavements*. Indian Road Congress, New Delhi.

Izaks, R., Haritonovs, V., Klasa, I., Zaumanis, M. 2015. Hot mix asphalt with high RAP content. *Procedia Engineering*, 114: 676–684.

Kandhal, P. S., Foo, K. Y. 1997. *Hot Mix Recycling Design Using Superpave Technology*. ASTM, Special Technical Publication, 1322.

Kennedy, T. W., Perez, I. 1978. Preliminary mixture design procedure for recycled asphalt materials. In *Recycling of Bituminous Pavements, ASTM STP 662*, ASTM International, pp. 47–67.

Khosla, N. P., Nair, H. K., Visintine, B., Malpass, G. 2009. Material characterization of different recycled asphalt pavement sources. *International Journal of Pavements*, 8: 13–24.

Khosla, N. P., Nair, H., Visintine, B., Malpass, G. 2012. Effect of reclaimed asphalt and virgin binder on rheological properties of binder blends. *International Journal of Pavement Research and Technology*, 5(5): 317–325.

Kim, S., Byron, T., Sholar, G. A., Kim, J. 2007. *Evaluation of Use of High Percentage of Reclaimed Asphalt Pavement (RAP) for Superpave Mixtures*. FDOT/SMO/07-507, Florida Department of Transportation, Florida.

Ma, T., Huang, X., Bahia, H. U., Zhao, Y. 2010. Estimation of rheological properties of RAP binder. *Journal of Wuhan University of Technology, Materials Science Edition*, 25(5), 866–870.

Ma, T., Zhao, Y., Huang, X., Zhang, Y. 2016. Using RAP material in high modulus asphalt mixture. *Journal of Testing and Evaluation*, 44(2): 781–787.

Mannan, U. A., Islam, M. R., Tarefder, R. A. 2015. Effects of recycled asphalt pavements on the fatigue life of asphalt under different strain levels and loading frequencies. *International Journal of Fatigue*, 78: 72–80.

Matolia, S., Guduru, G., Gottumukkala, B., Kuna, K. K. 2020. An investigation into the influence of aging and rejuvenation on surface free energy components and chemical composition of bitumen. *Construction and Building Materials*, 245: 118378.

McDaniel, R. S., Soleymani, H. R, Anderson, R. M., Turner, P., Peterson, R. 2000. *Recommended Use of Reclaimed Asphalt Pavement in the SuperPave Mixture Design Method*. NCHRP Final Report (9–12), Transportation Research Record. Washington, DC: National Research Council.

McDaniel, R. S., Anderson, R. M. 2001. *Recommended Use of Reclaimed Asphalt Pavement in the Superpave Mix Design Method: Technician's Manual (No. Project D9-12 FY'97)*. National Research Council (US), Transportation Research Board.

McDaniel, R. S., Soleymani, H., Shah, A. 2002. *Use of Reclaimed Asphalt Pavement (RAP) Under Superpave Specifications*. Washington, DC: Federal Highway Administration (FHWA/IN/JTRP-2002/6).

MoRTH. 2013. *Specifications for Road and Bridge Works*. Ministry of Road Transport and Highways, Indian Roads Congress, New Delhi.

Mogawer, W., Austerman, A., Mohammad, L., Kutay, E. M. 2013. Evaluation of high RAP-WMA asphalt rubber mixtures. *Road Materials and Pavement Design*, 14(2): 129–147.

Montanez, J., Caro, S., Carrizosa, D., Calvo, A., Sanchez, X. 2020. Variability of the mechanical properties of Reclaimed Asphalt Pavement (RAP) obtained from different sources. *Construction and Building Materials*, 230: 116968.

MS-2. 2015. *Asphalt Mix Design Methods*. Asphalt Institute, 7th Edition, Manual Series No. 2, Asphalt Institute, USA.

MS-20. 1986. *Asphalt Hot Mix Recycling*. Asphalt Institute, 2nd Edition, Manual Series No. 20, Asphalt Institute, USA.

Mullapudi, R. S., Sudhakar Reddy, K. 2020. An investigation on the relationship between FTIR indices and surface free energy of RAP binders. *Road Materials and Pavement Design*, 21(5): 1326–1340.

Mullapudi, R. S., Deepika, K. G., Reddy, K. S. 2019. Relationship between chemistry and mechanical properties of RAP binder blends. *Journal of Materials in Civil Engineering*, 31(7): 04019124.

Mullapudi, R. S., Aparna Noojilla, S. L., Reddy, K. S. 2020a. Fatigue and healing characteristics of RAP mixtures. *Journal of Materials in Civil Engineering*, 32(12): 04020390.

Mullapudi, R. S., Chowdhury, P. S., Reddy, K. S. 2020b. Fatigue and healing characteristics of RAP binder blends. *Journal of Materials in Civil Engineering*, 32(8): 04020214.

Mullapudi, R. S., Sudhakar Reddy, K. 2020c. Relationship between rheological properties of RAP binders and cohesive surface free energy. *Journal of Materials in Civil Engineering*, 32(6), 04020137.

Mullapudi, R. S., Noojilla, S. L. A., Kusam, S. R. 2020d. Effect of initial damage on healing characteristics of bituminous mixtures containing reclaimed asphalt material (RAP). *Construction and Building Materials*, 262: 120808.

NAPA. 1996. *Recycling Hot Mix Asphalt Pavements*. Lanham, MD: NAPA.

Newcomb, D. E., Brown, E. R., Epps, J. A. 2007. *Designing HMA Mixtures with High RAP Content: A Practical Guide*. National Asphalt Pavement Association.

Noureldin A. S., Wood L. E. 1987. Rejuvenator diffusion in binder film for hot-mix recycled asphalt pavement. *Journal of the Transportation Research Board*, 1115: 51–61.

Oliveira, J. R. M., Silva, H. M. R. D., Abreu, L. P. F., Gonzalez-Leon, J. A. 2012. The role of a surfactant based additive on the production of recycled warm mix asphalts – Less is more. *Construction and Building Materials*, 35: 693–700.

Reddy, P. S., Reddy, N. G., Serjun, V. Z., Mohanty, B., Das, S. K., Reddy, K. R., Rao, B. H. 2021. Properties and assessment of applications of red mud (bauxite residue): Current status and research needs. *Waste and Biomass Valorization*, 12: 1185–1217.

Pasetto, M., Baldo, N. 2017. Fatigue performance of recycled hot mix asphalt: A laboratory study. *Advances in Materials Science and Engineering*, 1–10.

Sabahfer, N., Hossain, M. 2015. Effect of fractionation of reclaimed asphalt pavement on properties of Superpave mixtures with reclaimed asphalt pavement. *Advances in Civil Engineering Materials*, 4(1): 47–60.

Shannon, C. P. 2012. Fractionation of recycled asphalt pavement materials: Improvement of volumetric mix design criteria for High-RAP content surface mixtures.

Shirodkar, P., Mehta, Y., Nolan, A., Sonpal, K., Norton, A., Tomlinson, C., Sauber, R. 2011. A study to determine the degree of partial blending of reclaimed asphalt pavement (RAP) binder for high RAP hot mix asphalt. *Construction and Building Materials*, 25(1): 150–155.

Shirodkar, P., Mehta, Y., Nolan, A., Dubois, E., Reger, D., McCarthy, L. 2013. Development of blending chart for different degrees of blending of RAP binder and virgin binder. *Resources, Conservation and Recycling*, 73: 156–161.

Shiva Kumar, G., Suresha, S. N. 2019. State of the art review on mix design and mechanical properties of warm mix asphalt. *Road Materials and Pavement Design*, 20(7): 1501–1524.

Shu, X., Huang, B., Vukosavljevic, D. 2008. Laboratory evaluation of fatigue characteristics of recycled asphalt mixture. *Construction and Building Materials*, 22(7): 1323–1330.

Tabaković, A., Gibney, A., McNally, C., Gilchrist, M. D. 2010. Influence of recycled asphalt pavement on fatigue performance of asphalt concrete base courses. *Journal of Materials in Civil Engineering*, 22(6): 643–650.

Taylor, N. H. 1978. Life expectancy of recycled asphalt paving. In *Recycling of Bituminous Pavements, ASTM STP 662*, ASTM International, pp. 3–15.

Valdés, G., Pérez-Jiménez, F., Miró, R., Martínez, A., Botella, R. 2011. Experimental study of recycled asphalt mixtures with high percentages of reclaimed asphalt pavement (RAP). *Construction and Building Materials*, 25(3): 1289–1297.

West, R. 2010. *Reclaimed Asphalt Pavement Management: Best Practices*. Auburn, AL: National Center for Asphalt Technology, NCAT Draft Report.

West, R. C. 2015. Best practices for RAP and RAS management (No. QIP 129).

Widyatmoko, I. 2008. Mechanistic-empirical mixture design for hot mix asphalt pavement recycling. *Construction and Building Materials*, 22(2): 77–87.

Willis, J. R., Turner, P., de Goes Padula, F., Tran, N., Julian, G. 2013. Effects of changing virgin binder grade and content on high reclaimed asphalt pavement mixture properties. *Transportation Research Record*, 2371(1): 66–73.

Xiao, F., Amirkhanian, S., Juang, C. H. 2007. Rutting resistance of rubberized asphalt concrete pavements containing reclaimed asphalt pavement mixtures. *Journal of Materials in Civil Engineering*, 19(6): 475–483.

Zaumanis, M., Mallick, R. B., Frank, R. 2013. Use of rejuvenators for production of sustainable high content RAP hot mix asphalt. In The XXVIII International Baltic Road Conference, pp. 1–10.

Zaumanis, M., Oga, J., Haritonovs, V. 2018. How to reduce reclaimed asphalt variability: A full-scale study. *Construction and Building Materials*, 188: 546–554.

Zhao, S., Huang, B., Shu, X., Woods, M. 2013. Comparative evaluation of warm mix asphalt containing high percentages of reclaimed asphalt pavement. *Construction and Building Materials*, 44: 92–100.

Zhou, F., Hu, S., Das, G., Scullion, T. 2011. *High RAP mixes design methodology with balanced performance*. Texas Transportation Institute, FHWA/TX-11/0-6092-2.

8 Valorization of Solid and Liquid Wastes Generated from Agro-Industries

Lopa Pattanaik,[1] *Pritam Kumar Dikshit,*[2]
Vikalp Saxena,[3] *and Susant Kumar Padhi*[3]
[1]Department of Biotechnology and Bioinformatics,
NIIT University, Neemrana, Rajasthan, India
[2]Department of Biotechnology, Koneru
Lakshmaiah Education Foundation, Vaddeswaram,
Guntur District, Andhra Pradesh, India
[3]Department of Civil Engineering, Shiv Nadar
University, Greater Noida, Uttar Pradesh, India

CONTENTS

DOI: 10.1201/9781003201076-8

8.1   INTRODUCTION

In today's global lignocellulosic waste scenario, agricultural waste remains a significant contributor, generating around 1.5×10^{11} tons of waste annually (Gupta and Verma, 2015; Pattanaik et al., 2019). As a part of agricultural waste, agro-industrial waste, which is generally referred to as the waste derived during the industrial processing of agricultural products, is quite dominant. This agro-industrial processing waste is broadly categorized into either solid or liquid waste, which includes residues remaining after processing of cereal crops (husk, bran, and straw, etc.), by-products generated from fruits and vegetable processing industries (peels, pulp, pomace, waste from fruit and vegetable processing, etc.), sugar and oil industries waste (molasses, sugar cane bagasse, and de-oiled cake, etc.), dairy industries waste (whey and buttermilk, etc.), slaughterhouse waste (skin, feather, meat, scales, and slaughterhouse processing wastewater, etc.), and other food or non-food-based miscellaneous waste (corn cob, soybean meals, natural dye industry waste, and oil cake from Jatropha, etc.) (Figure 8.1). The wastes

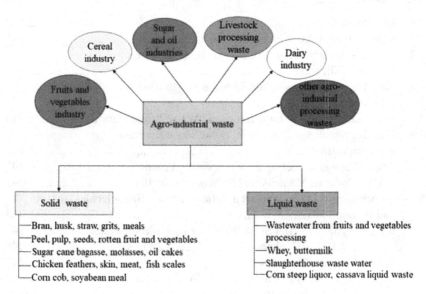

FIGURE 8.1   Types of agro-industrial wastes.

generated from these agro-industries are huge and unutilized. For instance, the global sugarcane bagasse production was estimated to be 180.73 million tons (Saini et al., 2015). Similarly, the waste generated from the oil palm industry is nearly 40% of the fresh palm fruit bunch (Mohanty et al., 2021; Sukiran et al., 2017). The reported wastes from natural indigo industries is almost 99% of the total biomass utilized for dye production (Pattanaik et al., 2020a, 2020b). These processed wastes are generally rich in organic nutrients and cause environmental damage and economic loss due to improper management and non-utilization (Beltrán-Ramírez et al., 2019). So far, mostly these agro-industrial wastes are either burned, landfilled, or disposed of in an uncontrolled manner, resulting in environmental pollution due to the release of harmful gases (CH_4, NH_3, and CO_2) and leachates. Additionally, the financial burden during collection, treatment, and disposal are also unavoidable (Panesar et al., 2015). Therefore, sustainable management of these wastes is of utmost necessity. Exploring various agro-industrial wastes potential and valorization toward biofuels/bioenergy, biomaterials, and bioactive compounds could result in a promising solution. Additionally, integrated waste management by adopting a biorefinery approach and addressing every residue/by-product generated from an agro-industry (solid and liquid waste) into multiple value-added products can also establish the concept of circular economy in the management of agro-industrial wastes (Beltrán-Ramírez et al., 2019). In this chapter, various aspects of agro-industrial waste valorization have been addressed.

8.2   VALORIZATION OF AGRO-INDUSTRIAL WASTES

Generally, agro-industrial wastes are rich in nutrients such as complex polysaccharide/proteins, carbohydrates, and polyphenolic constituents, etc. These nutrient-rich substrates can be exploited for either low-value high-volume products (bioethanol, biogas, compost, and biodiesel, etc.) or high-value low-volume products (biochemicals, bionutrients, biopharmaceuticals, and biomaterials, etc.) (Table 8.1). Even, in a bio-refinery approach, not single but multiple products can be targeted simultaneously from these wastes. However, the route for utilizing these agro-industrial wastes for single or multiple products generally depends on the source, type, and characteristics of the agro-industrial waste and its end-product utilization. For instance, based on the waste characteristics, Miranda et al. (2019) showed the possible utilization of spent olive pomace for lignin, glucose, and functional sugars. Whereas, Federici et al. (2009) utilized olive mill wastewater for the isolation of enzymes and antioxidants. Similarly, Devesa-Rey et al. (2011) proposed different routes for utilization of winery waste like trimming wastes, wine less, and grape marc for production of lactic acid, biosurfactants, and compost. Figure 8.2 depicts various directions of agro-industrial waste valorization.

TABLE 8.1

Single or Multiple Product Generation from Various Agro-Industrial Wastes

Agro-Industrial Waste	Value-Added Products	References
Olive mill wastewater	Antioxidants and enzymes	Federici et al. (2009)
Winery waste	Lactic acid, biosurfactants, and compost	Devesa-Rey et al. (2011)
Spent coffee grounds oil	Biodiesel, biohydrogen, biogas, glycerin, and hydrocarbon fuel, etc.	Atabani et al. (2019)
Defatted spent coffee ground	Bioethanol, compost, bio-oil, polymer, biochar, biogas, and bioactive compounds	Atabani et al. (2019)
Olive oil mill wastewater	Rhamnolipid	Gudiña et al. (2016)
Olive pomace	Lignin, glucose, and functional sugar	Miranda et al. (2019)
Moringa seeds	Biodiesel and bioethanol	Boulal et al. (2019)
Date palm wastes	Bioethanol	Boulal et al. (2019)
Orange peel	Pectin and activated carbon or adsorbent	Tovar et al. (2019) Kebaili et al. (2018)
Apple and grape waste from cider and wine industries	Bioactive compounds and biochar	Sette et al. (2020)
Mango waste	Cellulose nanocrystals	Henrique et al. (2013)
Egg plant	Pectin, phenolics, and pullulan	Kazemi et al. (2019)
Indigofera plant biomass waste	Indigo dye, compost, and nutrients	Pattanaik et al. (2020a)
Liquid waste from the natural indigo dye industry	Act as a source of nutrients for algae and bacterial growth	Pattanaik et al. (2020a)

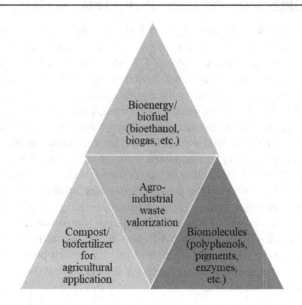

FIGURE 8.2 Agro-industrial waste valorization.

8.3   CHARACTERIZATION OF AGRO-INDUSTRIAL WASTE BIOMASS FOR ITS POSSIBLE UTILIZATION

Before utilization, characterization of the wastes and its potential toward different value-added products is essential (Devappa et al., 2015; Mishra and Mohanty, 2018; Sánchez-Acosta et al., 2019). Various studies have been conducted on characterization of different waste biomass, i.e., rice straw, wheat straw, sugar cane bagasse, oil palm fruit bunch, coffee husk, cattle manure, and orange-peel, etc. to value-added products such as biofuels and compost (Tables 8.2 and 8.3).

Majorly, biochemical and thermochemical are the two possible routes for utilizing different waste biomass by converting them into various value-added products. The biochemical route includes the use of various microorganisms and/ or enzymes to breakdown the biomass into intermediates such as sugars, amino acids, and short-chain fatty acids. These intermediates further converted into biofuels (biogas and bioethanol) and compost. In thermochemical process heat and chemicals are used either individually or in combination to produce syngas, biochar, and bio-oil. In the thermochemical route, a broad range of wastes has been utilized as compared to the biochemical route. However, biochemical process is preferable than the thermochemical from the perspective of fossil fuel consumption and greenhouse gas emission (Mu et al., 2010).

Prior to conversion, the physical and chemical characterization of the biomass is very much important. The biomass compositional analysis includes proximate, ultimate, and lignocellulose composition. The proximate composition of the biomass includes moisture, fixed carbon, volatile solids, and ash content. Moisture is an essential parameter in the biomass, and the moisture content in the different waste biomass generally varies between 3% and 80%, depending on the type of biomass and type of measurement such as oven-dried or air-dried (Vassilev et al., 2010). The biomass with low moisture content (10–15%) is preferred for thermochemical combustion or direct combustion, as higher moisture content would make the process less energy efficient by lowering the heating value (García et al., 2012). However, biomass having high moisture (>30%) content is generally preferable for biochemical conversion, as the optimum moisture content is suitable for the growth and proliferation of micro organisms. So, before utilizing these wastes for any biochemical processes such as biogas, bioethanol, and compost, additional water is added to maintain the optimum moisture content in the process (Chandra et al., 2012).

Similar to the moisture, ash content of the biomass is also important which helps in analyzing mostly the bulk inorganic matter present in the biomass, such as oxide of silica (Si), sodium (Na), potassium (K), calcium (Ca), magnesium (Mg), aluminum (Al), and iron (Fe). The ash contents in the agro-industrial waste (2–18%) are found to be much lower as compared to the other wastes (Pattanaik et al., 2019). However, the inorganic elements (P, K, Na, Ca, Mg, and Fe, etc.) present in the ash generated from the plant biomass provides nutrient for plant growth (Belyaeva and Haynes, 2012). Therefore, plant biomass containing inorganic elements are not suitable for thermochemical conversion, which can be used for composting. The biomass containing higher fixed carbon provides more

TABLE 8.2

Proximate and Ultimate Analyses of Different Agro-Industrial Waste Biomass

Agro-Industrial Waste Biomass	Ultimate Analysis (wt.%)				Proximate Analysis (wt.%)				Heating Value (MJ/kg)	References
	C	H	N	O	Moisture	Ash	Volatile Content	Fixed Carbon		
Rice husk	38.6	5.4	0.7	–	7.7	18.8	64.3	9.2	13.4	Lu et al. (2008); Ji-Lu (2007)
Sugarcane bagasse	45.5	6.0	0.2	45.2	16.0	3.2–4.3	79–83.7	–	18.6–18.7	Tsai et al. (2006)
Orange peel	46.4	5.7	1.5	46.3	7.05	4.6	77.1	18.73	18.4	Aguiar et al. (2008)
Coffee husk	47.50	6.40	–	43.7	–	2.4	78.50	19.1	19.80	Suarez et al. (2000)
Olive deoiled cake	53.7	6.7	0.6	36.2	–	2.8	62.1	34.6	21.6	Yin (2011); Demirbas and Ilten (2004)
Indigofera waste biomass	44.2	6.6	2.2	37.9	6.8	8.9	82.1	1.9	17.4	Pattanaik et al. (2019)

TABLE 8.3
Lignocellulosic Composition and Its Potential of Different Agro-Industrial Wastes

Agro-Industrial Waste	Lignocellulosic Composition			Potential			References
	Cellulose	Hemicellulose	Lignin	Biogas (mL/g VS)	Bioethanol (g/L)	Compost	
Rice husk	39	31	4	584	15.63	Co-composting with chicken manure	Nachaiwieng et al. (2015); Abdul-Hamid and Luan (2000)
Sugarcane bagasse	43.6–45.8	31.3–33.5	18.1–22.9	239	7.27	Enrichment with rock phosphate	Vassilev et al. (2012); Zayed and Abdel-Motaal (2005)
Orange peel	26.0	11.9	18.5	217	41.0	Co-composting with the organic fraction of municipal solid waste	Wikandari et al. (2015); Santi et al. (2014)
Coffee husk	24.5–43	7–29.7	9–23.7	239.7	–	Co-composting with coffee pulp and cow manure	Gouvea et al. (2009); Vassilev et al. (2012),
Olive deoiled cake	22.0	18.2	50.0	419	41.8	Composting with biochar as additives	López-Cano et al. (2016); Messineo et al. (2019)
Indigofera waste biomass	41.15	28.9	12.4	–	–	Composting using Jeevamrutha as an inoculum	Pattanaik et al. (2019); Pattanaik et al. (2020a)

calorific value, which is suitable for combustion (Nanda et al., 2013). Whereas the biomass with higher volatile content is usually preferred for the production of biogas and compost (Jain and Kalamdhad, 2018).

The ultimate analysis of the biomass generally represents five elements, such as C (Carbon), H (Hydrogen), N (Nitrogen), and O (Oxygen), and S (Sulfur). The C, H, N, and O content in different agro-industry waste biomass varies between 34–53%, 4.6–7.3%, 0.1–3.3%, and 33–51%, respectively (Table 8.2), which represents the efficacy of fuel production from the waste biomass. A higher C (lower H/C) and lower O (lower O/C) content of the biomass represents an increase in high heating value (HHV) (Sheng and Azevedo, 2005). Further, the C/N ratio derived from the ultimate analysis can also be helpful in predicting the potential of waste biomass for production of biofuel and compost. In addition, the compositional analysis mainly determines the lignocellulosic content (cellulose, hemicellulose, and lignin), which is required before processing the waste biomass for biofuels production. It is observed that the agro-industry waste majorly contains lignocellulosic contents (80–85%), i.e., cellulose, hemicellulose, and lignin. The waste biomass with high cellulose and hemicellulose contents preferable for biofuel production. The agro-industry wastes contain 21–45% cellulose, 15–33% hemicellulose, and 5–24% lignin, which vary depending on the originating source (Table 8.3). The rice husk usually contains low lignin (4%), for which it is preferable for bioethanol production. On the contrary, sugarcane bagasse contains high lignin (18.1–22.9%) that needs pretreatment to enhance the production of ethanol (Sun and Cheng, 2002).

8.4 VALORIZATION OF AGRO-INDUSTRIAL SOLID WASTE BIOMASS FOR BIOETHANOL

Among various biofuels (biodiesel, biobutanol, and ethanol), ethanol is a major alternative to fossil-derived fuels and a predominant biofuel in transportation sector (Domínguez-Bocanegra et al., 2015). Additionally, bioethanol is renewable, non-toxic and biodegradable in nature. The second-generation biofuels are produced while utilizing non-edible organic matter, such as sugarcane bagasse, rice straw, wood, and agricultural residues, etc. This lignocellulosic biomass is mainly composed of biological polymers such as cellulose, hemicellulose, and lignin, which need to be fractionated and hydrolyzed before utilizing it in the biological processes. Several pretreatment methods such as physical (ultrasonication and microwave, etc.), chemical (acid hydrolysis, alkaline hydrolysis, and ionic liquid, etc.), physico-chemical (steam explosion, ammonia fiber explosion, and CO_2 explosion, etc.), and biological methods (enzymatic and whole cell pretreatment) are exploited for the pretreatment of lignocellulosic biomass (Baruah et al., 2018). Among all these pretreatment methods, acid pretreatment using dilute sulfuric acid and alkaline pretreatment using sodium hydroxide are the most commonly applied techniques for pretreatment of lignocellulosic biomass. The main motives behind the pretreatment process are reducing the degree of polymerization and crystallinity index, removal of hemicellulose and lignin, breakage of lignin-carbohydrate linkage, and increase in porosity for increased enzyme access for enzymatic hydrolysis of biomass in the subsequent step. Enzymatic hydrolysis is the important step in biochemical conversion process that liberates simple sugars (glucose, xylose, mannose, galactose, rhamnose, and arabinose) from the lignocellulosic biopolymer, cellulose, and hemicellulose. These monomeric sugars are further converted by diverse groups of microorganisms for ethanol production. These microorganisms include several bacteria (*Escherichia coli, Klebsiella oxytoca, Erwinia* spp., *Enterobacter asburiae, Corynebacterium glutamicum, Zymomonas mobilis,* and *Zymobacter palmae*), yeast (*Saccharomyces cerevisiae* and *Schizosaccharomyces pombe*), and fungi (*Fusarium oxysporum*) (Bušić et al., 2018). Strains like *S. cerevisiae* and *S. pombe* are industrially important microorganisms for the production of ethanol due to their ability to utilize a wide range of sugars under anaerobic conditions.

Table 8.4 highlights utilization of various agro-industrial wastes for the production of bioethanol. Several agro-industrial residues such as rice straw, rice husk, rice bran, sugarcane bagasse, wheat bran, saw dust, corn stalks, and fruit waste have been explored for their potential in bioethanol production process.

Rice bran, one of the low-cost and abundant by-product from rice production process, can be used as potential substrates for bioethanol production (Okamoto et al., 2011; Watanabe et al., 2009). Todhanakasem et al. (2014) reported the development and application of *Zymomonas mobilis* biofilm for ethanol production from diluted acid pretreated and enzymatic saccharified rice bran. The *Z. mobilis* illustrated significant increase in ethanol yield compared to free cell, which was due to the high metabolic rate and higher survival with exposure to toxic inhibitors. In a similar study, Todhanakasem et al. (2015) used two different strains of *Z. mobilis* (ZM4 and TISTR 551) for immobilization in the form of biofilm on DEAE cellulose, which was used for enhanced ethanol production from rice bran hydrolysate. The fermentation experiments were carried out in both batch and repeated mode. A maximum ethanol yield 0.43 g/g (theoretic yield 83.89%) was produced by *Z. mobilis* TISTR 551 in batch process, while the highest yield 0.354 g/g was observed in fed-batch process with *Z. mobilis* ZM4. The efficiency of ethanol production was mainly limited by the detachment of biofilm from the surface that reduced the biomass concentration leading to reduction in ethanol concentration. A large amount of fruit waste are generated from agricultural processing and food industry worldwide, which end up in the land filling site causing serious threats to the environment (Choi et al., 2015). In contrast, these fruit wastes possesses high levels of fermentable sugars (glucose, sucrose, and fructose), which can be used as substrate in ethanol production. To evaluate the efficacy of this process, Choi et al. (2015) developed a process for conversion of citrus peel waste (CPW) (i.e., orange, mandarin, grapefruit, and lemon) or CPW along with other fruit waste (i.e., banana peel, apple pomace, and pear waste) were used for bioethanol production. The ethanol concentration of 14.4–29.5 g/L and yields reached 90.2–93.1% in an immobilized cell reactor after the removal of D-limonene, an inhibitory compound prior to the fermentation. The removal of D-limonene increased the ethanol production by 12-fold compared to the immobilized cell reactor alone. Other studies also reported the use of orange peel, mandarin peel (Boluda-Aguilar et al., 2010), and lemon peel

TABLE 8.4

Agro-Industrial Wastes and Their Pre-Treatment Conditions for the Production of Bioethanol

Raw Materials	Pre-Treatment/Hydrolysis	Microorganisms	Fermentation Condition	Ethanol Yield	Reference
Rice bran hydrolysate	KOH (0.2 M, 4 h) and H_2SO_4 (2% v/v, 121 °C, 30 min)	Z. mobilis TISTR 551	Batch fermentation; temperature: 30 °C; duration: 5 days	0.43 g/g	Todhanakasem et al. (2015)
	KOH (0.2 M, 4 h) and H_2SO_4 (2% v/v, 121 °C, 30 min)	Z. mobilis ZM4	Batch fermentation; temperature: 30 °C; duration: 5 days	0.18 g/g	Todhanakasem et al. (2015)
Rice byproduct and whey	Ammonia fiber expansion	S. cerevisiae Y904	SFS and SHF: temperature: 35 °C; rotation: 150 rpm; pH – 4.5, duration: 12 h	SFS: 11.7 g/L	Rocha et al. (2013)
Rice husk	NaOH (2.0% w/v), 130°C, 30 min	Kluyveromyces marxianus CK8	SSF; temperature: 43 °C; rotation: 150 rpm; pH: 4.2: duration: 96 h	15.63 g/L	Nachaiwieng et al. (2015)
Corncob	Alkalic salt ($Na_3PO_4 \cdot 12H_2O$, 121°C, 15 min) and dilute acid (H_2SO_4 1.04% v/v, 121 °C, 15 min)	S. cerevisiae BY4743	OSSF and PSSF, temperature: 35 °C; rotation: 120 rpm; duration: 2 days	OSSF: 35.04 g/L PSSF: 36.92 g/L	Sewsynker-Sukai and Kana (2018)
Corncob hydrolysate	Ultrasound (33 kHz, 37°C)	S. cerevisiae	Batch fermentation; temperature: 30 °C; rotation: 50 rpm; duration: 3 days	0.29 mL/g	Rekha and Saravanathamizhan (2021)
Corn stover	Dilute acid (H_2SO_4 0.5% w/w), 160 °C, 20 min	S. cerevisiae 424A	–	0.14 g/g	Uppugundla et al. (2014)
	Ionic Liquid, 140 °C, 3 h	S. cerevisiae 424A	–	0.21 g/g	Uppugundla et al. (2014)
	Ammonia fiber expansion (AFEX™), 140°C, 15 min	S. cerevisiae 424A	–	0.20 g/g	Uppugundla et al. (2014)
Sugarcane bagasse	Steam explosion and NaOH (1% w/v, 100 °C, 1 h)	S. cerevisiae UFPEDA 1238	Fed-batch fermentation, temperature: 34 °C; rotation: none ; pH: 5.5, duration: 12 h	0.39 g/g	de Albuquerque Wanderley et al. (2013)
	Acid-alkali	S. cerevisiae PE-2	SSF; duration: 18 h	0.34 g/g	de Araujo Guilherme et al. (2019)

(Continued)

TABLE 8.4   (Continued)
Agro-Industrial Wastes and Their Pre-Treatment Conditions for the Production of Bioethanol

Raw Materials	Pre-Treatment/Hydrolysis	Microorganisms	Fermentation Condition	Ethanol Yield	Reference
Wheat bran with recombinant cellulase cocktail	Acid pre-treatment (H_2SO_4: 1%), 121 °C, 30 min	*S. cerevisiae* MEL2	SSF	0.50 g/g	Cripwell et al. (2015)
Potato peel waste	Acid hydrolysis, 121 °C, 15 min	*S. cerevisiae* var. *bayanus*	Batch fermentation; temperature: 32 °C; rotation: 100 rpm ; pH – 5.0, duration: 48 h	0.46 g/g	Arapoglou et al. (2010)
Mandarin peel	Acid hydrolysis (H_2SO_4: 4%), 121 °C, 1 h	Immobilized *S. cerevisiae* KCTC 7906	SHF; Continuous fermentation; temperature: 30 °C; pH: 4.8; duration: 10 days	0.48 g/g	Choi et al. (2015)
Grapefruit peel	Acid hydrolysis (H_2SO_4: 4%), 121 °C, 1 h	Immobilized *S. cerevisiae* KCTC 7906	SHF; Continuous fermentation; temperature: 30 °C; pH: 4.8; duration: 10 days	0.46 g/g	Choi et al. (2015)
Lemon peel	Acid hydrolysis (H_2SO_4: 4%), 121 °C, 1 h	Immobilized *S. cerevisiae* KCTC 7906	SHF; Continuous fermentation; temperature: 30 °C; pH: 4.8; duration: 10 days	0.47 g/g	Choi et al. (2015)
Lime peel	Acid hydrolysis (H_2SO_4: 4%), 121 °C, 1 h	Immobilized *S. cerevisiae* KCTC 7906	SHF; Continuous fermentation; temperature: 30 °C; pH: 4.8; duration: 10 days	0.46 g/g	Choi et al. (2015)
Pineapple fruit peel	Alkali pre-treatment (NaOH: 1%, 2%, 3%), 30 °C, 48 h	*S. cerevisiae*	SHF; temperature: 36 °C; pH: 5.6; duration: 72 h	5.98 g/L	Casabar et al. (2019)
Pineapple waste (core and peel)	–	*S. cerevisiae* 11020	DF, CSF, SSF; temperature: 28 °C; pH: 5.0; duration: 72 h	DF – 0.43 g/g	Gil and Maupoey (2018)

Abbreviations: PSSF – simultaneous saccharification and fermentation with prehydrolysis; OSSF – simultaneous saccharification and fermentation without prehydrolysis; SFS/ SSF – simultaneous fermentation and saccharification; SHF – separated hydrolysis and fermentation.

(Boluda-Aguilar and López-Gómez, 2013) for bioethanol production after adopting suitable pretreatment method.

Gil and Maupoey (2018) utilized pineapple waste (core and peel) in three different fermentation processes, *viz.* direct fermentation, consecutive saccharification and fermentation, and simultaneous saccharification and fermentation, for the production of bioethanol. Prior to fermentation experiment, the solid biomass was hydrolyzed using enzyme cellulase and hemicellulase obtained from *Aspergillus niger*. The experiments for ethanol production were carried by using three different industrial strains such as *Saccharomyces bayanus* 1926, *Saccharomyces cerevisiae* 11020, and *Saccharomyces cerevisiae* 1319. Maximum ethanol yield of 0.418, 0.430, and 0.428 $g_{ethanol}/g_{glucose}$ were obtained from *S. bayanus* 1926, *S. cerevisiae* 11020, and *S. cerevisiae* 13190, respectively, in simultaneous saccharification and fermentation. In a similar study, Casabar et al. (2019) reported a highest ethanol concentration of 5.98 g/L using *Saccharomyces cerevisiae* from pineapple peel at 48 h of fermentation.

The production cost, technology, and environmental problems are the major hindrance in successful commercialization of the process, which mainly depends on ethanol productivity. Optimization of various process parameters, utilization of low-cost substrates, isolation of novel high-yield microbial strains, and adaptation of genetic engineering techniques could lead to increasing ethanol productivity.

8.5 VALORIZATION OF AGRO-INDUSTRIAL SOLID WASTE BIOMASS FOR BIOGAS

Agro-industry wastes have high nutritional value, therefore gaining more attention to be utilized for various bio-based products mainly biogas (Adámez et al., 2012; Katalinic et al., 2010). These waste residues mostly constitute organic content relatively higher than the inorganic content, which is a good source of energy in the form of biogas (Valijanian et al., 2018). It is reported that about 20% of the vegetables and fruit are wasted from different agro-industry every year in India (Rudra et al., 2015). Biogas majorly contains methane and carbon dioxide generally produced in anaerobic reactor from the digestion of organic waste. Biogas from agro-industrial wastes offers great opportunities because of its widely availability, economical and also reduces the greenhouse gases (GHGs), which is serving as a clean source of energy as compared to fossil fuels (Akinbami et al., 2001; Meyer-Aurich et al., 2012). Various anaerobic bioreactors are used for biogas production, which are classified based on the types of wastes used, processing mechanisms and environmental conditions (Caruso et al., 2019). The generation of biogas from various agro-industrial wastes can protect our environment and make the industry more sustainable (Weiland, 2010).

Different agro-industrial wastes such as sisal fiber, orange peel, coffee husk, cotton gin, olive pomace, rice straw, and sugarcane, etc. have been used for biogas production. Table 8.5 summarizes the various compositional characteristics such as pH, total solid, volatile solids, and C/N ratio of agro-industrial wastes and their biogas yield reported by various researchers using anaerobic digester. Results show that neutral pH range (6.5–8.5) favors the yield of biogas. On the contrary,

TABLE 8.5

Biogas Yield from Various Agro-Industrial Wastes

Agro-Industry Waste	pH	Total Solid (TS) (%)	Volatile Solid (VS) (%)	C/N	Methane Yield	Biogas Yield	References
Sisal fiber waste	8.8	91.6	92	65	216 L/kg VS added	–	Mshandete et al. (2006)
Orange peels waste	7–7.7	19.72	18.83	–	366 L/kg VS added (SMP)	771 L/kg VS added (SBP)	Suhartini et al. (2020)
Apple peels	7–7.7	21.63	21.29	–	407 L/kg VS added (SMP)	702 L/kg VS added (SBP)	Suhartini et al. (2020)
Olive pomace	6.1	53.5	52.3	–	213 (SMY) L/kg VS substrate	–	Pellera and Gidarakos (2016)
Rice straw	7–7.7	42.02	20.8	–	211 L/kg VS added (SMP)	433 L/kg VS added (SBP)	Suhartini et al. (2020)
Juice industry waste	4.6	16.2	15.7	–	445 (SMY) L/kg VS substrate	–	Pellera and Gidarakos (2016)
Brewery grains	5.08	20–30	84–96	11–25	180–270 L/kg VS	390–490 L/kg VS	Aliyu et al. (2011) Okoye et al. (2016)
Maize straw	7–7.7	24.93	21.97	–	224 L/kg VS added (SMP)	462 L/kg VS added (SBP)	Suhartini et al. (2020)
Coffee husk	7–7.7	86.36	81.24	–	181 L/kg VS added (SMP)	366 L/kg VS added (SBP)	Suhartini et al. (2020)
Sugarcane bagasse	–	93	83–90	20	–	400 L/kg VS	Nzila et al. (2010)
Barley straw	7.9	89–90	94	–	229 L/kg VS	417 L/kg VS	Dinuccio et al. (2010) Nzila et al. (2010)
Tea residue	–	91	87	–	–	424 L/kg VS	Nzila et al. (2010)
Corn residue	5.05	81–93	94–98	51–57	229 L/kg VS	417 L/kg VS	Nzila et al. (2010) Zhou et al. (2017)

Abbreviations: SMY – specific methane yield; SMP – specific methane potential; SBP – specific biogas potential.

some agro-industry wastes generated from juice industry also produces biogas at acidic pH range. But, at low pH the yield of biogas is less. The high VS degradation observed in various studies under anaerobic condition for agro-industry waste residues found to be feasible alternative for bioenergy production. High volatile content offered high rate of degradation, indicating good potential for high methane yield such as sisal fiber attributing about 92% VS resulting about 216 L CH_4/kg VS (Mshandete et al., 2006). It has also been observed that out of the total biogas produced from waste, 40–50% consisted of methane showing its potential of agro-industry wastes.

8.6 VALORIZATION OF AGRO-INDUSTRIAL SOLID WASTE BIOMASS FOR COMPOSTING

Composting is the most preferred options for organic mass degradation in many developing countries due to the lower investment and operation costs, less technical complexity and requirement of scientific expertise (Wang et al., 2019). Composting is not only a sustainable approach for reducing a large quantity of organic waste but also a recycling technique for the conversion of organic waste into a stable and nutrient-rich product, which is beneficial for soil health and plant growth (Onwosi et al., 2017).

8.6.1 Factors Affecting the Composting Process

Composting is a natural microbial mediated decomposing process. However, the effectiveness of the composting process is influenced and regulated by many factors such as nutrient balance (C/N ratio), pH, moisture, temperature, microorganism, oxygen concentration, and aeration, etc. (Bernal et al., 2009). Among all the factors, nutrient balance (which majorly defined as C/N ratio in the composting process) of the initial composting mixture is quite an essential factor. C represents an energy source, while N is for building cell structure. So, in the composting mix, if the N is limiting, it resulted in the slow growth of the microbial population, which directly affects the C degradation, whereas the excessive N content of the composting mixture resulted in N loss in the form of ammonia. In order to balance the optimum C/N ratio in the composting mixture, addition of a C rich material such as sawdust, wood chips, and cardboard or N rich materials, such as soybean residue, livestock manure is preferred depending on the characteristics of the initial raw material (Wang et al., 2019). A preferable C/N ratio for the composting process is 25–35 (Bernal et al., 2009; Wang et al., 2019). However, successful composting studies have been performed using a substrate containing a low initial C/N ratio than the desirable range. Awasthi et al. (2016) reported composting of gelatin industry sludge mixed with municipal solid waste, poultry waste, pig manure, and sawdust with an initial C/N ratio of 17.80. Similarly, Rich et al. (2018) used segregated vegetable waste mixed with water hyacinth, garden prune, sawdust, and cow dung with an initial C/N ratio of 22.01 for successful composting (Table 8.6). On the contrary, an excellent composting

TABLE 8.6

The Composting Conditions and the Process Parameters of Some Agro-Industrial Waste Reported in Literature

Raw Materials	Experimental Condition	Temperature (°C)[a]	Change in pH		Change in EC (mS/cm)		TC Reduction (%)		Change in TN (%)		TOM Reduction (%)		Change in C/N Ratio		GI (%)	References
			I	F	I	F	I	F	I	F	I	F	I	F		
Hydrolyzed spent grape marc mixed with vinification lees	Composting in a plastic bucket (10 L) for a period of 150 days	30	3.0	7.0	8.5	12	52	48	2.0	2.3	90	84	26	21	–	Paradelo et al. (2013)
Distilled grain waste at different initial pH values adjusted by CaO	Composting in a computer-controlled 28 L lab-scale reactor for a period of 65 days	67.5	5.0	7.4	1.9	0.9	–	–	2.0	1.6	–	–	–	–	65.4	Wang et al. (2017b)
Indigofera plant biomass waste from the natural indigo dye industry mixed with cow dung and Jeevamrutha	Composting in an earthen pot for a period of 42 days	38.8	7.0	6.8	1.14	2.28	44.28–38.29		2.24	2.64	83.30	72.41	19.8	14.07	129	Pattanaik et al. (2020a)
Flower waste mixed with saw dust and cow dung in different combinations	Rotary drum composting for a period of 60 days	45	5.4	7.4	4.0	4.2	46	36	1.6	2.2	–	–	28	17	–	Sharma et al. (2018)

(Continued)

TABLE 8.6 (Continued)
The Composting Conditions and the Process Parameters of Some Agro-Industrial Waste Reported in Literature

Raw Materials	Experimental Condition	Temperature (°C)[a]	Change in pH		Change in EC (mS/cm)		TC Reduction (%)		Change in TN (%)		TOM Reduction (%)		Change in C/N Ratio		GI (%)	References
			I	F	I	F	I	F	I	F	I	F	I	F		
Gelatin industry sludge mixed with OFMSW, poultry waste, 10% zeolite, and enriched nitrifying bacteria consortium	Windrows composting for a duration of 42 days	68	8.2	7.3	1.3	2.2	47	35	2.6	2.7	–		17	13	123.50[a]	Awasthi et al. (2016)
Segregated vegetable waste mixed with water hyacinth, garden prune, sawdust, and cow dung in different combinations	Rotary drum composting for a period of 30 days	52	5.6	7.5	–	–	46	29	2.0	2.4	86	66	22	12	110[a]	Rich et al. (2018)

[a] Maximum temperature reached during the composting process, I – initial value and F – final value of composting parameter.

was also reported in the initial C/N ratio between 25 and 40, even as high as 50 (Petric et al., 2015). The pH is also an essential parameter, the optimum pH for a good composting process should be in the range of 7–8 (Chan et al., 2016). However, in case of raw material containing low pH such as food waste (Chan et al., 2016) or distillery grain waste (Wang et al., 2017a), the initial pH has been adjusted by addition of alkali agents.

Appropriate moisture content is essential in the composting process, as it influences the oxygen uptake rate, temperature, and microbial activity of the process. The moisture content of the feedstock generally varies depending on the type and source but should be between 50 to 60% (Bernal et al., 2009). Whereas, Luangwilai et al. (2011) observed the optimum moisture content of 40–70% is required for microbial activity. The higher moisture content in the composting mass inhibits the gas diffusion and creates an anaerobic condition in the composting pile, which results in slower microbial degradation due to insufficient oxygen concentration (Mohammad et al., 2012). In such case, co-composting of high moisture containing raw material with low moisture containing or dry material has been carried out. Sharma et al. (2018) used flower waste with sawdust with a moisture content of 80.05% and 15.84%, respectively, for co-composting, to enhance the process. On the contrary, desiccated or low moisture containing composting mixture is the indicator of slow microbial growth due to insufficient water content, and in such cases, additional water must be provided to the composting mass to achieve the optimum moisture content (Wang et al., 2019).

In the composting process, the primary source of micro organisms, such as fungi, bacteria, archaea, and protozoa, generally resides in the composting mass itself and has rapid decomposition activity (Onwosi et al., 2017). However, in some instances, the microbial population has been added from external sources to promote organic matter degradation rate and enhance humification. The most simplified and traditional source of the microbial community used in the composting process is cow dung (Jain and Kalamdhad, 2018; Rich et al., 2018; Sharma et al., 2018). Further, a specific microbe (cellulose-degrading or lignin-degrading bacteria) or microbial inoculation has also been used for the effective degradation of composting mass (Onwosi et al., 2017). Influence of the fungal consortium (*Trichoderma viride*, *Aspergillus niger*, and *Aspergillus flavus*) on the organic fraction of municipal solid waste was investigated by Awasthi et al. (2014), and a higher organic mass decomposition was observed as compared to a control sample. Manu et al. (2017) used commercial microbial inoculums (EM1), which is a mixture of lactic acid bacteria, phototrophic bacteria, and yeast for composting food waste and noticed a reduction in the total composting period.

Temperature is the real-time indicator of the composting process, which regulates the microbial population and metabolism (Awasthi et al., 2014). Microbial breakdown of organic matter is an exothermic process, which generates heat due to microbial metabolic activity during the degradation of organic matter; as a result, the temperature of the composting pile rises (Tang et al., 2007). Microbial heat generation or temperature evolution during composting typically follows four phases (Onwosi et al., 2017):

i. Mesophilic (10–45°C, which lasts for a few days)
ii. Thermophilic (45–65°C, which lasts for few days to few weeks)
iii. Cooling (slightly higher than the ambient temperature, lasts for few weeks)
iv. Maturation

In the initial phase (mesophilic phase) the growth of diverse microbes (bacteria, fungi, actinomycetes, and protozoa) occur, which degrade the simple and easily degradable organic matter such as sugars and amino acids. In the later phase, the degradation of complex organic matter, such as protein, fat, and lignin takes place by consumption of a considerable amount of oxygen and releasing a large amount of energy as heat (Wang et al., 2019). As a result, the temperature of the composting pile increases resulting in the thermophilic phase with the dominance of thermophilic micro organisms. After rapid degradation of organic matter at the thermophilic phase, the microbial growth slows down due to exhaust of readily available organic matter, and the temperature drops down to the ambient temperature or slightly higher than that (Wang et al., 2019). Hence, the microbial heat generation is one of the crucial indicators to ascertain the organic matter degradation and thermophilic stage during the composting process. However, this concept does not hold good for all the cases. On the contrary, in certain instances, higher decomposition rates have been reported in mesophilic temperatures than the thermophilic temperature range (Liang et al., 2003; Paradelo et al., 2013; Semitela et al., 2019). The reason could be due to the presence of more diverse microbial communities in the mesophilic temperature range and higher metabolic rates associated with lower thermal degradation (Narihiro et al., 2004; Tang et al., 2007).

Aeration influence the compost stability by maintaining the O_2 concentration in the composting mass. An adequate amount of O_2 helps the oxidation of organic material, as well as evaporates excess moisture from the substrate (Onwosi et al., 2017). However, excessive aeration can cause rapid cooling of the composting pile and prevent the thermophilic conditions, whereas less aeration can lead to an anaerobic condition in the composting pile (Gao et al., 2010).

Maintaining optimum turning frequency is a crucial solution to ensure enough aeration in the composting pile. Kalamdhad and Kazmi (2009) optimized the turning frequency once a day for composting of cow manure mixed with green vegetable and sawdust in a rotary drum composter. They opined that turning the composting pile once a day in the early composting period would lead to nutrient-rich stable compost. Both forced and natural aeration have been proven to enhance the composting process. Chowdhury et al. (2014) observed that an airflow rate of 0.21 L/min/kg TS (Total solid) was quite effective in the reduction of NH_3 losses and increased CH_4 losses, by a pressurized aeration system, where compressed air was applied beneath the reactor through a perforated plate. On the contrary, Oudart et al. (2015) opined that natural aeration is a better alternative (cost-effective and less energy-intensive process) than forced aeration in the use of composting animal excreta under static pile conditions.

8.6.2 COMPOST MATURITY AND STABILITY EVALUATION

Compost maturity and stability are two essential parameters in evaluating compost quality. Most often, compost maturity and stability are used interchangeably. However, they individually refer to two specific properties of the composting material. Compost stability generally refers to the resistance of composted organic matter to further rapid degradation. It typically refers to the microbial activity and the composting process that can be measured either by the degradation of different chemical species in the compost organic matter or by the respiration index (Guo et al., 2012; Hachicha et al., 2009).

On the other hand, maturity refers to the degree or level of completeness of composting and implies improved qualities resulting from "aging" or "curing" of a product (Bernal et al., 2009). In other words, maturity also refers to the extent of phytotoxic organic matter transformation and generally measured by germination index (GI) or plant bioassay (Onwosi et al., 2017). However, both stability and maturity usually go parallelly, as the micro organisms present in the unstable composts generally produce phytotoxic compounds (Zucconi et al., 1985). The maturity and stability of the final compost sample are of utmost importance, as immature and poorly stabilized compost leads to the development of anaerobic pockets, which results in the production of odors and toxic compounds. Further, immature compost may have negative impacts on plant growth, due to the presence of phytotoxic compounds, or the presence of low oxygen and nitrogen content (Bernal et al., 2009).

Several chemical and biological parameters have been proposed to evaluate compost maturity and stability, such as C/N ratio, total organic carbon, volatile solids content, pH, EC, inorganic elements concentration (especially heavy metals), self-healing capacity, oxygen uptake rate, production of humic and fulvic acids, GI to measure the phytotoxicity, humification, microbial activity, and plant phytotoxicity, etc. (Onwosi et al., 2017). Further, certain physical parameters, such as color, odor, humidity, and temperature, etc., are also used as onsite parameters to assess compost maturity (Oviedo-Ocaña et al., 2015).

The maturity and stability characteristics of the end product or final compost are verified by comparing with the national and international standards, such as FAI (2007) (Indian Fertilizer Control Order, 1985), SWM (2016) (Solid Waste Management Rules), TMECC (2002) (Test Methods for the Examination of Composts and Composting, United States Composting Council), and CCME (2005) (Candian Council of the Ministers of the Environment), etc. (Awasthi et al., 2014; Manu et al., 2017). Among other physiochemical parameters, C/N ratio is a critical parameter and considered as an indicator to evaluate compost maturity (Awasthi et al., 2016, 2014; Chen et al., 2015; Iqbal et al., 2015; Manu et al., 2017). A C/N ratio equal to or less than 25 has been recommended by Awasthi et al. (2014). However, Iqbal et al. (2015) pointed out a C/N ratio of less than 20 is an indicator of acceptable compost maturity, and a ratio of 15 or less is preferable. pH and EC are also two crucial indicators of compost maturity. A relatively stable pH in the neutral range (6.5–7.5) at the end of composting is an indicator of mature compost (Manu et al., 2017).

Similarly, the desirable range of electrical conductivity (EC) in a compost sample must be <4 mS/cm for its utilization as fertilizer. The EC determines the total salt content of compost and high EC (>4 mS/cm) indicates the presence of more soluble products and affect adversely to the plant growth (low germination rate and withering, etc.) (Chan et al., 2016). The total organic matter (TOM) helps in assessing the amount of biologically degradable matter and biologically inert materials (lignin) present in the compost sample (Jain and Kalamdhad, 2018). Higher the degradation of organic matter signifies better composting process, but the organic matter should not be less than 30% in the final compost (Fertilizer Association of India, 2007). In evaluating compost maturity and phytotoxicity, GI acts as a sensitive indictor (Guo et al., 2012). In several studies, the recommended GI values for phytotoxic free compost were varied. Wang et al. (2016) has recommended a GI value of 50%, as an indicator of compost maturity during pig manure composting and suggested that it can be further used for agricultural application. Whereas, Rawoteea et al. (2017) opined that compost of GI value above 50% can be considered as phytotoxic free and above 80% as mature compost during vegetable waste composting. On the other hand, Rich et al. (2018) pointed out that the GI value of greater than 90% indicates the compost samples to be phytotoxic free, but not mature enough to help in germination or plant growth. And, GI value is greater than 100–110% was proposed as maturity index for animal manure compost.

8.7 VALORIZATION OF AGRO-INDUSTRIAL SOLID WASTE FOR BIOMOLECULES AND BIOMATERIALS

Other than biofuels and compost application, the scope of agro-industrial waste has also been widened to various biomolecules or bioactive compounds (polyphenols, pigments, and enzymes, etc.) and biomaterials (bioplastic and biocomposite, etc.).

8.7.1 AGRO-INDUSTRIAL SOLID WASTE FOR BIOMOLECULES

8.7.1.1 Polyphenols

Polyphenols are the phenolic group compounds, which are widely found in most of the plant species and some microbial species. Polyphenols are highly demanded molecules in the therapeutic market due to their antioxidant activity. Certain polyphenols, such as tannins, flavonoids, anthocyanins, and alkaloids, have industrial significance and present abundantly in most fruits, vegetables, and exotic plant species (Joshi et al., 2012). Recently, agro-industrial wastes have been widely used as a potential source for production of bioactive compound through various extraction routes. For instance, wastes (peels, seeds, and pruning) generated from tomato industries accounts for 12,874 kilo tons globally and rich in bioactive compounds, such as polyphenols, carotenes, terpenes, tocopherols, and sterols. These compounds are reported to exhibit excellent antioxidant and antimicrobial activities (Beltrán-Ramírez et al., 2019; Sharma et al., 2020). Similarly, other vegetable processing wastes such as potato waste, onion skin, carrot waste, etc., along with fruit wastes, such as the grape processing industry, apple juice processing industries, are also reported to contain various polyphenols (Sharma et al., 2020). Other than fruits and vegetable processing industries, the wastes generated from beverage industries (coffee and tea) are also enormous and have been explored for polyphenols extraction (Sharma et al., 2020). The global annual production of coffee industry processing waste is around 105 kilo tons. Exhausted and spent coffee ground wastes from various industries, restaurants, and domestics are a valuable source of polyphenols and tannins (Beltrán-Ramírez et al., 2019).

8.7.1.2 Pigments

Agro-industrial wastes have been an essential source for natural pigment extraction through direct extraction methods or microbial fermentation. Bio-pigments have immense market value in the food and cosmetic sectors due to their eco-friendliness, biodegradability, zero or less toxicity than their synthetic counterpart. Further, natural pigments extracted from fruit by-products are considered important due to their high color stability and purity (Yusuf, 2017). By direct extraction method, natural pigments, such as chlorophylls, carotenoids, anthocyanins, and melanin can be derived from various plant materials. By fermentative production route, certain micro organisms such as bacteria, fungi, yeast, and algae have been utilized for bio-pigment production, such as xanthophylls, phycocyanins, and melanins (Mishra et al., 2019). Specific genera of microbial strains renowned for bio-pigment production from waste are *Monascus*, *Aspergillus*, *Rhodotorula*, and *Penicillium*.

8.7.2 AGRO-INDUSTRIAL SOLID WASTES FOR BIOMATERIALS

8.7.2.1 Bioplastics

Polyhydroxybutyrates (PHBs) are considered bioplastics due to their non-toxic, water-soluble, biodegradable, biocompatible, thermoplastic, and elastomeric nature (Shivakumar et al., 2012). Prokaryotes or bacteria generally synthesize PHB as intracellular granules, which act as an energy storage facility. The PHB production by micro organisms occurs under unbalanced or unfavorable conditions, such as high carbon concentration and limited concentration of oxygen, nitrogen, phosphorous, and other trace elements (magnesium, calcium, and iron, etc.)

Among various solid agro-industrial wastes, by-products obtained from potato processing industries are considered a good source for PHB production. The major by-products obtained from potato industries are potato peel, pulp, and slurry, consisting of 80% of the original potato. Apart from water, the major constituent of potato industry waste is starch, which is a good source for fermentation. Haas et al. (2008) reported the conversion of waste potato starch into PHB with a productivity of 1.47 g PHB/L/h and a yield of 0.22 g PHB/g starch. Other agro-industrial wastes such as soya flour, bagasse, molasses, wheat bran, wheat germ, rice bran, and ragi bran were also evaluated by Shivakumar (2012) using the microbe *Bacillus thuringiensis*.

8.7.2.2 Biocomposites

Biocomposites, which act as reinforcement, are made up of natural fibers and polymer matrix. The natural fibers are majorly composed of cellulose, hemicellulose, and lignin. In biocomposite material preparation, cellulose is the major component of plant material, provides stability and strength. Whereas lignin, which has a highly cross-linked structure, influences the biocomposite material by providing the structure, morphology, properties, hydrolysis rate, and fibers' flexibility. The natural fibers can be used in both thermoplastics [polypropylene (PP), high-density polyethylene (HDPE), polystyrene (PS), polyvinyl chloride (PVC)], and thermosets (epoxy, polyester, and phenolics resins) matrices. The major applications of biocomposites are in automotive parts, packaging, medical sector, military industry, and aerospace, etc.

For biocomposite materials, agro-industrial wastes such as *Jatropha curcas* fiber, castor plant, and agave are widely used. In the case of *Jatropha curcas* and castor plants, oil has been extracted only from seeds, where the rest of the plant biomass is discarded, which is further utilized for natural fiber production. Whereas, agave plant biomass is used for tequila production, resulting in a large amount of waste generation, which is mostly fiber in nature (Beltrán-Ramírez et al., 2019; Khalil et al., 2013).

8.8 VALORIZATION OF AGRO-INDUSTRIAL LIQUID WASTE

The liquid waste generated from different agro-industries such as, livestock, soybean processing, dairy industry, and slaughterhouse are generally rich in carbon (C), nitrogen (N) and phosphorus (P), which is the primary cause of eutrophication while being discharged into nearby water bodies (Cai et al., 2013; Hongyang et al., 2011; Li et al., 2011; Pittman et al., 2011). These high nutrient-containing wastewaters are either directly used for biofuel, biofertilizer, and biomolecule production or are being utilized by biological agents for efficient removal of nutrients and these biological agents are further being utilized for biofuel or biofertilizer production.

8.8.1 AGRO-INDUSTRIAL WASTEWATER FOR BIOFUEL, BIOFERTILIZER, AND OTHER VALUE-ADDED PRODUCTS

Like other agro-industrial solid waste, various wastewater obtained from different agro-industries can also be utilized as a substrate for biofuel or biofertilizer as long as they are rich in carbohydrates, lipid, protein, cellulose, and hemicellulose (del Real Olvera and Lopez-Lopez 2012). Biomolecules have also remained an essential part of liquid waste, which is in continuous exploration.

TABLE 8.7

Biogas Production from Different Agro-Industrial Liquid Waste

Type of Agro-Industrial Wastewater	Mode of Operation	Temperature (°C)	Hydraulic Retention Time (d)	Productivity	Reference
Sewage sludge co-digestion of dried mixture of food waste, cheese whey and olive mill wastewater	Continuous mode	37	24	815 mL CH$_4$/L$_{reactor}$/d	Maragkaki et al. (2018)
Poultry manure and cheese whey wastewater	Batch mode	37	35	223 mL/g VS	Carlini et al. (2015)
Slaughter house wastewater	Tubular digester in continuous mode	27	3.2–87.4	0.25 kg COD m^3/d	Martí-Herrero et al. (2018)

Maragkaki et al. (2018) have reported the co-digestion of dried mixture of cheese whey, olive mill wastewater, food waste, and sewage sludge for the production of biogas in a continuous mode of operation, resulting in biogas productivity of 815 mL CH$_4$/L$_{reactor}$/d. Similarly, Carlini et al. (2015) reported with a biogas production of 223 mL/g VS by using cheese whey wastewater and poultry manure as substrate. Martí-Herrero et al. (2018) reported biogas productivity of 0.25 kg COD m^3/d using a low-cost tubular digester (Table 8.7).

Like biogas production, scientific studies on bioethanol production by using agro-industrial wastewater have also been reported. Domínguez-Bocanegra et al. (2015) investigated the use of raw agro-industrial wastes such as coconut milk, tuna juice, and pineapple juice to produce bioethanol using *Saccharomyces cerevisiae* CDBB 790 strain. The ethanol yields per sugar consumption (YP/S) were observed to be 0.43, 0.29, and 0.47 g/g while using coconut milk, tuna juice, and pineapple juice in the fermentation medium, respectively. The increase in ethanol yield for coconut milk and pineapple juice could be due to high nutrient composition (potassium and phosphorous) (Table 8.8).

Agro-industrial wastewater is used as a biofertilizer in several studies with the purpose of reuse for irrigation and crop cultivation. However, the direct application of raw agro-industrial wastewater is not recommended due to the presence of excess nutrients (above the irrigation discharge standard), which might affect plant growth and soil quality. Therefore, in most of the studies, agro-industrial wastewater after certain treatment is considered as "liquid fertilizer" (Kokkora et al., 2015; Libutti et al., 2018). Libutti et al. (2018) have shown the effect of

TABLE 8.8

Bioethanol Production from Different Agro-Industrial Liquid Waste

Type of Agro-Industrial Wastewater	Pre-Treatment/ Hydrolysis	Microorganisms	Fermentation Condition	Ethanol Yield	Reference
Coconut milk	–	S. cerevisiae CDBB 790	Batch fermentation; temperature: 28 °C; rotation: 150 rpm; duration: 5 days	0.43 g/g	Domínguez-Bocanegra et al. (2015)
Tuna juice	–	S. cerevisiae CDBB 790	Batch fermentation; temperature: 28 °C; rotation: 150 rpm; duration: 5 days	0.29 g/g	Domínguez-Bocanegra et al. (2015)
Pineapple juice	–	S. cerevisiae CDBB 790	Batch fermentation; temperature: 28 °C; rotation: 150 rpm; duration: 5 days	0.47 g/g	Domínguez-Bocanegra et al. (2015)

secondary and tertiary treated agro-industrial wastewater compared to groundwater on the cultivation of tomato, broccoli in southern Italy and found the secondary treated water followed by the tertiary treated wastewater in improving the soil quality and plant productivity. Torr (2009) has proposed utilization of wastewater derived from dairy farms and manure could be utilized as alternate fertilizers for the cultivation of crops. With the aim of zero discharge, Kokkora et al. (2015) have utilized, microfiltered olive milled wastewater (OMWW) as liquid fertilizer for the production of maize crop. The soil quality and plant productivity by microfiltered-OMWW (MF-OMWW) have been compared with the application of mineral N fertilizer. The results obtained in both cases were similar and further recommended the complete substitution of mineral N fertilizer by MF-OMWW.

On the contrary, Galliou et al. (2018), instead of using OMWW as liquid fertilizer, used solar-dried OMW with swine manure (as a bulking agent) for composting. The dried OMWW material contains high nutrients and phenols (N: 27.8 g/kg, P: 7.3 g/kg, K: 81.6 g/kg, and phenols: 18.4 g/kg). To detoxify the final dried OMWW, the composting process was applied by incorporating grape marc as a bulking agent. The final compost obtained has the characteristics equivalent to organic fertilizer, which is rich in nutrient (C: 57%, N: 3.5%, P: 1%, and K: 6.5%) and low phenolic contents (2.9 g/kg). Further, the fertility of compost was compared with commercial NPK fertilizer and was found to be equally effective in improving the soil fertility.

Like, agro-industrial solid waste, biomolecules from liquid waste are also explored for the application in food and fine chemicals (Alves et al., 2021; Cañadas et al., 2021; Dammak et al., 2016). Cañadas et al. (2021) recovered phenolic compounds from winery wastewater by using hydrophobic eutectic solvents. A satisfactory recovery of 64.14–84.10% w/w of the phenolic compound was obtained from this liquid effluent. Similarly, Dammak et al. (2016) targeted OMWW for extraction of polyphenols to have the simultaneous benefit of obtaining bioactive compounds and reduce its phytotoxicity. The method adopted for the extraction of polyphenols from OMWW is the drowning-out crystallization-based separation process, which is depending on the solubility behavior of targeted compound. This method resulted in a highly concentrated (75% w/w) polyphenols isolate. Slaughterhouse waste, such as chicken blood meal, rich in iron and protein and often used as fertilizer and animal feed, has also been explored for antioxidant protein hydrolysates.

Dairy whey, a rich source of carbon and has been explored for multiple value-added product formation, such as lactic acid and bioethanol, etc. It has also been investigated for bioactive compounds production such as isoflavones. These compounds have recently received much attention from the research communities due to their health benefits (Chua and Liu, 2019). It is not only limited to biomolecules, dairy whey or cheese whey have been successfully explored for PHB production by genetically engineered *Escherichia coli* with the highest productivity of 4.6 g PHB/L/h (Ahn et al., 2000; Elain et al., 2016).

8.8.2 Nutrient Recovery from Agro-Industrial Wastewater and Its Further Value Addition

Nutrient removal from such high strength wastewater generated from agro-industries, using biological agents such as algae or algal-bacterial consortia have been extensively reported to be efficient, less energy-intensive, and environmentally friendly approaches (Choudhary et al., 2016). Several algal strains, such as *Scenedesmus*, *Chlorella*, *Chlamydomonas* species, etc., have been effectively used for nutrient (N and P) removal from wastewater (Pittman et al., 2011). Abou-Shanab et al. (2013) have tested and compared the effectiveness of six algal species (*Chlamydomonas mexicana*, *Chlorella vulgaris*, *Ourococcus multisporus*, *Nitzschia* cf. *pusilla*, *Scenedesmus obliquus*, and *Micractinium reisseri*) in nutrient removal from piggery wastewater, in which *Chlamydomonas* species was found to be most efficient in N (62%) and P (28%) removal. In case of algal nutrient removal, N and P are majorly consumed for growth and energy metabolism, whereas the uptake of carbon either from atmosphere (CO_2) or from the culture medium (organic or inorganic carbon from wastewater) would lead to autotrophic or heterotrophic mode of growth, respectively (Cai et al., 2013).

Other than individual algal species, microbial consortia are also quite effective in treating various municipal and agro-industrial wastewater (Choudhary et al., 2016; Hernández et al., 2013; Lee et al., 2016). Choudhary et al. (2016) studied the nutrient removal potential of nine native algal consortia (isolated from the sewage treatment plant and slaughterhouse effluent) in unsterilized and untreated

TABLE 8.9

Nutrient Removal from Various Agro-Industrial Wastewaters and Their Further Value Addition

Type of Wastewater	Microorganisms	Cultivation Time (days)	Nutrient Composition (mg/L)	Nutrient Removal Efficiency (%)	Value-Added Product	References
Algae						
Piggery wastewater	Chlorella vulgaris	30	COD: 168; NH_3-N: 53.6 NO_3-N: 0.34; PO_4-P: 11.4	COD:58; NH_3-N: 84 NO_3-N: 41; PO_4-P: 83	Biodiesel Lipid production (0.07 g/L)	Ji et al. (2013)
Aquaculture wastewater	Chlorella sorokiniana	7	COD: 96; NH_3-N: 5.32 NO_3-N: 40.67; PO_4-P: 8.82	COD:72; NH_3-N: 76 NO_3-N: 85; PO_4-P: 73	Lipid (150.19 mg/L/d), carbohydrate (172.91 mg/L/d) and protein productivity (141.57 mg/L/d)	Guldhe et al. (2017)
Piggery wastewater	Chlamydomonas mexicana	20	TIC: 336; TN: 56 TP: 13.5	TIC: 29; TN: 63 TP: 28	Biodiesel Lipid content (0.31 g/L)	Abou-Shanab et al. (2013)
Livestock wastewater	Chroococcus sp.	12	COD: 2965; NH_3-N: 161; NO_3-N: 75	COD: 693; NH_3-N: 3.14; NO_3-N: 12.17; TP: 31	Biogas (291.83 ± 3.904 mL CH_4/g VS_{fed})	Choudhary et al. (2016)
Mixed microbial consortia						
Potato processing wastewater	Chlorella sorokiniana + aerobic sludge	10	COD: 1536; TP: 4.2 NH_3-N: 12.1	COD: 84.8; TP: 80.7; NH_3-N: >95	Biogas (518 mL CH_4/g COD)	Hernández et al. (2013)
Livestock wastewater	Chlorella + Phormidium	12	COD: 2940; NH_3-N: 161; NO_3-N: 75 PO_4-P: 200	COD: 80; NH_3-N: 99; NO_3-N: 87 PO_4-P: 83	Theoretical biogas (0.79 m³/kg VS)	Choudhary et al. (2016)

Abbreviations: COD – chemical oxygen demand; TIC – total inorganic carbon; TN – total nitrogen; TP – total phosphorus.

high strength livestock wastewater. The authors screened and identified the algal consortia as *Chlorella* and *Phormidium*, which are efficient for N (NH_3-N: 99%, NO_3-N: 87%) and P (PO_4^{-3}-P:83%) removal as compared to other algal consortia. Further, a high C (COD: 80%) reduction was also noticed during this study. Along with nutrient removal efficiency, the authors observed the same algal consortia's growth potential in terms of the highest chlorophyll concentration (32.3 g/mL) and biomass productivity (1.9 g dry weight/L). From the above study, the authors opined that the simultaneous reduction in C (reduction in COD) from the culture medium and enhancement of growth (biomass and chlorophyll content) could be due to either heterotrophic growth of algae under dark conditions or mixotrophic growth in the presence of light over a range of organic C substrates. These algal species or microbial consortia not only limited to nutrient removal from agro-industrial wastewater, but also been utilized effectively for biofuel production. Prajapati et al. (2014) have evaluated four algal species for nutrient removal from livestock wastewater and found *Chroococcus* sp. to be the best in terms of nutrient removal (>80%) and biomass production (4.44 g/L). The authors further tested this algal species for biogas production along with cattle dung as co-substrate and found a biogas productivity of 291.83 ± 3.904 mL CH_4/g VS_{fed} (Table 8.9).

8.9 CONCLUSIONS

Recycling and reuse of agro-industrial waste and its subsequent valorization can serve as a significant contributor to agricultural waste management. The utilization of agro-industrial waste can not only help in the reduction of ecological damage, global environmental pollution, and its associated health diseases but also boost the economics of the industries by incorporating the value-added products obtained from waste in the mainstream. Therefore, research interests have been multifold to inspect various agricultural processing wastes and their possible valorization. Exploring the potential of residual wastes through characterization can act as a preliminary route for further investigation into multiple products. Channelizing agro-industrial wastes for fuel or energy and compost or biofertilizer can bestow relief to the renewable energy and organic fertilizer sector. At the same time, obtaining bioactive compounds and making biocomposite from wastes will definitely boost the economy of the commercial market. Undoubtedly, addressing both solid and liquid sectors of agro-industrial processed wastes can cater to the criteria of the zero-waste approach and enable the processing industries to achieve the sustainability goals.

REFERENCES

Abdul-Hamid, A., & Luan, Y.S. 2000. Functional properties of dietary fibre prepared from defatted rice bran. *Food Chemistry*, 68: 15–19.

Abou-Shanab, R.A.I., Ji, M.-K., Kim, H.C., Paeng, K.J., & Jeon, B.H. 2013. Microalgal species growing on piggery wastewater as a valuable candidate for nutrient removal and biodiesel production. *Journal of Environmental Management* 115: 257–264.

Adámez, J.D., Samino, E.G., Sánchez, E.V., & González, G.D. 2012. In vitro estimation of the antibacterial activity and antioxidant capacity of aqueous extracts from grape-seeds (*Vitis vinifera L.*). *Food Control*, 24: 136–141.

Aguiar, L., Márquez-Montesinos, F., Gonzalo, A., Sánchez, J.L., & Arauzo, J. 2008. Influence of temperature and particle size on the fixed bed pyrolysis of orange peel residues. *Journal of Analytical and Applied Pyrolysis*, 83: 124–130.

Ahn, W.S., Park, S.J., & Lee, S.Y. 2000. Production of poly(3-hydroxybutyrate) by fed-batch culture of recombinant *Escherichia coli* with a highly concentrated whey solution. *Applied and Environmental Microbiology*, 66: 3624–3627.

Akinbami, J.F.K., Ilori, M.O., Oyebisi, T.O., Akinwumi, I.O., & Adeoti, I.O.O. 2001. Biogas energy use in Nigeria: Current status. Future prospects and policy implication. *Renewable and Sustainable Energy Reviews*, 5: 97–112.

Aliyu, S., & Bala, M. 2011. Brewer's spent grain: A review of its potential applications. *African Journal of Biotechnology* 10: 324–331.

Alves, F.E.D.S.B., Carpiné, D., Teixeira, G.L., Goedert, A.C., de Paula Scheer, A., & Ribani, R.H. 2021. Valorization of an abundant slaughterhouse by-product as a source of highly technofunctional and antioxidant protein hydrolysates. *Waste and Biomass Valorization*, 12: 263–279.

Arapoglou, D., Varzakas, T., Vlyssides, A., & Israilides, C. 2010. Ethanol production from potato peel waste (PPW). *Waste Management*, 30: 1898–1902.

Atabani, A.E., Ala'a, H., Kumar, G., Saratale, G.D., Aslam, M., Khan, H.A., Said, Z., & Mahmoud, E. 2019. Valorization of spent coffee grounds into biofuels and value-added products: Pathway towards integrated bio-refinery. *Fuel*, 254: 115640.

Awasthi, M.K., Pandey, A.K., Bundela, P.S., Wong, J.W.C., Li, R., & Zhang, Z. 2016. Co-composting of gelatin industry sludge combined with organic fraction of municipal solid waste and poultry waste employing zeolite mixed with enriched nitrifying bacterial consortium. *Bioresource Technology* 213: 181–189.

Awasthi, M.K., Pandey, A.K., Khan, J., Bundela, P.S., Wong, J.W.C., & Selvam, A. 2014. Evaluation of thermophilic fungal consortium for organic municipal solid waste composting. *Bioresource Technology* 168: 214–221.

Baruah, J., Nath, B.K., Sharma, R., Kumar, S., Deka, R.C., Baruah, D.C., & Kalita, E. 2018. Recent trends in the pretreatment of lignocellulosic biomass for value-added products. *Frontiers in Energy Research*, 6: 141.

Beltrán-Ramírez, F., Orona-Tamayo, D., Cornejo-Corona, I., González-Cervantes, J.L.N., de Jesús Esparza-Claudio, J., & Quintana-Rodríguez, E. 2019. Agro-industrial waste revalorization: The growing biorefinery. In *Biomass for Bioenergy-Recent Trends and Future Challenges*. IntechOpen.

Belyaeva, O.N., & Haynes, R.J. 2012. Comparison of the effects of conventional organic amendments and biochar on the chemical, physical and microbial properties of coal fly ash as a plant growth medium. *Environmental Earth Sciences*, 66: 1987–1997.

Bernal, M.P., Alburquerque, J.A., & Moral, R. 2009. Composting of animal manures and chemical criteria for compost maturity assessment. A review. *Bioresource Technology*, 100: 5444–5453.

Boluda-Aguilar, M., & López-Gómez, A. 2013. Production of bioethanol by fermentation of lemon (*Citrus limon L.*) peel wastes pretreated with steam explosion. *Industrial Crops and Products*, 41: 188–197.

Boluda-Aguilar, M., García-Vidal, L., delPilar González-Castañeda, F., & López-Gómez, A. 2010. Mandarin peel wastes pretreatment with steam explosion for bioethanol production. *Bioresource Technology*, 101: 3506–3513.

Boulal, A., Atabani, A.E., Mohammed, M.N., Khelafi, M., Uguz, G., Shobana, S., Bokhari, A., & Kumar, G. 2019. Integrated valorization of *Moringa oleifera* and waste *Phoenix dactylifera* L. dates as potential feedstocks for biofuels production from Algerian Sahara: An experimental perspective. *Biocatalysis and Agricultural Biotechnology*, 20: 101234.

Bušić, A., Marđetko, N., Kundas, S., Morzak, G., Belskaya, H., IvančićŠantek, M., Komes, D., Novak, S., & Šantek, B. 2018. Bioethanol production from renewable raw materials and its separation and purification: A review. *Food Technology and Biotechnology*, 56: 289–311.

Cai, T., Park, S.Y., & Li, Y. 2013. Nutrient recovery from wastewater streams by microalgae: Status and prospects. *Renewable & Sustainable Energy Review*, 19: 360–369.

Cañadas, R., González-Miquel, M., González, E.J., Díaz, I., & Rodríguez, M. 2021. Hydrophobic eutectic solvents for extraction of natural phenolic antioxidants from winery wastewater. *Separation and Purification Technology*, 254: 117590.

Carlini, M., Castellucci, S., & Moneti, M. 2015. Biogas production from poultry manure and cheese whey wastewater under mesophilic conditions in batch reactor. *Energy Procedia*, 82: 811–818.

Caruso, M.C., Braghieri, A., Capece, A., Napolitano, F., Romano, P., Galgano, F., Altieri, G., & Genovese, F. 2019. Recent updates on the use of agro-food waste for biogas production. *Applied Sciences (Switzerland)*, 9: 1217.

Casabar, J.T., Unpaprom, Y., & Ramaraj, R. 2019. Fermentation of pineapple fruit peel wastes for bioethanol production. *Biomass Conversion and Biorefinery*, 9: 761–765.

CCME, 2005. Canadian Council of the Ministers of the Environment, Guidelines for Compost Quality, Ministry of Public Works and Government Services Canada. Cat. No. PN1341.

Chan, M.T., Selvam, A., & Wong, J.W.C. 2016. Reducing nitrogen loss and salinity during 'struvite' food waste composting by zeolite amendment. *Bioresource Technology*, 200: 838–844.

Chandra, R., Takeuchi, H., & Hasegawa, T. 2012. Methane production from lignocellulosic agricultural crop wastes: A review in context to second generation of biofuel production. *Renewable & Sustainable Energy Review*, 16: 1462–1476.

Chen, Z., Zhang, S., Wen, Q., & Zheng, J. 2015. Effect of aeration rate on composting of penicillin mycelial dreg. *Journal of Environmental Science*, 37: 172–178.

Choi, I.S., Lee, Y.G., Khanal, S.K., Park, B.J., & Bae, H.J. 2015. A low-energy, cost-effective approach to fruit and citrus peel waste processing for bioethanol production. *Applied Energy*, 140: 65–74.

Choudhary, P., Prajapati, S.K., & Malik, A. 2016. Screening native microalgal consortia for biomass production and nutrient removal from rural wastewaters for bioenergy applications. *Ecological Engineering*, 91: 221–230.

Chowdhury, M.A., de Neergaard, A., & Jensen, L.S. 2014. Potential of aeration flow rate and bio-char addition to reduce greenhouse gas and ammonia emissions during manure composting. *Chemosphere*, 97: 16–25.

Chua, J.Y., & Liu, S.Q. 2019. Soy whey: More than just wastewater from tofu and soy protein isolate industry. *Trends in Food Science & Technology*, 91: 24–32.

Cripwell, R., Favaro, L., Rose, S.H., Basaglia, M., Cagnin, L., Casella, S., & van Zyl, W. 2015. Utilisation of wheat bran as a substrate for bioethanol production using recombinant cellulases and amylolytic yeast. *Applied Energy*, 160: 610–617.

Dammak, I., Neves, M., Isoda, H., Sayadi, S., & Nakajima, M. 2016. Recovery of polyphenols from olive mill wastewater using drowning-out crystallization based separation process. *Innovative Food Science & Emerging Technologies*, 34: 326–335.

de Albuquerque Wanderley, M.C., Martín, C., de Moraes Rocha, G.J., & Gouveia, E.R. 2013. Increase in ethanol production from sugarcane bagasse based on combined pretreatments and fed-batch enzymatic hydrolysis. *Bioresource Technology*, 28: 448–453.

de Araujo Guilherme, A., Dantas, P.V.F., de AraújoPadilha, C.E., Dos Santos, E.S., & de Macedo, G.R. 2019. Ethanol production from sugarcane bagasse: Use of different fermentation strategies to enhance an environmental-friendly process. *Journal of Environmental Management*, 234: 44–51.

del Real Olvera, J., & Lopez-Lopez, A. 2012. Biogas production from anaerobic treatment of agro-industrial wastewater. In: *Biogas*, Rijeka: InTech, pp. 91–112.

Demirbas, A., & Ilten, N. 2004. Fuel analyses and thermochemical processing of olive residues. *Energy Sources*, 26: 731–738.

Devappa, R.K., Rakshit, S.K., & Dekker, R.F.H. 2015. Potential of poplar bark phyto-chemicals as value-added co-products from the wood and cellulosic bioethanol industry. *Bio Energy Research*, 8: 1235–1251.

Devesa-Rey, R., Vecino, X., Varela-Alende, J.L., Barral, M.T., Cruz, J.M., & Moldes, A.B. 2011. Valorization of winery waste vs. the costs of not recycling. *Waste Management*, 31: 2327–2335.

Dinuccio, E., Balsari, P., Gioelli, F., & Menardo, S. 2010. Evaluation of the biogas produc-tivity potential of some Italian agro-industrial biomasses. *Bioresource Technology*, 101: 3780–3783.

Domínguez-Bocanegra, A.R., Torres-Muñoz, J.A., & López, R.A. 2015. Production of bioethanol from agro-industrial wastes. *Fuel*, 149: 85–89.

Elain, A., Le Grand, A., Corre, Y.M., Le Fellic, M., Hachet, N., Le Tilly, V., & Bruzaud, S. 2016. Valorisation of local agro-industrial processing waters as growth media for polyhydroxyalkanoates (PHA) production. *Industrial Crops and Products*, 80: 1–5.

Fertilizer Association of India. 2007. The Fertilizer (Control) Order 1985. The Fertilizer Association of India 10.

Federici, F., Fava, F., Kalogerakis, N., & Mantzavinos, D. 2009. Valorisation of agro-indus-trial by-products, effluents and waste: Concept, opportunities and the case of olive mill wastewaters. *Journal of Chemical Technology & Biotechnology: International Research in Process, Environmental & Clean Technology*, 84: 895–900.

Galliou, F., Markakis, N., Fountoulakis, M.S., Nikolaidis, N., & Manios, T. 2018. Production of organic fertilizer from olive mill wastewater by combining solar greenhouse drying and composting. *Waste Management*, 75: 305–311.

Gao, M., Li, B., Yu, A., Liang, F., Yang, L., & Sun, Y. 2010. The effect of aeration rate on forced-aeration composting of chicken manure and sawdust. *Bioresource Technology*, 101: 1899–1903.

García, R., Pizarro, C., Lavín, A.G., & Bueno, J.L. 2012. Characterization of Spanish biomass wastes for energy use. *Bioresource Technology*, 103: 249–258.

Gil, L.S., & Maupoey, P.F. 2018. An integrated approach for pineapple waste valorisation. Bioethanol production and bromelain extraction from pineapple residues. *Journal of Cleaner Production*, 172: 1224–1231.

Gouvea, B.M., Torres, C., Franca, A.S., Oliveira, L.S., & Oliveira, E.S. 2009. Feasibility of ethanol production from coffee husks. *Biotechnology Letter*, 31: 1315–1319.

Gudiña, E.J., Rodrigues, A.I., de Freitas, V., Azevedo, Z., Teixeira, J.A., & Rodrigues, L.R. 2016. Valorization of agro-industrial wastes towards the production of rham-nolipids. *Bioresource Technology*, 212: 144–150.

Guldhe, A., Ansari, F.A., Singh, P., & Bux, F. 2017. Heterotrophic cultivation of micro-algae using aquaculture wastewater: A biorefinery concept for biomass production and nutrient remediation. *Ecological Engineering*, 99: 47–53.

Guo, R., Li, G., Jiang, T., Schuchardt, F., Chen, T., Zhao, Y., & Shen, Y. 2012. Effect of aeration rate, C/N ratio and moisture content on the stability and maturity of compost. *Bioresource Technology*, 112: 171–178.

Gupta, A., & Verma, J.P. 2015. Sustainable bio-ethanol production from agro-residues: A review. *Renewable and Sustainable Energy Reviews*, 41: 550–567.

Haas, R., Jin, B., & Zepf, F.T. 2008. Production of poly (3-hydroxybutyrate) from waste potato starch. *Bioscience, Biotechnology, and Biochemistry*, 72: 253–6.

Hachicha, S., Sellami, F., Cegarra, J., Hachicha, R., Drira, N., Medhioub, K., & Ammar, E. 2009. Biological activity during co-composting of sludge issued from the OMW evaporation ponds with poultry manure-physico-chemical characterization of the processed organic matter. *Journal of Hazardous Materials*, 162: 402–409.

Henrique, M.A., Silvério, H.A., Neto, W.P.F., & Pasquini, D. 2013. Valorization of an agro-industrial waste, mango seed, by the extraction and characterization of its cellulose nanocrystals. *Journal of Environmental Management*, 121: 202–209.

Hernández, D., Riaño, B., Coca, M., & García-González, M.C. 2013. Treatment of agro-industrial wastewater using microalgae-bacteria consortium combined with anaerobic digestion of the produced biomass. *Bioresource Technology*, 135: 598–603.

Hongyang, S., Yalei, Z., Chunmin, Z., Xuefei, Z., & Jinpeng, L. 2011. Cultivation of *Chlorella pyrenoidosa* in soybean processing wastewater. *Bioresource Technology*, 102: 9884–9890.

Iqbal, M.K., Nadeem, A., Sherazi, F., & Khan, R.A. 2015. Optimization of process parameters for kitchen waste composting by response surface methodology. *International Journal of Environmental Science and Technology*, 12: 1759–1768.

Jain, M.S., & Kalamdhad, A.S. 2018. Efficacy of batch mode rotary drum composter for management of aquatic weed (*Hydrilla verticillata* (Lf) Royle). *Journal of Environmental Management*, 221: 20–27.

Ji, M.K., Kim, H.C., Sapireddy, V.R., Yun, H.S., Abou-Shanab, R.A., Choi, J., Lee, W., Timmes, T.C., Inamuddin, & Jeon, B.H. 2013. Simultaneous nutrient removal and lipid production from pretreated piggery wastewater by *Chlorella vulgaris* YSW-04. *Applied Microbiology and Biotechnology*, 97: 2701–2710.

Ji-Lu, Z. 2007. Bio-oil from fast pyrolysis of rice husk: Yields and related properties and improvement of the pyrolysis system. *Journal of Analytical and Applied Pyrolysis*, 80: 30–35.

Joshi, V.K., Kumar, A., & Kumar, V. 2012. Antimicrobial, antioxidant and phyto-chemicals from fruit and vegetable wastes: A review. *International Journal of Food and Fermentation Technology*, 2: 123.

Kalamdhad, A.S., & Kazmi, A.A. 2009. Effects of turning frequency on compost stability and some chemical characteristics in a rotary drum composter. *Chemosphere*, 74: 1327–1334.

Katalinic, V., Mozina, S.S., Skroza, D., Generalic, I., Abramovic, H., Milos, M., Ljubenkov, I., Piskernik, S., Pezo, I., Terpinc, P., & Boban, M. 2010. Polyphenolic profile, antioxidant properties and antimicrobial activity of grape skin extracts of 14 *Vitis vinifera* varieties grown in Dalmatia (Croatia). *Food Chemistry*, 119: 715–723.

Kazemi, M., Khodaiyan, F., Hosseini, S.S., & Najari, Z. 2019. An integrated valorization of industrial waste of eggplant: Simultaneous recovery of pectin, phenolics and sequential production of pullulan. *Waste Management*, 100: 101–111.

Kebaili, M., Djellali, S., Radjai, M., Drouiche, N., & Lounici, H. 2018. Valorization of orange industry residues to form a natural coagulant and adsorbent. *Journal of Industrial and Engineering Chemistry*, 64: 292–299.

Khalil, H.A., Aprilia, N.S., Bhat, A.H., Jawaid, M., Paridah, M.T., & Rudi, D. 2013. A Jatropha biomass as renewable materials for biocomposites and its applications. *Renewable and Sustainable Energy Reviews*, 22: 667–685.

Kokkora, M., Vyrlas, P., Papaioannou, C., Petrotos, K., Gkoutsidis, P., Leontopoulos, S., & Makridis, C. 2015. Agricultural use of microfiltered olive mill wastewater: Effects on maize production and soil properties. *Agriculture and Agricultural Science Procedia*, 4: 416–424.

Lee, C.S., Oh, H.-S., Oh, H.M., Kim, H.S., & Ahn, C.Y. 2016. Two-phase photoperiodic cultivation of algal–bacterial consortia for high biomass production and efficient nutrient removal from municipal wastewater. *Bioresource Technology*, 200: 867–875.

Liang, C., Das, K.C., & McClendon, R.W. 2003. The influence of temperature and moisture contents regimes on the aerobic microbial activity of a biosolids composting blend. *Bioresource Technology*, 86: 131–137.

Li, Y., Chen, Y.F., Chen, P., Min, M., Zhou, W., Martinez, B., Zhu, J., & Ruan, R. 2011. Characterization of a microalga *Chlorella* sp. well adapted to highly concentrated municipal wastewater for nutrient removal and biodiesel production. *Bioresource Technology*, 102: 5138–5144.

Libutti, A., Gatta, G., Gagliardi, A., Vergine, P., Pollice, A., Beneduce, L., & Tarantino, E. 2018. Agro-industrial wastewater reuse for irrigation of a vegetable crop succession under Mediterranean conditions. *Agricultural Water Management*, 196: 1–14.

López-Cano, I., Roig, A., Cayuela, M.L., Alburquerque, J.A., & Sánchez-Monedero, M.A., 2016. Biochar improves N cycling during composting of olive mill wastes and sheep manure. *Waste Management*, 49: 553–559.

Lu, Q., Yang, X., & Zhu, X. 2008. Analysis on chemical and physical properties of bio-oil pyrolyzed from rice husk. *Journal of Analytical and Applied Pyrolysis*, 82: 191–198.

Luangwilai, T., Sidhu, H.S., Nelson, M.I., & Chen, X.D. 2011. Modelling the effects of moisture content in compost piles. In: Australian Chemical Engineering Conference Australia: Engineers Australia, pp. 2–12.

Manu, M.K., Kumar, R., & Garg, A. 2017. Performance assessment of improved composting system for food waste with varying aeration and use of microbial inoculum. *Bioresource Technology*, 234: 167–177.

Maragkaki, A.E., Vasileiadis, I., Fountoulakis, M., Kyriakou, A., Lasaridi, K., & Manios, T. 2018. Improving biogas production from anaerobic co-digestion of sewage sludge with a thermal dried mixture of food waste, cheese whey and olive mill wastewater. *Waste Management*, 71: 644–651.

Martí-Herrero, J., Alvarez, R., & Flores, T. 2018. Evaluation of the low technology tubular digesters in the production of biogas from slaughterhouse wastewater treatment. *Journal of Cleaner Production*, 199: 633–642.

Messineo, A., Maniscalco, M.P., & Volpe, R. 2019. Biomethane recovery from olive mill residues through anaerobic digestion: A review of the state of the art technology. *Science of Total Environment*, 135508.

Meyer-Aurich, A., Schattauer, A., Hellebrand, H.J., Klauss, H., Plochl, M., & Berg, W. 2012. Impact of uncertainties on greenhouse gas mitigation potential of biogas production from agricultural resources. *Renewable Energy* 37: 277–284.

Miranda, I., Simões, R., Medeiros, B., Nampoothiri, K.M., Sukumaran, R.K., Rajan, D., Pereira, H., & Ferreira-Dias, S. 2019. Valorization of lignocellulosic residues from the olive oil industry by production of lignin, glucose and functional sugars. *Bioresource Technology*, 292: 121936.

Mishra, B., Varjani, S., & Varma, G.K.S. 2019. Agro-industrial by-products in the synthesis of food grade microbial pigments: An eco-friendly alternative. In: *Green Bio-Processes*, Springer, Singapore, pp. 245–265.

Mishra, R.K., & Mohanty, K. 2018. Characterization of non-edible lignocellulosic biomass in terms of their candidacy towards alternative renewable fuels. *Biomass Conversion and Biorefinery*, 8: 799–812.

Mohammad, N., Alam, M.Z., Kabbashi, N.A., & Ahsan, A. 2012. Effective composting of oil palm industrial waste by filamentous fungi: A review. *Resources, Conservation and Recycling*, 58: 69–78.

Mohanty, A., Rout, P.R., Dubey, B., Meena, S.S., Pal, P., & Goel, M. 2021. A critical review on biogas production from edible and non-edible oil cakes. *Biomass Conversion and Biorefinery*, 1–18.

Mshandete, A., Björnsson, L., Kivaisi, A.K., Rubindamayugi, M.S.T., & Mattiasson, B. 2006. Effect of particle size on biogas yield from sisal fibre waste. *Renewable Energy*, 31: 2385–2392.

Mu, D., Seager, T., Rao, P.S., & Zhao, F. 2010. Comparative life cycle assessment of lignocellulosic ethanol production: Biochemical versus thermochemical conversion. *Environmental Management*, 46: 565–578.

Nachaiwieng, W., Lumyong, S., Yoshioka, K., Watanabe, T., & Khanongnuch, C. 2015. Bioethanol production from rice husk under elevated temperature simultaneous saccharification and fermentation using *Kluyveromyces marxianus* CK8. *Biocatalysis and Agricultural Biotechnology*, 4: 543–549.

Nanda, S., Mohanty, P., Pant, K.K., Naik, S., Kozinski, J.A., & Dalai, A.K. 2013. Characterization of North American lignocellulosic biomass and biochars in terms of their candidacy for alternate renewable fuels. *Bioenergy Research*, 6: 663–677.

Narihiro, T., Abe, T., Yamanaka, Y., & Hiraishi, A. 2004. Microbial population dynamics during fed-batch operation of commercially available garbage composters. *Applied Microbiology and Biotechnology*, 65: 488–495.

Nzila, C., Dewulf, J., Spanjers, H., Kiriamiti, H., & van Lagenhove, H. 2010. Biowaste energy potential in Kenya. *Renewable Energy*, 35: 2698–2704.

Okamoto, K., Nitta, Y., Maekawa, N., & Yanase, H. 2011. Direct ethanol production from starch, wheat bran and rice straw by the white rot fungus *Trametes hirsuta*. *Enzyme and Microbial Technology*, 48: 273–277.

Okoye, B.O., Igbokwe, P.K., & Ude, C.N. 2016. Comparative study of biogas production from cow dung and brewer's spent grain. *International Journal of Research in Advanced Engineering and Technology*, 2: 19–21.

Onwosi, C.O., Igbokwe, V.C., Odimba, J.N., Eke, I.E., Nwankwoala, M.O., Iroh, I.N., & Ezeogu, L.I. 2017. Composting technology in waste stabilization: On the methods, challenges and future prospects. *Journal of Environmental Management*, 190: 140–157.

Oudart, D., Robin, P., Paillat, J.M., & Paul, E. 2015. Modelling nitrogen and carbon interactions in composting of animal manure in naturally aerated piles. *Waste Management*, 46: 588–598.

Oviedo-Ocaña, E.R., Torres-Lozada, P., Marmolejo-Rebellon, L.F., Hoyos, L. V, Gonzales, S., Barrena, R., Komilis, D., & Sanchez, A. 2015. Stability and maturity of biowaste composts derived by small municipalities: Correlation among physical, chemical and biological indices. *Waste Management*, 44: 63–71.

Panesar, R., Kaur, S., & Panesar, P.S. 2015. Production of microbial pigments utilizing agro-industrial waste: A review. *Current Opinion in Food Science*, 1: 70–76.

Paradelo, R., Moldes, A.B., & Barral, M.T. 2013. Evolution of organic matter during the mesophilic composting of lignocellulosic winery wastes. *Journal of Environmental Management*, 116: 18–26.

Pattanaik, L., Duraivadivel, P., Hariprasad, P., & Naik, S.N. 2020a. Utilization and re-use of solid and liquid waste generated from the natural indigo dye production process – A zero waste approach. *Bioresource Technology*, 301: 122721.

Pattanaik, L., Naik, S.N., & Hariprasad, P. 2019. Valorization of waste *Indigofera tinctoria* L. biomass generated from indigo dye extraction process—potential towards biofuels and compost. *Biomass Conversion and Biorefinery*, 9: 445–457.

Pattanaik, L., Padhi, S.K., Hariprasad, P., & Naik, S.N. 2020b. Life cycle cost analysis of natural indigo dye production from *Indigofera tinctoria* L. plant biomass: A case study of India. *Clean Technologies and Environmental Policy*, 22: 1639–1654.

Pellera, F.M., & Gidarakos, E. 2016. Effect of substrate to inoculum ratio and inoculum type on the biochemical methane potential of solid agroindustrial waste. *Journal of Environmental Chemical Engineering*, 4:3217–3229.

Petric, I., Avdihodžić, E., & Ibrić, N. 2015. Numerical simulation of composting process for mixture of organic fraction of municipal solid waste and poultry manure. *Ecological Engineering*, 75: 242–249.

Pittman, J.K., Dean, A.P., & Osundeko, O. 2011. The potential of sustainable algal biofuel production using wastewater resources. *Bioresource Technology*, 102: 17–25.

Prajapati, S.K., Kumar, P., Malik, A., & Vijay, V.K. 2014. Bioconversion of algae to methane and subsequent utilization of digestate for algae cultivation: A closed loop bioenergy generation process. *Bioresource Technology*, 158: 174–180.

Rawoteea, S.A., Mudhoo, A., & Kumar, S. 2017. Co-composting of vegetable wastes and carton: Effect of carton composition and parameter variations. *Bioresource Technology*, 227: 171–178.

Rekha, B., & Saravanathamizhan, R. 2021. Catalytic conversion of corncob biomass into bioethanol. *International Journal of Energy Research*, 45(3): 4508–4518.

Rich, N., Bharti, A., & Kumar, S. 2018. Effect of bulking agents and cow dung as inoculant on vegetable waste compost quality. *Bioresource Technology*, 252: 83–90.

Rocha, N.R.D.A.F., Barros, M.A., Fischer, J., CoutinhoFilho, U., & Cardoso, V.L. 2013. Ethanol production from agroindustrial biomass using a crude enzyme complex produced by *Aspergillus niger*. *Renewable Energy*, 57: 432–435.

Rudra, S.G., Nishad, J., Jakhar, N., & Kaur, C. 2015. Food industry waste: Mine of nutraceuticals. *International Journal of Science Environment and Technology*, 4: 205–229.

Saini, J.K., Saini, R., & Tewari, L. 2015. Lignocellulosic agriculture wastes as biomass feedstocks for second-generation bioethanol production: Concepts and recent developments. *3 Biotech*, 5: 337–353.

Sánchez-Acosta, D., Rodriguez-Uribe, A., Álvarez-Chávez, C.R., Mohanty, A.K., Misra, M., López-Cervantes, J., & Madera-Santana, T.J. 2019. Physicochemical characterization and evaluation of pecan nutshell as biofiller in a matrix of poly (lactic acid). *Journal of Polymers and the Environment*, 27, 521–532.

Santi, G., Crognale, S., D'Annibale, A., Petruccioli, M., Ruzzi, M., Valentini, R., & Moresi, M. 2014. Orange peel pretreatment in a novel lab-scale direct steam-injection apparatus for ethanol production. *Biomass and Bioenergy*, 61: 146–156.

Semitela, S., Pirra, A., & Braga, F.G. 2019. Impact of mesophilic co-composting conditions on the quality of substrates produced from winery waste activated sludge and grape stalks: Lab-scale and pilot-scale studies. *Bioresource Technology*, 289: 121622.

Sette, P., Fernandez, A., Soria, J., Rodriguez, R., Salvatori, D., & Mazza, G. 2020. Integral valorization of fruit waste from wine and cider industries. *Journal of Cleaner Production*, 242: 118486.

Sewsynker-Sukai, Y., & Kana, E.G. 2018. Simultaneous saccharification and bioethanol production from corn cobs: Process optimization and kinetic studies. *Bioresource Technology*, 262: 32–41.

Sharma, D., Yadav, K.D., & Kumar, S. 2018. Role of sawdust and cow dung on compost maturity during rotary drum composting of flower waste. *Bioresource Technology*, 264: 285–289.

Sharma, P., Gaur, V.K., Kim, S.H., & Pandey, A. 2020. Microbial strategies for bio-transforming food waste into resources. *Bioresource Technology*, 299: 122580.

Sheng, C., & Azevedo, J.L.T. 2005. Estimating the higher heating value of biomass fuels from basic analysis data. *Biomass and Bioenergy*, 28: 499–507.

Shivakumar, S. 2012. Polyhydroxybutyrate (PHB) production using agro-industrial residue as substrate by *Bacillus thuringiensis* IAM 12077. *International Journal of ChemTech Research*, 4: 1158–1162.

Solid Waste Management (SWM) Rules, 2016. http://www.moef.gov.in/sites/default/files/SWM%202016.pdf. Accessed May 2019.

Suarez, J.A., Luengo, C.A., Felfli, F.F., Bezzon, G., & Beatón, P.A. 2000. Thermochemical properties of Cuban biomass. *Energy Sources*, 22: 851–857.

Suhartini, S., Nurika, I., Paul, R., & Melville, L. 2020. Estimation of biogas production and the emission savings from anaerobic digestion of fruit-based agro-industrial waste and agricultural crops residues. *Bioenergy Research*, 1–16.

Sukiran, M.A., Abnisa, F., Daud, W.M.A.W., Bakar, N.A., & Loh, S.K. 2017. A review of torrefaction of oil palm solid wastes for biofuel production. *Energy Conversion and Management*, 149: 101–120.

Sun, Y., & Cheng, J. 2002. Hydrolysis of lignocellulosic materials for ethanol production: A review. *Bioresource Technology*, 83: 1–11.

Tang, J.-C., Shibata, A., Zhou, Q., & Katayama, A. 2007. Effect of temperature on reaction rate and microbial community in composting of cattle manure with rice straw. *Journal of Bioscience and Bioengineering*, 104: 321–328.

TMECC, 2002. Test methods for the examination of composts and composting. In: Thompson, W., Leege, P., Millner, P., Watson, M.E. (Eds.), *The US Composting Council*, US Government Printing Office.

Todhanakasem, T., Narkmit, T., Areerat, K., & Thanonkeo, P. 2015. Fermentation of rice bran hydrolysate to ethanol using *Zymomonas mobilis* biofilm immobilization on DEAE-cellulose. *Electronic Journal of Biotechnology*, 18(3): 196–201.

Todhanakasem, T., Sangsutthiseree, A., Areerat, K., Young, G.M., & Thanonkeo, P. 2014. Biofilm production by *Zymomonas mobilis* enhances ethanol production and tolerance to toxic inhibitors from rice bran hydrolysate. *New Biotechnology*, 31: 451–459.

Torr, L.C. 2009. Applications of dairy wastewater as a fertilizer to agricultural land: An environmental management perspective: Stellenbosch, South Africa, University of Stellenbosch Department of Geology, Geography and Environmental Studies, 133.

Tovar, A.K., Godínez, L.A., Espejel, F., Ramírez-Zamora, R.-M., & Robles, I. 2019. Optimization of the integral valorization process for orange peel waste using a design of experiments approach: Production of high-quality pectin and activated carbon. *Waste Management*, 85: 202–213.

Tsai, W.T., Lee, M.K., & Chang, D.Y. 2006. Fast pyrolysis of rice straw, sugarcane bagasse and coconut shell in an induction-heating reactor. *Journal of Analytical and Applied Pyrolysis*, 76: 230–237.

Uppugundla, N., da Costa Sousa, L., Chundawat, S.P., Yu, X., Simmons, B., Singh, S., Gao, X., Kumar, R., Wyman, C., Dale, B., & Balan, V. 2014. A comparative study of ethanol production using dilute acid, ionic liquid and AFEX™ pretreated corn stover. *Biotechnology for Biofuels*, 7: 1–14.

Valijanian, E., Tabatabaei, M., Aghbashlo, M., Sulaiman, A., & Chisti, Y. 2018. Biogas production systems. In: *Biogas*, Springer, pp. 95–116.

Vassilev, S. V, Baxter, D., Andersen, L.K., & Vassileva, C.G. 2010. An overview of the chemical composition of biomass. *Fuel*, 89: 913–933.

Vassilev, S. V, Baxter, D., Andersen, L.K., Vassileva, C.G., & Morgan, T.J. 2012. An overview of the organic and inorganic phase composition of biomass. *Fuel*, 94: 1–33.

Wang, Q., Awasthi, M.K., Zhang, Z., & Wong, J.W. 2019. Sustainable composting and its environmental implications. In: Taherzadeh, M.J., Bolton, K., Wong, J., Pandey, A. (Eds.), *Sustainable Resource Recovery and Zero Waste Approaches*, Elsevier, pp. 115–132.

Wang, Q., Li, R., Cai, H., Awasthi, M.K., Zhang, Z., Wang, J.J., Ali, A., & Amanullah, M. 2016. Improving pig manure composting efficiency employing Ca-bentonite. *Ecological Engineering*, 87: 157–161.

Wang, S.P., Zhong, X.Z., Wang, T.T., Sun, Z.Y., Tang, Y.Q., & Kida, K. 2017a. Aerobic composting of distilled grain waste eluted from a Chinese spirit-making process: The effects of initial pH adjustment. *Bioresource Technology*, 245: 778–785.

Wang, X., Zhao, Y., Wang, H., Zhao, X., Cui, H., & Wei, Z. 2017b. Reducing nitrogen loss and phytotoxicity during beer vinasse composting with biochar addition. *Waste Management*, 61: 150–156.

Watanabe, M., Takahashi, M., Sasano, K., Kashiwamura, T., Ozaki, Y., Tsuiki, T., Hidaka, H., & Kanemoto, S. 2009. Bioethanol production from rice washing drainage and rice bran. *Journal of Bioscience and Bioengineering*, 108: 524–526.

Weiland, P. 2010. Biogas production: Current state and perspectives. *Applied Microbiology and Biotechnology* 85: 849–860.

Wikandari, R., Nguyen, H., Millati, R., Niklasson, C., & Taherzadeh, M.J. 2015. Improvement of biogas production from orange peel waste by leaching of limonene. *BioMed Research International*, 2015: 494182.

Yin, C.Y. 2011. Prediction of higher heating values of biomass from proximate and ultimate analyses. *Fuel*, 90: 1128–1132.

Yusuf, M. 2017. Agro-industrial waste materials and their recycled value-added applications. In: *Handbook of Ecomaterials*, pp. 1–11.

Zayed, G., & Abdel-Motaal, H. 2005. Bio-active composts from rice straw enriched with rock phosphate and their effect on the phosphorous nutrition and microbial community in rhizosphere of cowpea. *Bioresource Technology*, 96: 929–935.

Zhou, J., Yang, J., Yu, Q., Yong, X., Xie, X., Zhang, L., Wei, P., & Jia, H. 2017. Different organic loading rates on the biogas production during the anaerobic digestion of rice straw: A pilot study. *Bioresource Technology*, 244: 865–871.

Zucconi, F., Monaco, A, Forte, M., & de Bertoldi, M. 1985. Phytotoxins during the stabilization of organic matter. In: Gasser, J.K.R. (Ed.), *Composting of Agricultural and Other Wastes*, Elsevier Applied Science Publ., London, pp. 73–85.

9 Management of Emerging Contaminants in Wastewater
Detection, Treatment, and Challenges

Aryama Raychaudhuri and Manaswini Behera
School of Infrastructure, Indian Institute of
Technology Bhubaneswar, Khordha, Odisha, India

CONTENTS

DOI: 10.1201/9781003201076-9

9.1 INTRODUCTION

Emerging contaminants (ECs) are a comparatively new group of unregulated substances of anthropogenic origin that can cause detrimental effects on the marine ecosystem and human health. ECs in the environmental samples have not been investigated widely before the late 1990s owing to the absence of sensitive analytical detection techniques to measure their comparatively low concentration. In general, EC concentration varies from a few ng/L to a few hundred µg/L in the aquatic environment. However, even at low levels, many of these contaminants are capable of posing severe environmental and human health hazards; thus, they are also identified as Contaminants of Emerging Concern (CEC). ECs in the biosphere may have adverse implications on aquatic and human life via disruption of the endocrine system of living organisms, antimicrobial resistance, and soil accumulation. Another instrumental problem for the ecosystem and human health is the uptake of ECs by plants from polluted soils and its accumulation in the food chain. Inadequate knowledge on the transport and fate of emerging compounds as well as their possible toxic responses makes it difficult for policymakers to formulate the regulatory guidelines that could support their environmental management (Naidu et al., 2016; Noguera-Oviedo et al., 2016). The broad category of ECs includes pharmaceutical drugs, personal care products (PCPs), natural and synthetic hormones, antibiotics, microplastics, artificial sweeteners, pesticides, surfactants, plasticizers, endocrine-disrupting compounds (EDCs), etc. (Ahmed et al., 2017). Abnormal physiological responses, reproductive dysfunction, increased incidences of cancer, antimicrobial resistance in bacteria, and increased toxicity of spent chemicals are possible environmental and health issues related to these emerging pollutants (Gogoi et al., 2018). Pesticides present in the surface water bodies can cause immune reactions in mammals and fish, altering the anterior kidney's hemopoietic tissue. EDCs, which modify the endocrine systems' functions, trigger many detrimental health problems for an organism or its descendants. The acute and chronic effects of EDCs include reproductive impairments including polycystic ovaries, human sperm reduction, decreased fertility, breakage of egg (for fishes, birds, and turtles), prostate cancer, endometriosis, and testicular or breast cancer (Noguera-Oviedo et al., 2016; Rout et al., 2021).

Most of these ECs were detected at concentrations of µg/L in water supplies and originate from industrial discharge, domestic and hospital wastewater, landfill leachates, or agricultural and livestock runoffs. ECs can have more lasting effects than commonly believed, despite being detected at trace levels, often lower than their toxic concentrations. Few may be harmful or cancerous and very persistent, with a propensity to bioaccumulate. They are transformed or remain similar as they enter the traditional wastewater treatment plant (WWTP) and, in certain instances, discharged into surface waters in their native form. From that viewpoint, the development of strategies for their removal is of utmost importance (Pesqueira et al., 2020).

Although various treatment techniques have been explored to remove ECs from wastewater, most of them demonstrate inadequate removal efficiency. Owing to its high solubility in water, most pesticides and pharmaceuticals remain partitioned in the aqueous phase. Their removal by physical techniques such as flocculation and sedimentation had limited success and documented to be less than 10% (Ahmed et al., 2017). Conventional WWTPs are capable of removing ECs but not up to the desired level; thus modification of traditional treatment techniques is necessary to minimize their environmental release. Different biological (activated sludge process (ASP), constructed wetland, microalgae and fungal treatment, membrane bioreactor (MBR), trickling filter, rotating biological reactor, biosorption, enzymatic treatment, bioelectrochemical system, etc.), physical (activated carbon adsorption, membrane filtration, etc.), and chemical (advanced oxidation process (AOP), ozonation, photocatalysis, photo-Fenton, electro-Fenton, etc.) treatment technologies have shown effective EC removal (Ahmed et al., 2017; Rout et al., 2021). Activated carbon has shown more favorable results due to its ability to adsorb most ECs. However, pore blocking, competition for adsorption sites, and inefficient bacterial inactivation have undermined its long-term usage. The membrane filtration techniques (ultrafiltration and nanofiltration) face challenges because the membrane's average pore size is higher than the target EC. On the other hand, reverse osmosis and nanofiltration suffer from excessive membrane fouling and concentrate disposal issues. Besides being expensive, both UV irradiation and AOPs produce harmful disinfection by-products (Shah et al., 2020). Therefore, new approaches and technologies are required to efficiently eliminate ECs and meet conventional treatment standards at minimal capital and operating costs.

Although several studies have addressed the toxic effects of ECs and recommended treatment technologies for their management, some critical knowledge gap still exists, requiring further investigation. Appropriate analytical methods and sophisticated tools for the detection of ECs and their metabolites are still under development. The suitable reference materials for several ECs are still unknown. The fate, toxicity, and behavior of ECs in the environment and their long-term (chronic) health effects are still unknown (Naidu et al., 2016). Future research should be focused on a more in-depth understanding of ECs and their interaction in the environment.

9.2 EMERGING CONTAMINANTS IN WASTEWATER

9.2.1 Types of Emerging Contaminants and Their Effects

As new compounds that belong to this category are being detected, the wide range of chemical groups that comprise ECs continues to expand. A wide variety of compounds can be included in the EC group consists of pesticides, pharmaceuticals, food additives, artificial sweeteners, disinfection by-products, flame/fire retardants, and PCPs. The primary sources and pathways of the ECs are depicted in Figure 9.1. The different ECs present in water and wastewater samples are tabulated in Table 9.1. The broad spectrum of ECs can be classified into the following groups.

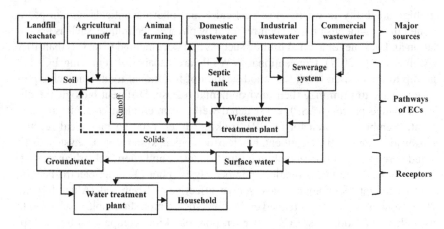

FIGURE 9.1   Emerging contaminants in the environment: major sources, pathways, and receptors.

TABLE 9.1
The Presence of Different ECs in Various Wastewater Samples

Emerging Contaminant	Sample	Concentration (ng/L)	Reference
Personal care products			
Triclocarban	Effluent of WWTP	29.6 ± 2.9 (wet weather)	Lu et al. (2018)
		12.5 ± 1.2 (dry weather)	
Benzotriazole	River water	39 ± 6 to 526 ± 127	Williams et al. (2019)
Bisphenol A	River water	95 ± 4 to 299 ± 100	Williams et al. (2019)
Triclosan	Effluent of WWTP	1.5 ± 0.1 (wet weather)	He et al. (2018)
		2.0 ± 0.2 (dry weather)	
Pharmaceuticals			
Sulfamethoxazole	Effluent of WWTP	374.3–1,309.5	Archundia et al. (2017)
Erythromycin	Effluent of WWTP	4–133	Prabhasankar et al. (2016)
Ciprofloxacin	Tidal brackish water	0.4–28.8	Panthi et al. (2019)
Tetracycline	Fresh water creek	3.1–15.1	Panthi et al. (2019)
Ibuprofen	River water	4.75	Li et al. (2013)
Atenolol	Public supply wells	8	Schaider et al. (2014)
Ofloxacin	River water	33.1–37.0	Zuccato et al. (2005)
Estrogen			
Estrone	River water	26 ± 0.3 to 124 ± 15	Williams et al. (2019)
17 β-estradiol	River water	4 ± 1 to 28 ± 5	Williams et al. (2019)

9.2.1.1 Analgesics and Antibiotics

Non-steroidal anti-inflammatory medications such as diclofenac, paracetamol (acetaminophen), ibuprofen, and naproxen are widely used and readily available. Ibuprofen rapidly converts into its two primary metabolites: carboxyibuprofen and hydroxyibuprofen prior to sewage treatment works (STW) discharge. In STW effluents and seawater, parent-ibuprofen concentrations of 24,600 ng/L (median 4,000 ng/L) and 700 ng/L were detected. One of the oldest, most effective, and most widely used pharmaceuticals are antibiotics, commonly found in wastewater. Antibiotic metabolites such as erythromycin and penicillin were also identified in ground and surface water at concentrations up to 30 ng/L (Focazio et al., 2008). Even in relatively low concentrations, they can trigger antibiotic resistance in microbes (Wilkinson et al., 2017).

9.2.1.2 Pharmaceuticals

Pharmaceutics is an important EC category that poses serious concerns about the risks of estrogenic and other harmful impacts on humans and wildlife. Around 3000 different chemical compounds, including painkillers (ibuprofen, paracetamol, naproxen, codeine, diclofenac, etc.), antibiotics (amoxicillin, erythromycin, triclosan, sulfamethoxazole, trimethoprim, etc.), antidepressants (amitriptyline, fluoxetine, venlafaxine, etc.), beta-blockers (metoprolol, atenolol), lipid regulators (bezafibrate, gemfibrozil), hormones/steroids (ethynylestradiol, estradiol, norethisterone, coprostanol), and antidiabetics are known to be used as pharmaceutical ingredients (Rodriguez-Narvaez et al., 2017; Wilkinson et al., 2017). Pharmaceuticals have been identified in WWTP effluents, sludge deposits, surface, and groundwater. They are perceived as pseudo-persistent contaminants that continually enter the atmosphere at trace levels and bioaccumulate, causing severe ecological impacts (Gogoi et al., 2018).

9.2.1.3 Pesticides and Biocides

One of the major issues with pesticides is related to the biodegradation of agricultural crop material resulting in disruption of the food chain. The term "pesticide" refers to agricultural chemicals, whereas "biocide" refers to urban-environmental chemicals (Margot et al., 2015). Biocides and pesticides are introduced to the ground and surface water by storm-water runoff during events of precipitation. The metabolites of pesticides are extremely toxic and biologically active, so some of the pesticides considered allowable have become less acceptable and are now categorized as ECs. These metabolites include 3-hydroxycarbofuran, clethodim, acephate, captan, acetochlor ethanesulfonic acid, acetochlor oxanilic acid, etc. (Naidu et al., 2016; Rodriguez-Narvaez et al., 2017).

9.2.1.4 Personal Care Products (PCPs)

PCPs mainly incorporate perfumes and deodorants (musk compounds), cosmetic products, personal hygiene products, cleaning agents (detergent, soaps, and shampoos), and ultraviolet (UV) filters (sunscreens), which are usually released through domestic wastewater. Synthetic musk compounds (SMCs) are extensively utilized in PCPs (fragrances, cosmetics, ointments, detergents, and soaps) and are usually found in various chemical structures in the environment. Tonalide, galaxolide, musk xylene, and musk ketone are some of the most commonly found SMCs present in PCPs such as creams, scents, and deodorants at concentrations up to 8,000 mg/g, 22,000 mg/g, 26 mg/g, and 0.5 mg/g, respectively (Roosens et al., 2007). Cleaning products containing non-ion surfactants, such as alkylphenol ethoxylates, used in domestic and industrial applications, have been identified as an emerging pollutant. UV filters (benzophenone and 2-Phenylbenzimidazole-5-sulphonic acid), having estrogenic activity, have recently been found in WWTP effluents and surface waters entered via bathing, swimming, and other activities. The effects of such PCPs on marine environments are somewhat ambiguous. Further research is necessary, primarily to assess possible endocrine disruptions associated with UV-filter exposure and PCP bioaccumulation in aquatic ecosystems (Wilkinson et al., 2017).

9.2.1.5 Endocrine-Disrupting Compounds (EDCs)

EDCs are synthetic chemicals that copy or block the hormones and affect the normal functioning of the body. EDCs are characterized by the Environmental Protection Agency (EPA) as foreign substances which interfere with the production, release, transportation, attachment, function, or relocation of natural hormones in the body that regulate homeostasis, growth, reproduction, and behavioral patterns. Three main groups of ECs that are widely recognized are estrogenic (mimics or changes the activity of natural estrogens), androgenic (copies or hinders natural testosterone), and thyroidal (causes instant or oblique effects to the thyroid). The long-term toxicity of these substances and their harmful effects on human beings remain uncertain (Gogoi et al., 2018). A variety of epidemiological studies have found that exposure to bisphenol A (also an EDC with widespread human exposure) is directly associated with diseases, including diabetes, cardiovascular system disorders, irregular neural responses, and even obesity (Vandenberg et al., 2013).

9.2.1.6 Flame/Fire Retardants

Polybrominated diphenyl ether and phosphate-based flame retardants are two kinds of flame retardants used in domestic and industrial applications. Perfluorinated compounds (PFCs) such as perfluorooctane sulfonate and perfluorooctanoic acid have been used in fire-fighting foams. The presence of PFCs in WWTPs effluents and surface water was documented in several studies. PFCs and their metabolites demonstrate persistence and bioaccumulation in the aquatic ecosystem and are resistant to biological or chemical degradation (Naidu et al., 2016; Wilkinson et al., 2017).

9.2.1.7 Food Additives

Some studies have found that sweeteners, such as acesulfame, saccharin, and sucralose, occur in domestic sewage through human excretion. Other food preservatives, for example, parabens, were found in groundwater (Naidu et al., 2016).

9.2.1.8   Disinfection By-Products

During wastewater chlorination, N-nitrosodimethylamine can be formed, which is found in potable water. It has recently raised concern because of its carcinogen effects.

9.2.1.9   Microplastics

Microplastics are defined as plastic waste debris with a diameter ranging between 1 μm and 5 mm, whereas nanoplastics have a diameter of <1 μm. Microplastics enter into the marine ecosystem as an ingredient in PCPs and cosmetics (facial scrubs and toothpaste) as well as through the decay of larger plastic items over centuries or fragmentation of macro plastics. Physicochemical properties of microplastics, such as high mobility, longevity, and slow decomposition, facilitate their dispersion by natural hydraulic migration in water and sediments around the world. Microplastics can serve as a carrier material for transportation of few pharmaceuticals, PCPs, and other ECs as they bind to them. Marine species can swallow plastic particles, possibly mistaken for larvae, unveiling a potential infection route as PCPs/ECs become attached to the particles (Wilkinson et al., 2017; Vieira et al., 2020).

9.2.2   Quantification of Emerging Contaminants

The detection and measurement of ECs in water or wastewater have emerged as a significant research interest, requiring highly advanced analytical tools to detect concentrations ranging from micrograms to nanograms per liter. From an analytical chemistry viewpoint, there is a vital requirement for in-depth knowledge on ECs, including information on the implementation of wide-range monitoring methods and the development of quick and successful screening methods to determine these substances. For the quantification of ECs, it is of utmost importance to follow the sampling guidelines, which include determining sampling objectives and sample quantity and identifying the sampling method. The collection of samples is carried out in apparatus made of glass, stainless steel, aluminum, or fluorocarbon polymer. Plastic objects such as polyethylene, rubber, Tygon, etc., are avoided because of their ability to adsorb targeted contaminants from the sample. The benefit the grab sampling method is that they are easily obtainable and relatively cheap (Alvarez and Jones-Lepp, 2010). Passive sampling methods are preferred in monitoring ECs due to their capacity to concentrate trace amounts of ECs, capture EC pulses into the atmosphere, and precisely extract soluble chemicals (not attached to particulate matter). An increasing number of passive samplers have been developed for sampling organic chemicals in the water. These samplers include polar organic chemical integrative samplers (POCISs), semi-permeable membrane devices (SPMDs), chemcatchers, polyethylene strips, glass polymers, and solid-phase microextraction (SPME) devices (Namieśnik et al., 2005).

As concentrations of ECs are usually under µg/L range, sample pretreatment (extraction, pre-concentration, and cleanup) is a crucial step to consider for precise measurement. The most commonly documented method for extracting ECs from aqueous environmental samples is solid-phase extraction (SPE), which was introduced as a substitute for liquid-liquid extraction (LLE). LLE method is less advantageous as it is resource-intensive, challenging to automate, and requires a massive amount of solvents. Nonetheless, LLE has been used to extract ECs containing hydroxyl groups (e.g., bisphenol A, nonylphenol ethoxylates, alkylphenol ethoxylates, and most steroids and hormones) from water. Due to the lack of partitioning of many hydrophilic ECs into organic solvents, which results in poor extraction performance, SPE should be employed rather than LLE. Other advantages of SPE include lower solvent usage, shorter processing cycles, and automation. SPE is commercially available in three standard formats: thin flat discs, small cylindrical cartridges, and 96-well plates. Other pre-concentration techniques reported include automated SPE, on-line SPE, molecular imprinted polymers, solid-phase microextraction, and magnetic SPE. The use of nanomaterials is an interesting option for further research in this area. Recent studies have shown that SPE applications focus on using magnetic iron nanoparticles (Moliner-Martínez et al., 2011).

Chromatography (gas/liquid) along with mass spectrometry (MS) is the primary analytical technique available for ECs due to their inherent complexity. Cutting-edge MS techniques such as triple quadrupole and ion trap permit the detection of ECs in the ng/L level, and other groundbreaking innovations such as linear ion traps quadrupole, triple quadrupole, quadrupole time of flight, and quadrupole-linear ion trap have been used to elucidate the structure of transformation products (Nikolaou, 2013). Environmental samples can be incredibly complex, particularly surface water samples comprising WWTP effluents. Correct detection of chemicals can be challenging to almost unattainable, even with advanced mass spectrometers. Co-eluting compounds, chemicals with comparable mass-to-charge ratios, shifting of retention time, and matrix properties like ion suppression can lead to misinterpretation of compounds. Besides, the accessibility to pure standard chemicals is also limited. It is also essential to elucidate the degradation pathway of several ECs to understand the parent compound in the environmental sample (Alvarez and Jones-Lepp, 2010; Rodriguez-Narvaez et al., 2017).

9.3 TREATMENT METHODS

It is necessary to improve the existing treatment practices and develop novel, cost-effective technologies, which can attain comprehensive removal of a wide range of emerging pollutants (Schwarzenbach et al., 2006). These treatments can be broadly classified into adsorption, membrane separation, biological treatment, and AOPs. The advantages and limitations of different available treatment techniques are listed in Table 9.2.

TABLE 9.2

Merits and Drawbacks of Various Existing Treatment Technologies for Removing Emerging Contaminants

Treatment Techniques	Advantages	Drawbacks
Physical treatment techniques		
Adsorption	• Several adsorbents are available for the effective removal of targeted pollutants • Waste biomass and other waste materials can be converted to adsorbents	• Disposal issues for the concentrated contaminants • The efficiency of the adsorbents are influenced by the presence of suspended organic particles
Micro/Ultra-Filtration	• Effective pathogen removal can be achieved • Efficient heavy metal removal can be achieved	• Partial EC removal due to pore size restriction • Expensive operation • Performance limitation due to membrane fouling • Disposal of discarded materials
Nano-Filtration	• Saline water and wastewater can be effectively treated • Effective heavy metal removal can be achieved • Effective for removal of pesticides and dye chemicals	• High energy consumption • Inefficient pharmaceuticals removal
Reverse Osmosis	• Wastewater and saline water can be effectively treated • PCPS, EDCs, and pharmaceuticals can be effectively removed	• Performance limitation due to membrane fouling • Disposal of discarded materials • High energy consumption • Inefficient pharmaceuticals removal and discharge of corrosive effluent
Chemical treatment techniques		
Advanced Oxidation Process	• PCPs, EDCs, and pesticides can be effectively removed • Pollutants can be degraded in a short operational time	• Formation of toxic intermediates and by-products • High energy consumption • High operating and maintenance cost • Intervention of radical scavengers

(Continued)

TABLE 9.2 (Continued)

Merits and Drawbacks of Various Existing Treatment Technologies for Removing Emerging Contaminants

Treatment Techniques	Advantages	Drawbacks
Fenton and Photo-Fenton	• Enhanced degradation and mineralization of ECs can be achieved • Solar irradiation can be used instead of UV light	• Scavenge of OH· radical when exposed to chloride and sulfate ion for the production of chloro and sulfato-iron (III) complexes
Sonochemical	• Enhanced degradation and mineralization of ECs can be achieved • Pollutants can be degraded in a short operational time	• Difficulties in the optimization of ultrasonic frequency • OH· radicals and the size of bubbles influence the removal efficiency
Photocatalytic Oxidation Using TiO₂	• Persistent organic pollutants and ECs can be degraded • Increased reaction rate with catalyst usage • TiO₂ catalyst is cheap, chemically stable, and easy to recover • Solar irradiation can be used instead of UV light	• Difficult to separate and recycle photocatalytic particles in slurry suspension • Cannot handle a large volume of wastewater • High power consumption and expensive artificial UV lamps elevate the treatment cost
Ozonation	• In the presence of peroxide, several ECs could be removed • Both pollutant removal and disinfection can be achieved	• High power consumption • Intervention of radical scavengers • Formation of toxic by-products
Biological treatment techniques		
Activated Sludge Process	• Lower investment and operating cost in comparison to chemical treatment processes • Environmentally sustainable process • Ease of operation	• Low removal efficiency for pharmaceuticals and beta-blockers • Sludge generation and disposal issue • Inability to handle high strength wastewater
Membrane Bioreactor	• High removal efficiency for various ECs • Lesser environmental footprint	• Low removal efficiency for pharmaceuticals • Membrane fouling issue • High energy requirement

(Continued)

TABLE 9.2 (Continued)
Merits and Drawbacks of Various Existing Treatment Technologies for Removing Emerging Contaminants

Treatment Techniques	Advantages	Drawbacks
Constructed Wetland	• High removal efficiency for pesticides, PCPs, estrogen, and pathogens • Inexpensive and ease of operation	• Large area requirement and high operational time • Performance can be varied with the change of season • Formation of sediment and problem of clogging
Waste Stabilization Ponds/Algal Reactors	• Recovered algal biomass can be recycled as a fertilizer • Removes ECs effectively and provide high-quality effluent	• EDCs cannot be removed entirely • Performance can be varied with the change of season (especially cold season)
Biological - Activated Carbon	• Effective for removing several ECs • Generation of toxic by-products can be avoided • Removes residuals from ozonation or disinfection	• Sludge generation and disposal issue • High operating and maintenance cost • Hard to regenerate

Source: Modified from Ahmed et al. (2017) and Dhangar and Kumar (2020).

9.3.1 ADSORPTION

Adsorption is a phase-changing technology in which pollutants (adsorbates) shift to solid phase (adsorbent) from the aqueous phase. Owing to its high porosity, large specific surface area, and surface interaction properties, activated carbon (AC) has most commonly been used to remove ECs (Rizzo et al., 2019). van der Waals interaction and electrostatic interactions are the dominant measures for removing antibiotics and other ECs (Snyder et al., 2003). Both powdered and granular activated carbon (PAC and GAC) have displayed promising performance for removing ECs. As high as 90% removal efficiency of EDCs were achieved by PAC. Similarly, 84–99% removal efficiency was reported for particular ECs like diclofenac using GAC during WWTPs effluent treatment (Grover et al., 2011). Nevertheless, the production of AC is an energy-consuming process. Thus from the sustainability point of view, biochars have been investigated for cost-effective removal of emerging pollutants. Pyrolysis settings are significant aspects of the biochar manufacturing process that affect its ability to adsorb ECs. A cane species known as *Arundo donax* L. was utilized as a raw material for biochar production during sulfamethoxazole removal. Without thermal activation, biochar could achieve 35% removal in the same experimental conditions, while the maximum removal obtained was <16% using thermal activation. These findings are associated with the impact of thermal activation, which can alter the properties of biochar (acid-base, hydrophilic-hydrophobic, particle size, porosity, etc.), disrupting its ability to remove ECs (Zheng et al., 2013). The use of carbon nanotubes as an adsorbent to ECs was also investigated, and 92% removal of tetracycline was obtained. Several other adsorbent materials, including clay minerals, zeolites, meso- and micro-porous materials, resins, and metal oxides, have been documented in the literature to remove ECs (Rodriguez-Narvaez et al., 2017). However, a crucial drawback for the realistic implementation of these research findings is the lack of knowledge of the feasibility of its full-scale application.

9.3.2 MEMBRANE TECHNOLOGY

Membranes are made of various materials that create unique filtering characteristics (e.g., pore size, surface charge, and hydrophobicity) that decide the kind of pollutant that can be retained. Hydrostatic pressure is the governing principle of membrane processes that allows water to pass while suspended solids and contaminants are retained. Different type of membrane filtration (MF) includes ultrafiltration (UF), nanofiltration (NF), microfiltration (MF), forward osmosis (FO), and reverse osmosis (RO) (Rodriguez-Narvaez et al., 2017). The key advantage of FO is the ability to produce high-quality permeate by removing various ECs while operating under osmotic driving force without a hydraulic pressure difference. The penetration of ECs through RO membranes requires adsorption of the ECs into the membrane surfaces, dissolution of the ECs into the membrane, and subsequently the diffusion of dissolved EC molecules through the membrane matrix. Although NF membranes can be used to remove a wide range of ECs, the retention of ECs by NF membranes is highly dependent on the physicochemical properties, which may be influenced by solution composition. UF membrane uses existing separation mechanisms (e.g., size/steric exclusion, hydrophobic adsorption, and electrostatic repulsion) to eliminate ECs from wastewater (Kim et al., 2018). Using polysulfone- and polyvinylidene-made UF membranes, the removal of bisphenol A from the water was examined. The former was able to eliminate 75%, while the latter could remove as high as 98% of the pollutant (Heo et al., 2012; Melo-Guimarães et al., 2013). The removal performance for ECs increases dramatically as the pore size decreases. For the treatment of water contaminated with a variety of ECs, FO and RO have been documented to have high removal efficacy. For example, the effective removal of carbamazepine (80–99%) and caffeine were achieved using FO membrane (Linares et al., 2011). In general, the EC removal efficiency of membranes follows the trend: RO > FO > NF > UF. Although UF alone does not remove ECs effectively, it can be used prior to FO and RO as a pretreatment step (Kim et al., 2018). Some researchers believe that ionic pollutants may have a higher affinity for membrane surfaces, resulting in higher removal efficiency than neutrally charged pollutants. This creates a knowledge gap and an opportunity to achieve a better understanding of the structure and reaction of different ECs with membranes.

9.3.3 ADVANCED OXIDATION PROCESS (AOP)

Degradation of ECs via AOP involves the production of extremely reactive oxygenated species with low selectivity (e.g., hydroxyl radicals (OH·), capable of oxidizing almost all organic compounds, generating either less toxic compounds, or mineralizing them entirely to CO_2, H_2O, and inorganic compounds. Ozone (O_3), hydrogen peroxide (H_2O_2), semiconductors, photolysis using UV, Fenton reagent, and ultrasound (sonolysis) are the most common AOPs that have been used on a bench scale and even on a pilot scale (Gogoi et al., 2018). Reactions between OH· and organic materials might consist of hydrogen abstraction to produce an organic radical, OH· addition in double bonds and aromatic rings, and electron transfer. The key advantages of these treatments in comparison to other treatment methods include the complete mineralization capability, oxidization of recalcitrant, and non-biodegradable substances, the possibility of using them as a pre- or post-treatment process, less reaction time, and the possibility of *in situ* remediations (de Oliveira et al., 2020). Photochemical AOP encompasses processes in which light radiation (UV light) is coupled with inorganic oxidants like ozone (O_3), hydrogen peroxide (H_2O_2), persulfate ($S_2O_8^{2-}$), sulfite ion (SO_3^{2-}), chlorine (Cl_2), or peroxymonosulfate (HSO_5^-) (Olatunde et al., 2020). Hydroxyl radical production can be increased using semiconductor photocatalyst (e.g., titanium oxide [TiO_2]), thereby improving the degradation process (Giraldo et al., 2015). The complete photocatalytic degradation of the EC was achieved using TiO_2 and UV radiation, while the sonochemical process attained just 80% degradation. Using the Fenton reaction, the overall degradation of oxacillin reached up to 90%, whereas the photoassisted Fenton process achieved complete degradation (Magureanu et al., 2011). Although ozonation constitutes chemical oxidation, the degradation rate is lower compared with free-radical techniques. Ozonation, O_3 along with H_2O_2 (O_3:H_2O_2), and O_3 with UV (O3:UV) were examined to remove pharmaceuticals from water as single oxidation, pre-oxidation, or a disinfection step prior to other treatments (Kıdak and Doğan, 2018). Electrochemical processes have developed as a suitable solution for removing ECs because reactive species are produced by electricity, without the need for chemicals. An electrochemical oxidation method achieved simultaneous degradation of sulfamethoxazole, propranolol, and carbamazepine, with 86%, 85%, and 82% degradation efficiency (Garcia-Segura et al., 2018; de Oliveira et al., 2020). The major bottleneck for scaling up this process is the economic constraints. Furthermore, the knowledge base for assessing the scaling up parameters is limited. A systematic framework is required to compare the research outcome documented to date accurately.

9.3.4 BIOLOGICAL PROCESS

Several aerobic and anaerobic biological processes have been investigated to remove ECs from wastewater. The applicability of aerobic or anaerobic treatment is associated with the prevailing conditions relating to the availability of terminal electron acceptors. The removal efficiency of EDCs has been documented to be higher in aerobic environments than in anaerobic conditions (Furuichi et al., 2006). To remove ECs, the ASP has been proven to be effective. The process comprises the biodegradation of EC occurring in different extents, ranging from full mineralization (to CO_2 and H_2O) to partial oxidation leading to the formation of transformed by-products. For example, biodegradation of methoxy-triclosan up to 81% can be achieved in ASP; but, it is converted to a lower degree (56.5%) by converting it into a comparatively recalcitrant version, 2,4-dichlorophenol in presence of laccase enzymes (Samaras et al., 2013; Rout et al., 2021). The unavailability of reliable analytical methodologies capable of defining and quantifying ECs in such a complex matrix is one of the key issues in implementing biological processes to remove ECs. Other biological processes such as waste stabilization ponds (WSP), constructed wetlands (CW), sequential batch reactors (SBR), and MBRs have also demonstrated efficient removal of ECs. Li et al. (2013) reported high removal efficiency ranging from 88 to 100% for pharmaceuticals and PCPs except for carbamazepine in WSP. Removal of certain ECs such as oxytetracycline HCl, nadolol, ciprofloxacin HCl, enrofloxacin, and cotinine in CWs was documented to be as high as 70% (Li et al., 2014). MBR is commonly regarded as an advanced technology that can effectively eliminate various ECs, including highly resistant compounds. The performance of the MBR system is generally governed by the retained sludge on the membrane surface, sludge age, presence of anoxic and anaerobic chambers, wastewater composition, operating temperature, pH, and conductivity. Due to the comparatively long sludge age, it is easier to remove certain slow degradable pharmaceuticals such as antibiotics and analgesics in MBRs, leading to the formation of distinct microbial communities. The general pattern of EC elimination by MBR can be presented as EDCs > PCPs > beta-blockers > pharmaceuticals > pesticides. Further research on this topic is imperative to address the efficacy of various microbial communities for the removal of selected ECs (in particular, pesticides and pharmaceuticals), as well as the optimization of operating and design parameters (Ahmed et al., 2016).

9.3.5 HYBRID TREATMENT SYSTEMS

Wastewater comprises a very complex matrix containing organic compounds, microbes and pathogens, humic substances, minerals, pharmaceuticals, etc. It is difficult for a single treatment technology to accomplish the necessary treatment standards for effluent discharge (Taheran et al., 2018). Several innovative and hybrid treatment units have been examined in recent years because the traditional treatment procedures are inadequate for removing a wide variety of ECs effectively. Integration of MBR and MF has been investigated to remove various ECs such as EDCs, PCPs, beta-blockers, and pharmaceuticals. MBRs are found to be efficient at eliminating hydrophilic compounds that are readily biodegradable but less efficient at removing hydrophobic and biologically persistent compounds. Whereas MF (RO and NF) has shown effective removal of hydrophilic compounds through steric and electrostatic interaction (Dhangar and Kumar, 2020). A hybrid treatment unit consisting of upflow anaerobic sludge blanket (UASB) reactor combining two sequential CWs was examined to remove several ECs. Due to the prevailing anaerobic conditions, ECs such as caffeine, ketoprofen, and triclosan were preferentially removed in UASB. However, due to its prevalent oxidation and photooxidation process, the CW system was observed to be most effective at removing pharmaceuticals and PCPs (Reyes-Contreras et al., 2011). The combination of a biological system and some physical/chemical treatment system has most commonly been explored. For example, the MBR system was combined with the UV oxidation process, in which the MBR removed the biodegradable and hydrophobic ECs and UV oxidation degraded biologically resistant compounds (Nguyen et al., 2013). A recent study investigated a novel hybrid treatment unit integrating ultrafiltration, adsorption through activated carbon, and ultrasound irradiation for the removal of ECs, particularly antibiotics such as amoxicillin, diclofenac, and carbamazepine (Secondes et al., 2014). The removal of ECs using a hybrid electrochemical process combining with MBR technology was investigated. It was observed that the eMBR process enhanced the removal of ECs and reduced the membrane fouling (Mameda et al., 2017). Some advanced treatment methods such as chitosan-based hydrogel application, ozone treatment coupled with electrocoagulation (Shah et al., 2020), and microbial fuel cell (MFC) (Bagchi and Behera, 2020) have also been studied for the removal of ECs.

9.4 IMPACT ASSESSMENT AND RISK MANAGEMENT OF EMERGING CONTAMINANTS

The treatment units must be evaluated to assess the most economical, environmentally friendly, and socially feasible process to be selected and implemented appropriately. During designing a treatment process to address a certain environmental problem, it must be understood that these processes will introduce additional environmental pressures by resource consumption and energy utilization. These need to be reduced to a minimum to avoid the emergence of further environmental issues from the initial problem. The Life Cycle Assessment (LCA) enables examining the possible effects of these processes throughout its life (starting from resource extraction, production, processing, utilization, reusing, recovering, and final disposal) on humans and the ecosystem. LCA has the potential to improve environmental sustainability and assist decision-makers in identifying goals and reformulating schemes and policies. It consists of four iterative phases: (i) goal/scope elucidation; (ii) inventory evaluation; (iii) impact assessment; and (iv) interpretation. Lack of knowledge regarding the long-term effect of ECs and their derivatives on living beings undermines the implementation of the LCA and contributes to the understatement of probable impacts. LCA is a comprehensive decision-making method that offers a thorough analysis of environmental effects. Strategic decisions should emerge from relevant consideration of environmental sustainability, expenses, time restraints, available capital and resources, social implications, and other related factors. Few conclusions can be drawn from the previous assessments. As MBR prevents more impact than it produces, it is considered a sustainable option. Under the analyzed conditions, reverse osmosis does not seem to be an environmentally friendly solution since it consumes a lot of electricity. The capacity for ozonation is highly dependent upon the source of energy. In adsorption-based approaches, it is often safer to regenerate and reuse activated carbon and other adsorbents. Adsorption is important because the EC removal efficiency from wastewater is high, and the formation of degradation by-products can be avoided; however, the contaminants are not mineralized but merely phase-shifted (Pesqueira et al., 2020).

9.5 CONCLUSIONS AND FUTURE PERSPECTIVES

Due to the difficulty in detecting and handling ECs, the concentration reduction at the source must be emphasized through policies and regulations, public awareness, and imposing discharge limit of ECs in the environment. There are many ways to reduce the overall environmental impact of ECs, including improvement in the handling of used chemicals, application of "green" alternatives, reduction in usage of ECs for the manufacturing of PCPs, reduction and proper management of waste, fewer chemical discharge, and improvement in water treatment systems. This poses a fundamental challenge because it involves improving the existing framework and implementing novel, cost-effective techniques that can permanently remove a wide range of pollutants. The most crucial step is the collection and preparation of wastewater samples for reliable analysis of targeted

EC. A suitable analytical technique should then be implemented to quantify ECs (range from µg to ng level). Detailed information about the degradation mechanism of ECs is still insufficient; consequently, the parent contaminant, its pathway, transport, and fate cannot be specified. Although treatment methods for removing ECs are evolving progressively, significant constraints remain which require future research. Hybrid technologies combining biological and chemical processes (for example, UV photolysis combining with H_2O_2 and subsequent MBR or biological activated carbon) demonstrated promising results for removing several ECs. For large-scale commercial applications, the robustness and feasibility of solar irradiation should be investigated as a potential substitute for UV. Membrane fouling is a common problem for MBRs, limiting their applicability to some degree. On the other hand, some ECs can be retained on the fouling layer and subsequently removed. It is therefore important to examine the positive and negative contributions of fouling mechanisms. Also, existing knowledge of genetic engineering must be implemented for selecting the most efficient EC degrading bacteria, decreasing hydraulic retention time, and saving investments in reactor design. In the AOP, the hydroxyl radicals may mineralize the ECs or produce intermediate by-products. Adequate information about the intermediates is therefore necessary in order to avoid further environmental hazards. For AC adsorption, future research should focus on the recyclability of the adsorbents and their proper disposal.

At present, a risk-based approach is followed to mitigate the issue of emerging pollutants. An integrative approach combining both risk and cost analysis is imperative for the management of ECs. To confirm that the environmental, social, and economic goals are established, the risk management process should incorporate an adequate communication level among risk assessors and other involved parties. Risk evaluation and management integrates activities that share a common goal and convey possible risks to stakeholders. Risk communication involves systematic information sharing between all stakeholders regarding health and environmental risks. That will lead to more effective, transparent, and strengthened processes, as all perspectives would have been acknowledged. Such information exchange often permits the establishment of short- and long-term management/remediation strategies, which will ensure an early selection of suitable remediation alternative.

REFERENCES

Ahmed, M. B., Zhou, J. L., Ngo, H. H., Guo, W., Thomaidis, N. S., Xu, J. 2017. Progress in the biological and chemical treatment technologies for emerging contaminant removal from wastewater: A critical review. *Journal of Hazardous Materials*, 323: 274–298.

Alvarez, D. A., Jones-Lepp, T. L. 2010. Sampling and analysis of emerging pollutants. *Water Quality Concepts, Sampling, and Chemical Analysis*, 199–226.

Archundia, D., Duwig, C., Lehembre, F., Chiron, S., Morel, M. C., Prado, B., Bourdat-Deschamps, M., Vince, E., Aviles, G. F., Martins, J. M. F. 2017. Antibiotic pollution in the Katari subcatchment of the Titicaca Lake: Major transformation products and occurrence of resistance genes. *Science of the Total Environment*, 576: 671–682.

Bagchi, S., Behera, M. 2020. Pharmaceutical wastewater treatment in microbial fuel cell. In *Integrated Microbial Fuel Cells for Wastewater Treatment*. Butterworth-Heinemann. 135–155.

de Oliveira, M., Frihling, B. E. F., Velasques, J., Magalhães Filho, F. J. C., Cavalheri, P. S., Migliolo, L. 2020. Pharmaceuticals residues and xenobiotics contaminants: Occurrence, analytical techniques and sustainable alternatives for wastewater treatment. *Science of the Total Environment*, 705: 135568.

Dhangar, K., Kumar, M. 2020. Tricks and tracks in removal of emerging contaminants from the wastewater through hybrid treatment systems: A review. *Science of the Total Environment*, 738: 140320.

Focazio, M. J., Kolpin, D. W., Barnes, K. K., Furlong, E. T., Meyer, M. T., Zaugg, S. D., Barber, L. B., Thurman, M. E. 2008. A national reconnaissance for pharmaceuticals and other organic wastewater contaminants in the United States—II. Untreated drinking water sources. *Science of the Total Environment*, 402: 201–216.

Furuichi, T., Kannan, K., Suzuki, K., Tanaka, S., Giesy, J. P., Masunaga, S. 2006. Occurrence of estrogenic compounds in and removal by a swine farm waste treatment plant. *Environmental Science & Technology*, 40(24): 7896–7902.

Garcia-Segura, S., Ocon, J. D., Chong, M. N. 2018. Electrochemical oxidation remediation of real wastewater effluents—A review. *Process Safety and Environmental Protection*, 113: 48–67.

Giraldo, A. L., Erazo-Erazo, E. D., Flórez-Acosta, O. A., Serna-Galvis, E. A., Torres-Palma, R. A. 2015. Degradation of the antibiotic oxacillin in water by anodic oxidation with Ti/IrO$_2$ anodes: Evaluation of degradation routes, organic by-products and effects of water matrix components. *Chemical Engineering Journal*, 279: 103–114.

Gogoi, A., Mazumder, P., Tyagi, V. K., Chaminda, G. T., An, A. K., Kumar, M. 2018. Occurrence and fate of emerging contaminants in water environment: A review. *Groundwater for Sustainable Development*, 6: 169–180.

Grover, D. P., Zhou, J. L., Frickers, P. E., Readman, J. W. 2011. Improved removal of estrogenic and pharmaceutical compounds in sewage effluent by full scale granular activated carbon: Impact on receiving river water. *Journal of Hazardous Materials*, 185: 1005–1011.

He, S., Dong, D., Zhang, X., Sun, C., Wang, C., Hua, X., Zhang, L., Guo, Z. 2018. Occurrence and ecological risk assessment of 22 emerging contaminants in the Jilin Songhua River (Northeast China). *Environmental Science and Pollution Research*, 25(24): 24003–24012.

Heo, J., Flora, J. R., Her, N., Park, Y. G., Cho, J., Son, A., Yoon, Y. 2012. Removal of bisphenol A and 17β-estradiol in single walled carbon nanotubes–ultrafiltration (SWNTs–UF) membrane systems. *Separation and Purification Technology*, 90: 39–52.

Kıdak, R., Doğan, Ş. 2018. Medium-high frequency ultrasound and ozone based advanced oxidation for amoxicillin removal in water. *Ultrasonics Sonochemistry*, 40: 131–139.

Kim, S., Chu, K. H., Al-Hamadani, Y. A., Park, C. M., Jang, M., Kim, D. H., Yu, M., Heo, J., Yoon, Y. 2018. Removal of contaminants of emerging concern by membranes in water and wastewater: A review. *Chemical Engineering Journal*, 335: 896–914.

Li, X., Zheng, W., Kelly, W. R. 2013. Occurrence and removal of pharmaceutical and hormone contaminants in rural wastewater treatment lagoons. *Science of the Total Environment*, 445: 22–28.

Li, Y., Zhu, G., Ng, W. J., Tan, S. K. 2014. A review on removing pharmaceutical contaminants from wastewater by constructed wetlands: Design, performance and mechanism. *Science of the Total Environment*, 468: 908–932.

Linares, R. V., Yangali-Quintanilla, V., Li, Z., Amy, G. 2011. Rejection of micropollutants by clean and fouled forward osmosis membrane. *Water Research*, 45(20): 6737–6744.

Lu, J. F., He, M. J., Yang, Z. H., Wei, S. Q. 2018. Occurrence of tetrabromobisphenol a (TBBPA) and hexabromocyclododecane (HBCD) in soil and road dust in Chongqing, western China, with emphasis on diastereoisomer profiles, particle size distribution, and human exposure. *Environmental pollution*, 242: 219–228.

Magureanu, M., Piroi, D., Mandache, N. B., David, V., Medvedovici, A., Bradu, C., Parvulescu, V. I. 2011. Degradation of antibiotics in water by non-thermal plasma treatment. *Water Research*, 45(11): 3407–3416.

Mameda, N., Park, H. J., Choo, K. H. 2017. Membrane electro-oxidizer: A new hybrid membrane system with electrochemical oxidation for enhanced organics and fouling control. *Water Research*, 126: 40–49.

Margot, J., Rossi, L., Barry, D. A., Holliger, C. 2015. A review of the fate of micropollutants in wastewater treatment plants. *Wiley Interdisciplinary Reviews: Water*, 2(5): 457–487.

Melo-Guimarães, A., Torner-Morales, F. J., Durán-Álvarez, J. C., Jiménez-Cisneros, B. E. 2013. Removal and fate of emerging contaminants combining biological, flocculation and membrane treatments. *Water Science & Technology*, 67(4): 877–885.

Moliner-Martínez, Y., Ribera, A., Coronado, E., Campíns-Falcó, P. 2011. Preconcentration of emerging contaminants in environmental water samples by using silica supported Fe_3O_4 magnetic nanoparticles for improving mass detection in capillary liquid chromatography. *Journal of Chromatography A*, 1218(16): 2276–2283.

Naidu, R., Espana, V. A. A., Liu, Y., Jit, J. 2016. Emerging contaminants in the environment: Risk-based analysis for better management. *Chemosphere*, 154: 350–357.

Namieśnik, J., Zabiegała, B., Kot-Wasik, A., Partyka, M., Wasik, A. 2005. Passive sampling and/or extraction techniques in environmental analysis: A review. *Analytical and Bioanalytical Chemistry*, 381(2): 279–301.

Nguyen, L. N., Hai, F. I., Kang, J., Price, W. E., Nghiem, L. D. 2013. Removal of emerging trace organic contaminants by MBR-based hybrid treatment processes. *International Biodeterioration & Biodegradation*, 85: 474–482.

Nikolaou, A. 2013. Pharmaceuticals and related compounds as emerging pollutants in water: Analytical aspects. *Global NEST Journal*, 15(1): 1–12.

Noguera-Oviedo, K., Aga, D. S. 2016. Lessons learned from more than two decades of research on emerging contaminants in the environment. *Journal of Hazardous Materials*, 316: 242–251.

Olatunde, O. C., Kuvarega, A. T., Onwudiwe, D. C. 2020. Photo enhanced degradation of contaminants of emerging concern in waste water. *Emerging Contaminants*, 6: 283–302.

Panthi, S., Sapkota, A. R., Raspanti, G., Allard, S. M., Bui, A., Craddock, H. A., Murray, R., Zhu, L., East, C., Handy, E., Callahan, M. T. 2019. Pharmaceuticals, herbicides, and disinfectants in agricultural water sources. *Environmental Research*, 174: 1–8.

Pesqueira, J. F., Pereira, M. F. R., Silva, A. M. 2020. Environmental impact assessment of advanced urban wastewater treatment technologies for the removal of priority substances and contaminants of emerging concern: A review. *Journal of Cleaner Production*, 261: 121078.

Prabhasankar, V. P., Joshua, D. I., Balakrishna, K., Siddiqui, I. F., Taniyasu, S., Yamashita, N., Kannan, K., Akiba, M., Praveenkumarreddy, Y., Guruge, K. S. 2016. Removal rates of antibiotics in four sewage treatment plants in South India. *Environmental Science and Pollution Research*, 23(9): 8679–8685.

Reyes-Contreras, C., Matamoros, V., Ruiz, I., Soto, M., Bayona, J. M. 2011. Evaluation of PPCPs removal in a combined anaerobic digester-constructed wetland pilot plant treating urban wastewater. *Chemosphere*, 84(9): 1200–1207.

Rizzo, L., Malato, S., Antakyali, D., Beretsou, V.G., Đolić, M.B., Gernjak, W., Heath, E., Ivancev-Tumbas, I., Karaolia, P., Ribeiro, A.R.L., Mascolo, G. 2019. Consolidated vs new advanced treatment methods for the removal of contaminants of emerging concern from urban wastewater. *Science of the Total Environment*, 655: 986–1008.

Rodriguez-Narvaez, O. M., Peralta-Hernandez, J. M., Goonetilleke, A., Bandala, E. R. 2017. Treatment technologies for emerging contaminants in water: A review. *Chemical Engineering Journal*, 323: 361–380.

Roosens, L., Covaci, A., Neels, H. 2007. Concentrations of synthetic musk compounds in personal care and sanitation products and human exposure profiles through dermal application. *Chemosphere*, 69(10): 1540–1547.

Rout, P. R., Zhang, T. C., Bhunia, P., Surampalli, R. Y. 2021. Treatment technologies for emerging contaminants in wastewater treatment plants: A review. *Science of the Total Environment*, 753: 141990.

Samaras, V. G., Stasinakis, A. S., Mamais, D., Thomaidis, N. S., Lekkas, T. D. 2013. Fate of selected pharmaceuticals and synthetic endocrine disrupting compounds during wastewater treatment and sludge anaerobic digestion. *Journal of Hazardous Materials*, 244: 259–267.

Schaider, L. A., Rudel, R. A., Ackerman, J. M., Dunagan, S. C., Brody, J. G. 2014. Pharmaceuticals, perfluorosurfactants, and other organic wastewater compounds in public drinking water wells in a shallow sand and gravel aquifer. *Science of the Total Environment*, 468: 384–393.

Schwarzenbach, R. P., Escher, B. I., Fenner, K., Hofstetter, T. B., Johnson, C. A., Von Gunten, U., Wehrli, B. 2006. The challenge of micropollutants in aquatic systems. *Science*, 313: 1072–1077.

Secondes, M. F. N., Naddeo, V., Belgiorno, V., Ballesteros Jr, F. 2014. Removal of emerging contaminants by simultaneous application of membrane ultrafiltration, activated carbon adsorption, and ultrasound irradiation. *Journal of Hazardous Materials*, 264: 342–349.

Shah, A. I., Dar, M. U. D., Bhat, R. A., Singh, J. P., Singh, K., Bhat, S. A. 2020. Prospectives and challenges of wastewater treatment technologies to combat contaminants of emerging concerns. *Ecological Engineering*, 152: 105882.

Snyder, S. A., Westerhoff, P., Yoon, Y., Sedlak, D. L. 2003. Pharmaceuticals, personal care products, and endocrine disruptors in water: Implications for the water industry. *Environmental Engineering Science*, 20(5): 449–469.

Taheran, M., Naghdi, M., Brar, S. K., Verma, M., Surampalli, R. Y. 2018. Emerging contaminants: Here today, there tomorrow!. *Environmental Nanotechnology, Monitoring & Management*, 10: 122–126.

Vandenberg, L. N., Hunt, P. A., Myers, J. P., Vom Saal, F. S. 2013. Human exposures to bisphenol A: Mismatches between data and assumptions. *Reviews on Environmental Health*, 28(1): 37–58.

Vieira, Y., Lima, E. C., Foletto, E. L., Dotto, G. L. 2020. Microplastics physicochemical properties, specific adsorption modeling and their interaction with pharmaceuticals and other emerging contaminants. *Science of the Total Environment*, 753: 141981.

Wilkinson, J., Hooda, P. S., Barker, J., Barton, S., Swinden, J. 2017. Occurrence, fate and transformation of emerging contaminants in water: An overarching review of the field. *Environmental Pollution*, 231: 954–970.

Williams, M., Kookana, R. S., Mehta, A., Yadav, S. K., Tailor, B. L., Maheshwari, B. 2019. Emerging contaminants in a river receiving untreated wastewater from an Indian urban centre. *Science of the Total Environment*, 647: 1256–1265.

Zheng, H., Wang, Z., Zhao, J., Herbert, S., Xing, B. 2013. Sorption of antibiotic sulfamethoxazole varies with biochars produced at different temperatures. *Environmental Pollution*, 181: 60–67.

Zuccato, E., Castiglioni, S., Fanelli, R. 2005. Identification of the pharmaceuticals for human use contaminating the Italian aquatic environment. *Journal of Hazardous Materials*, 122(3): 205–209.

10 Municipal Wastewater as a Potential Resource for Nutrient Recovery as Struvite

Nageshwari Krishnamoorthy,[1] Alisha Zaffar,[1] Thirugnanam Arunachalam,[1] Yuwalee Unpaprom,[2] Rameshprabu Ramaraj,[2] Gaanty Pragas Maniam,[3] Natanamurugaraj Govindan,[3] and Paramasivan Balasubramanian[1]
[1]Department of Biotechnology & Medical Engineering, National Institute of Technology Rourkela, Odisha, India
[2]Maejo University, Chiang Mai, Thailand
[3]Faculty of Industrial Sciences & Technology, Universiti Malaysia Pahang, Pahang, Malaysia

CONTENTS

DOI: 10.1201/9781003201076-10

10.1 INTRODUCTION

Rapid industrialization and urbanization have become remarkable trends worldwide that have caused negative insinuation on the environment. Due to the increasing population, notable issues such as land and environmental degradation, pollution, and climate change arise. One such problem is wastewater generation caused by several anthropogenic activities that demand major concern and treatment (Addagada, 2020, Krishnamoorthy et al., 2021a). Of the various sources of wastewater, municipal wastewater serves to be a chief contributor of nutrients and harmful pollutants. It mainly comprises household waste, medicinal waste, plastics, human feces, urine, heavy metals, organic, and inorganic waste. In this regard, it contains vital nutrients such as phosphorus, nitrogen, sodium, and potassium (Kataki et al., 2016; Luo et al., 2014; Mahy et al., 2019). Human urine alone contributes ~50% phosphorous and ~80% of ammonium nitrogen to municipal wastewater (Krishnamoorthy et al., 2021b; Wilsenach & Van Loosdrecht, 2004). Around 38.88 km^3 of wastewater is generally produced in various regions of southern parts of Asia. India alone produces ~109,598 tons of solid waste/day, of which only 7% is treated (Evans et al., 2012; Jasinski, 2014). This mineral-enriched water when let into ponds and natural streams will aid in algal growth and eutrophication leading to oxygen depletion in the water bodies. The hypoxic conditions prevalent will decrease life underwater (Gilbert, 2009). A feasible solution should be arrived at to establish a sustainable circular bio-economy.

    On the other hand, the alarming depletion of phosphorous resources has triggered a rising need for the recycling and recovery of this non-renewable nutrient. Phosphate reserves are scarred around the world, mentionable in Morocco, Iraq, China, and Algeria (Desmidt et al., 2012; Gurr, 2012; Zhang et al., 2013). According to an article published by International Fertilizer Development Center, the United States, phosphate rock mining for fertilizer production is increasing at a rate of 180–190 million tons per year as a result of the increase in population (Gurr, 2012; Van Kauwenbergh et al., 2013). The shooting up of per capita phosphorous demand due to an increase in population and a rise in phosphorus consumption suggests that the reserves will be depleted in the next few years (Gilbert, 2009). Such critical situations have urged the researchers to explore sustainable solutions not only to recover nutrients but also to utilize them as organic fertilizer for the enrichment of plant growth. Retrieval of P from municipal wastewater in the form of struvite ($MgNH_4PO_4 \cdot 6H_2O$) can be an alternative approach for the cessation of a leak in the phosphate cycle.

Struvite is a phosphate-rich mineral comprising mainly of phosphate, ammonium, and magnesium equimolar ratio). The research and commercialization of struvite are emerging in recent decades for utilization as a potential substitute to commercial fertilizers extracted from phosphate rocks. It has an NPK value of 6:29:0 (Barak & Stafford, 2006). Considering the nutritional composition of municipal wastewater, it is likely to be used for struvite precipitation. Domestic wastewater collected in China is estimated to contain approximately 5.5% of total phosphate as compared to the consumption of chemical fertilizer (Krishnamoorthy *et al.*, 2020). Solid waste accounts for up to 50% of the organic waste, comprising mainly of food scraps with a P concentration of ≤4 g/kg dry matter. An added advantage is that the flue gas produced by incineration of solid waste contains 10–15% of CO_2 which can reduce chemical cost by serving as a carbon source for calcium pre-treatment (Wu *et al.*, 2018).

The struvite precipitated can be directly marketed as a soil-conditioning slow-release fertilizer. The risk of heavy metal co-precipitation with struvite is low as the metal content is much lower than the ground itself, reducing the hazard of chemical contamination (Maaß *et al.*, 2014; Saerens *et al.*, 2021). This chapter converses the nutritional composition and aptness of municipal wastewater for struvite recovery and various principles that are employed for efficient struvite crystallization. Chemical and electrochemical precipitations are the common methods used for lab-scale and industry-scale production and several reactor designs have been discussed in detail. The chapter also highlights the most significant process parameters that need optimization for improved quality of struvite. In addition, it provides insights on the research gaps, challenges that should be overcome, and future perspectives of struvite research that require further attention.

10.2 NUTRITIONAL COMPOSITION OF MUNICIPAL WASTEWATER

Several works of literature have reported that municipal wastewater has a nutritional value equivalent to that of commercial fertilizers to be used in the agricultural field. It holds numerous macro and micronutrients, organic matter, and high NPK value that can be used for fertilizing applications (Saerens *et al.*, 2021; Srivastava *et al.*, 2018; Theregowda *et al.*, 2019). Among these nutrients, phosphorus, nitrogen, and magnesium are the most essential nutrients required by the plants for their growth and hence serve as a source of struvite. Animal waste in municipal wastewater alone constitutes up to 34% of the total nitrogen that enters the agricultural field (Matassa *et al.*, 2015). There are many physicochemical methods for recovering the nutrients, of which struvite precipitation has gained momentum in recent years.

P is a vital nutrient present in municipal wastewater that is required by the plants for various processes like photosynthesis, cell division, growth, energy transfer, and quality of fruits and crops. It can be incorporated into the food chain by the application of phosphate-rich fertilizers (Theregowda *et al.*, 2019). Every year approximately 27 million tons of P is used in agriculture, among which only

10–11% is actually utilized. The remaining proportion is dumped into the water bodies through diffusion resulting in eutrophication. Struvite precipitation can address this issue by recovering P in its most redeemable form, phosphate. In addition, crystallization of struvite is considered a biological way of water treatment as it recovers nitrogen in the form of ammonium, the most accessible form by the plants. Plants can easily take up ammonium and convert it to nitrate through nitrification and further alter nitrate to nitrogen by denitrification (Theregowda et al., 2019; Williams et al., 2015).

Potassium is another important essential nutrient present in municipal wastewater which is required for enrichment of soil and crop yield (Ryu et al., 2012; Zhang et al., 2012). It can be efficiently recovered as K-struvite by slightly varying the conditions of usual struvite precipitation. This isomer analog of struvite is formed by the replacement of ammonium by potassium and potentially used as a slow-release fertilizer. However, high concentrations of ammonium and sodium in municipal wastewater can inhibit the K-struvite formation. In order to avoid this, pre-treatment strategies such as ammonium stripping can be adopted. So far, there are no such approaches for preventing the effect of sodium, and further studies on understanding the variation in process and experimental conditions can provide insights on eliminating the inhibitory effect. Also, there is literature shred of evidence of co-precipitation of struvite, K-struvite, and other struvite family crystals which can serve as a value-added fertilizer (Huang et al., 2019; Li et al., 2020). Municipal wastewater also constitutes a high concentration of heavy metals such as copper and zinc and other micropollutants like pharmaceuticals, pesticides, organic, and inorganic pollutants (Decrey et al., 2011). Pre-treatment methods have been widely explored to reduce heavy metal amounts for efficient water treatment and reuse of irrigation. The presence of minuscule of such metal ions can help as co-factors of bacterial enzymes involved in anaerobic digestion. The list of nutrients, heavy metals, organic, and inorganic compounds present in municipal wastewater are shown in Table 10.1.

10.3   TECHNOLOGIES INVOLVED IN STRUVITE PRECIPITATION

Recovery of nutrients from wastewater is quite tedious concerning the concentration and form of ion species. In this regard, struvite recovery has bloomed to be an interesting approach not only for efficient recovery of P and nitrogen but also to serve as a sustainable alternative for chemical fertilizers. Several technologies such as chemical, electrochemical, and biological precipitation have been developed for this purpose to promote recycling, and reuse of nutrients and to avoid consumption of non-renewable natural resources.

10.3.1   CHEMICAL PRECIPITATION

A chemical method for the struvite precipitation is one of the most common processes by which nutrients can be managed to be removed from wastewater. Chemical precipitation requires the addition of magnesium as it is present

TABLE 10.1

Nutritional Composition and Constituents of Municipal Wastewater

	Constituents	Concentration (mg/L)
Nutrients	Nitrogen	20–35
	Phosphorus	5–30
	Potassium	13–20
Heavy metal (mg/kg)	Iron	20,000
	Cadmium	3
	Arsenic	3–230
	Chromium	1–3,410
	Copper	80–2,300
	Nickel	2–179
	Lead	13–465
	Zinc	101–49,000
	Sulfate	20–50
Organic and inorganic substances	Dioxins, polychlorinated biphenyls (PCBs), furans, organochlorine pesticides	8–35
	Inorganic substances	3–10
Physicochemical parameters	Alkalinity	50–200
	Total solids (TS)	350–1,200
	Total dissolved solids (TDS)	250–850
	Suspended solids (SS)	100–350
	Volatile compounds	80–275
	Volatile organic compounds	100–400
	Total organic carbon (TOC)	80–290
	Biological oxygen demand (BOD)	110–400
	Chemical oxygen demand (COD)	250–1,000

Sources: Esakku et al. (2005), Liu *et al.* (2013), and Werle and Dudziak (2012).

in restricted quantities in sludge liquors of municipal wastewater. The common magnesium salts used for precipitation include magnesium chloride, magnesium sulfate, magnesium oxide, and other cheap sources such as bitter, wood ash, and seawater (Borojovich *et al.*, 2010; Krishnamoorthy *et al.*, 2020; Sakthivel *et al.*, 2012). pH is reported to be the most influential parameter and hence, the addition of alkaline chemicals is required to increase the pH necessary for struvite crystallization. It has been reported in the literature that alkaline pH conditions and an appropriate molar ratio of external magnesium input are favorable for unintended precipitation of phosphate as struvite (Liu *et al.*, 2013). Figure 10.1a diagrammatically represents the processes taking place in chemical precipitation. In the case of anaerobic digestate, CO_2 stripping has been shown to improve pH (Hu *et al.*, 2020; Song *et al.*, 2011). Similarly, temperature conditions above 20°C with high constituent ion concentrations are shown to be ambient (Tilley *et al.*, 2008).

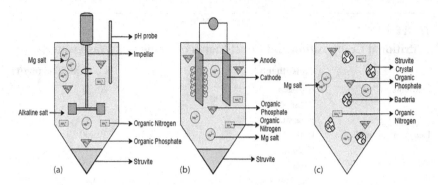

FIGURE 10.1   Diagrammatic representation of working principles of THE (a) chemical, (b) electrochemical, and (c) biological precipitation of struvite.

The optimal range of operational parameters is subject to differ with a source of wastewater and has been discussed in detail in Section 10.4. However, deviation from these conditions can lead to the formation of struvite analogs such as ferrous phosphate (vivianite), hydroxylapatite, and brushite (Capdevielle *et al.*, 2016).

In municipal wastewater, up to 90% of the phosphate is attached chemically or biologically to the surface of the solids. In order to make P available for struvite precipitation, \ has to undergo a series of steps. Initially, the bound phosphates have to be dissolved into the sludge with the help of thermal or chemical treatments. Sometimes, sulfuric acid is used to dissolve the P bound to heavy metals. Other means of improving phosphate dissolution include the wet oxidation process and metallurgical smelt-gasification process. The latter is a thermal treatment where the sludge will be smelted at 2,000°C with metallurgical coke for slag formation. Later, magnesium and alkaline salts are added to precipitate out struvite (Hester & Harrison, 2013).

Law and Pagilla (2018) have reported the highest P recovery potential up to 90% from municipal sludge influent. But, this is not the case with treated effluents, where recovery of only 10% was recorded. In addition, the process was claimed to be cost-intensive. P extraction from secondary effluents was performed by Hester and Harrison (2013) using a combination of ion exchange and precipitation techniques. It was concluded that the procedure demanded a high amount of chemicals making it economically unviable. The nitrogen and phosphate recovery as struvite from various municipal sources are shown in Table 10.2. Agitation of solution can improve nucleation and thus the crystallization. However, high stirring can produce unrecoverable struvite fines (Doino et al., 2011; Oliveira *et al.*, 2020). Studies have reported the use of such fines for seeding purposes when recycled (Liu *et al.*, 2011). Though the chemical process may sometimes be efficient, the high costs incurred by NaOH in addition to controlling pH and magnesium supply for struvite precipitation make the process impractical. The main benefit of the chemical method is its simplicity of operation.

TABLE 10.2

Various Technologies and Operational Parameters Involved for Struvite Precipitation from Municipal Wastewater

Source of Municipal Wastewater	Precipitation Technology Involved	P concentration (mg/L)	Mg:P	pH	Mixing Speed (rpm)	Mixing Time (min)	Recovery	Reference
Urine	Electrochemical	256±8.6	1.3	9.20	160	5	96% P	Liu et al. (2013)
Urine	Chemical	1,000	1.4	9.3	–	30	94% P	Morales et al. (2013)
Swine manure	Chemical	2,700	1.2	8–8.5	100	60	88.6% P	Lee et al. (2015)
Medicinal waste	Chemical	–	1	9.6	–	60	91% N	Li et al. (2018)
Digestate	Electrochemical	108.51	1.5	10.5	600	60	–	Corona et al. (2020)
Domestic sewage	Chemical	64.4	1.6	8.28–8.41	–	2 h	90% P	Quintana et al. (2004)
Poultry manure	Chemical	1,318	1	9.0	–	15	85.4% N	Yetilmezsoy and Sapci-zengin (2009)
Digested manure	Chemical	800	1.3	9.0	–	60	80% P	Siciliano et al. (2020)
Swine wastewater	Electrochemical	780	0.8	8.0–9.0	3,500	15	79–82% P	Ichihashi and Hirooka (2012)
Digested swine wastewater	Electrochemical	116.2	0.6	8.0–11.0	200	–	90% P	Ye et al. (2018)
Food waste	Chemical	–	–	7.5–9.0	–	–	70–100% P	Campos et al. (2019)
Human waste	Electrochemical	0.6 mM	2.1:0.55	8.3	600	–	80% P	Cid et al. (2018)
Domestic waste	Chemical	5	–	8.5	–	–	87.5% P 91.6% N	Zou and Wang (2016)

10.3.2 ELECTROCHEMICAL PRECIPITATION

In electrochemical precipitation of struvite, magnesium electrodes are used for the discharge of magnesium ions into the sludge. Magnesium acts as a sacrificial anode and releases divalent ions to serve the purpose of struvite formation and pH elevation (Figure 10.1b). The diffusion of oxygen increases during electrolysis leading to the generation of OH$^-$ ions. However, magnesium requirement is too high to precipitate all the phosphate which increases the overall cost of the process. Hence, alternate methods should be investigated for coupling pH adjustment with struvite precipitation. Aeration is one such technique reported to increase pH (Almatouq & Babatunde, 2016; Hug & Udert, 2013).

In the case of conventional wastewater treatment, several electrodes such as aluminum and iron are being used to supply necessary cations for the coagulation of particulate matter. However, magnesium electrodes are vital for struvite formation. Due to slower precipitation kinetics at high pH, phosphate removal was found to be slower in stored urine than fresh urine. But conditions prevailing in stored urine are relatively conducive for phosphate precipitation. Also, studies have proved that phosphate availability in struvite is significantly higher than aluminum or iron phosphates (Baierle et al., 2015; Jardin & Pöpel, 2001; Sengupta & Pandit, 2011). Struvite precipitation using magnesium electrodes has not been widely described in the literature.

Struvite is considered the best alternative for recycling nutrients and direct application into agricultural fields. Struvite crystals with high purity were obtained by Kruk et al. (2014) through the electrolytic release of magnesium. Though such dosage setup seems simpler to handle, it is comparatively more expensive than chemical dosing. Nonetheless, in chemical magnesium dosage, the risk of struvite powder incrustation in screw feeder and to prevent this, magnesium salts have to be packed in smaller bags of polyvinyl alcohol. The electrochemical method is advantageous in the fact that it involves the direct dissolution of magnesium without additional mechanical feed mechanisms. Such systems are particularly suitable for decentralized reactors considering a simple and scalable infrastructure (Hug & Udert, 2013; Ishii & Boyer, 2015, Krishnamoorthy et al., 2021a). The operational parameter conditions used in electrochemical struvite precipitation are shown in Table 10.2.

10.3.3 BIOLOGICAL PRECIPITATION

Biological precipitation is the process by which microorganisms mineralize the nutrients present in wastewater to form struvite like minerals on their surface for hardening the structural tissue. This process is also known as biomineralization (Figure 10.1c). Many bacterial species such as *Proteus mirabilis*, *Proteus vulgaris*, *Staphylococcus aureus*, *Enterobacter* sp., and *Ureaplasma urealyticum* exhibit this property. Such formation was initially observed in patients with urinary infection. Some researchers predict that the formation can be a result of dead bacterial cells present in the medium. The negatively charged surfaces of bacteria release molecules such as phospholipids and proteolipids that not only generate free surfaces but also attract positive ions like magnesium. However, the exact mechanism behind this process is yet to be explored. It is also assumed that such urease-producing strains develop a conducive physicochemical environment suitable for struvite crystallization. Due to this activity, ammonia is released leading to an elevation in pH. Later, the supersaturation of the solution takes place, followed by heterogeneous nucleation by cellular molecules with a low energy barrier. The mineralization can take place either intracellularly or extracellularly depending on the biological system. The process can take place actively by controlled mineralization (homogenous mineral) or passively by biological induction (heterogeneous mineral). The research on this aspect of struvite crystallization is still in the early stages and more studies have to be conducted in order to gain more insights on the actual mechanism, duration of struvite formation, reactor designing, and scale-up of the process (Bayuseno & Schmahl, 2020; Kataki *et al.*, 2016; Li *et al.*, 2015; Prywer & Torzewska, 2010).

10.4 OPERATIONAL PARAMETERS INFLUENCING STRUVITE PRECIPITATION

Research on the extraction of struvite from municipal wastewater has been going on for the last two decades. So far, several methods have been applied for struvite precipitation. Apart from the method, the most significant factors that govern struvite precipitation are the operational parameters that will guide the process and lead to a better yield. Various parameters that influence struvite precipitation are discussed below.

10.4.1 pH

The availability of the ions such as Mg^{2+}, NH_4^+, and PO_4^{3-} is the main component of struvite precipitation whose ionic forms are controlled by pH. Struvite precipitation occurs basically at a pH range of 8–11. During the process, H^+ ions are released decreasing the pH. In order to overcome this effect, NaOH is supplied to increase the pH and facilitate precipitation. The change in pH also results in the formation of phosphate ions like PO_4^{3-}, HPO_4^{2-}, and $H_2PO_4^-$ which may enable the formation of other compounds along with struvite. pH > 11 leads to the formation of $Mg(OH)_2$ and NH_3 reducing the availability of NH_4^+ and Mg^+ ions leading to a decrease in the efficiency of struvite recovered (Crutchik & Garrido, 2011, 2016; Crutchik *et al.*, 2017). The presence of the ions (Na^+, SO_4^{2-}, Cl^-, etc.) in lower concentrations tends to form $MgSO_4$ and $NaPO_4$, hindering the formation of struvite. Thus, to eliminate the formation of such precipitates an optimum pH should be maintained. Apart from this, the stability of struvite crystals is reported to be very minimum at pH 10 (Lahav *et al.*, 2013). Sparging can also be used in place of NaOH to maintain the pH. Hallas *et al.* (2019) through their experiments have shown that there is a slight increase in the amount of struvite crystal formation when air sparging is used.

TABLE 10.3

Solubility Trend of Ions in Municipal Wastewater with Increase in Temperature

Temperature (°C)	pK_{sp}	K_{sp} (×10^{-14})	Temperature (°C)	pK_{sp}	K_{sp} (×10^{-14})
10	14.36	0.436	30	13.15	7.12
15	14.04	0.916	35	13.23	5.920
15	13.28	5.3	35	13.10	7.90
20	13.69	2.050	35	13.08	8.32
20	13.22	6.03	40	13.40	4.000
25	13.36	4.330	45	13.60	2.530
25	13.26	5.51	50	13.68	2.110
25	13.17	6.76	55	03.84	1.460
30	13.17	6.840	60	14.01	0.973

Sources: Crutchik and Garrido (2016), Fang *et al.* (2016), and Hanhoun *et al.* (2011).

10.4.2 TEMPERATURE

Temperature plays a very important role in struvite precipitation as it has an effect on the solubility of ions. Several studies have concluded that the optimum temperature for struvite precipitation falls between 25 and 35°C. A study done by Salsabili *et al.* (2014) shows that the solubility of the crystals increases with an increase in temperature up to a certain limit, after which it declines. Fang *et al.* (2016) have concluded that the size of struvite crystals increases from 65 to 69 μm when temperature increases from 21 to 49°C. Hence, the solubility of the ion temperature may also influence the size and shape of the crystals. Table 10.3 below shows the trend of solubility with increasing temperature.

10.4.3 MAGNESIUM SOURCE AND DOSAGE

There are many sources of magnesium that can be used for struvite precipitation. According to studies, the most common magnesium sources are crystals of $Mg(OH)_2$, MgO, $MgCl_2$, and $MgSO_4$. $MgCl_2$ and $MgSO_4$ have the advantage of being easily soluble leading to the formation of struvite crystals with more purity. Among all the crystals, MgO is the least harmful and one of the most efficient and cheapest sources of Mg^{2+} for struvite precipitation. Though it provides better yield, its insoluble nature decreases the efficiency of struvite formation (Capdevielle *et al.*, 2014). Many new technologies like silicon-doped magnesium oxide (SMG) have been developed that helps in the production of struvite by the process of adsorption (Li *et al.*, 2019).

In recent years, various cheap alternatives of magnesium like seawater, magnesite, bittern, and ash powder are being explored (Heinonen-Tanski & Van Wijk-Sijbesma, 2005; Sakthivel *et al.*, 2012; Siciliano & De Rosa, 2014). Though the theoretical molar ratio of Mg:P is unity, it is suggested that an increase in magnesium concentration can improve struvite yield. The most obvious reason behind it is the more availability of magnesium. However, it may also lead to the formation of struvite analog resulting in the compromisation of purity. When Mg:P is in the range of 1–1.5, the yield and size of the crystals are reported to be optimum (Rahaman *et al.*, 2008).

10.4.4 PRESENCE OF POLLUTANTS

The struvite crystals that are precipitated from municipal wastewater are not pure due to the presence of pollutants in the form of ions, inorganic metals, organic compounds, and metalloids depending on the method of struvite formation and wastewater source. Ions like Na^+, Ca^{2+}, Cl^-, SO_4^{2-}, OH^-, Cu^{2+}, Zn^+, Cd^{2+}, Fe^{2+}, and Al^{3+} are formed at different phases of struvite precipitation which may or may not hinder the formation of struvite (Saidou *et al.*, 2015; Hallas *et al.*, 2019). Along with ions, heavy metals like As, Cd, Pd, Ni, and Hg are also present in municipal waste that can co-precipitate along with struvite crystals. Many organic pollutants like heptadecane, eicosane, tetra-tetracontane, 2-ethylacridine, 4-ethoxy-benzoic acid, PAH, and benzene have been detected in urban wastewater (Saerens *et al.*, 2021). Sometimes, the organic carbon content in the struvite crystal can exceed beyond 6% which can pose certain disadvantages (Huygens *et al.*, 2019). The amount of TSS influences the purity and stability of the struvite crystals. TSS levels above 153 mg/L can result in irregular-sized and broken crystals; however, a minimum of 20 mg/L TSS should be present to promote heterogenous precipitation, which can lead to formation of large crystals (Ping *et al.*, 2016). Table 10.4 depicts

TABLE 10.4
Literature References for Heavy Metal Concentration in Struvite below Maximum Permissible Addition (MPA)

Heavy Metal	Saerens *et al.* (2021) (mg/kg)	Maaß *et al.* (2014) (mg/kg)	Latifian *et al.* (2012) (mg/kg)	Antonini *et al.* (2011) (mg/kg)	MPA (mg/kg)
As	<2	–	0.5	<15	4.5
Cd	<0.2	0.3	0.02	<0.2	0.76
Cr	8.2	11.0	0.42	17.8	3.8
Cu	6.8	39.0	7.26	44.3	10
Hg	<0.007	–	0.01	<4.2	1.9
Pb	<10	5	0.30	<0.9	2
Ni	<3	2	–	<4.2	10
Zn	41	0.1	11	90.2	16
Fe	–	–	350	44.3	15
Mn	–	0.2	–	4.14	0.5

Sources: Kinuthia *et al.* (2020) and Vodyanitskii (2016).

the probable pollutants that can precipitate along with struvite, their concentration and permissible limits for application of struvite to the agricultural fields.

10.4.5 STIRRING SPEED

Struvite can be formed naturally without the help of stirring but the size of crystals remains small. In order to decrease the time of struvite crystal formation, promote nucleation and increase the size of the crystals, stirring is indispensable. Stirring can increase the mass transfer rate of ions in wastewater and enhance crystallization. The stirring speed for the precipitation of crystal of optimum size falls in the range of 100–200 rpm (Morales *et al.*, 2013). Stirring time also has to be taken into account for efficient struvite recovery. Stirring for a long time with a high mixing rate can lead to breakage of crystals resulting in the formation of unrecoverable struvite fines. An optimum of 15–60 minutes mixing time has been reported in literature followed by sufficient holding time (Liu *et al.*, 2008; Peng *et al.*, 2018; Rahaman *et al.*, 2008).

10.4.6 SEEDING

Struvite formation requires nucleation from the embryo, followed by crystal growth. In this context, the lag phase of crystallization is usually quite long. In order to speed up the precipitation process and enhance crystal size, many seed materials have been used by researchers. The struvite powder itself is sometimes used for seeding the next batch. As a result of spontaneous precipitation, struvite fines tend to settle at the bottom of wastewater. Such pre-formed crystals can be recovered and used for seeding purposes. They provide surfaces for other fine crystals to deposit on the active spots influencing the crystal size and morphology. Seeds can enhance ammonium removal through struvite formation; however, no significant effect of phosphate recovery was observed. Seed addition cannot induce spontaneous nucleation but promote the growth and formation of clusters. Other than struvite, several other seed materials are being used for enhanced struvite precipitation including biochar, quartz sand, kaolinite, and steel powder. Modifications in biochar such as Mg-laden biochar have been shown to be very effective in crystallization (Addagada, 2020; Ali & Schneider, 2006; Liu *et al.*, 2020; Liu *et al.*, 2013).

10.4.7 REACTOR DESIGN

There are several reactor designs used globally for the precipitation of struvite. Some of the most common designs are fluidized bed reactor (FBR), stirred tank reactor (STR), bioelectric systems such as microbial fuel cells (MFCs) and microbial electrolysis cells (MECs), ion exchange technology (IET), and membrane exchange technology (MBR). The operational conditions used in such reactors are shown in Table 10.5 and are discussed in detail below.

TABLE 10.5
Various Reactor Designs Used along with Operational Conditions for Struvite Precipitation

Reactor Technology Involved	Operational Conditions	References
Fluidized bed reactor	**Mg:P:** 1.3 **Alkaline reagent:** 0.5 M NaOH **pH:** 9.0–9.3 **Mg source:** $MgCl_2 \cdot 6H_2O$ **Flow rate:** 170–180 mL/min	Doino et al. (2011)
Fluidized bed reactor	**Mg:P:** 1.0 **Alkaline reagent:** 0.1 M NaOH **pH:** 8.5–9.5 **Mg source:** $MgCl_2 \cdot 6H_2O$ **Flow rate:** 0.01 m/s	Ye *et al.* (2018)
Stirred tank reactor	**Mg:P:** 1.0 **Alkaline reagent:** 1 M NaOH **pH:** 8.5–9.5 **Mg source:** $MgCl_2 \cdot 6H_2O$ **Stirring rate:** 500 rpm **Supersaturation range:** 1.83–3.44 **Temperature:** 25°C	Hanhoun *et al.* (2013)
Stirred tank reactor	**Mg:P:** 1.0 **Alkaline reagent:** 1 M NaOH **pH:** 8.0–8.5 **Mg source:** $MgCl_2 \cdot 6H_2O$ **Stirring rate:** 300 rpm **Temperature:** 25°C	Mehta and Batstone (2013)
Inverse fluidized bed reactor	**Mg:P:** 1.0 **Alkaline reagent:** 4 N NaOH **pH:** 8.0–9.0 **Mg source:** $MgCl_2 \cdot 6H_2O$ **Temperature:** 18.5–33.4°C	Sathiasivan *et al.* (2021)
Fluidized bed reactor	**Mg:P:** 1.3 **Alkaline reagent:** 1 M NaOH **pH:** 8.0–8.2 **Mg source:** $MgCl_2 \cdot 6H_2O$ **Supersaturation ratio:** 2–6	Iqbal *et al.* (2008)
Fluidized bed reactor	**Mg:P:** 1.3 **Alkaline reagent:** 1 M NaOH **pH:** 9.0–9.5 **Mg source:** $MgCl_2 \cdot 6H_2O$ **Temperature:** 20°C	Zamora et al. (2017)
Fluidized bed reactor	**Mg:P:** 1.25 **pH:** 9.0 **Mg source:** $MgCl_2 \cdot 6H_2O$	Guadie *et al.* (2014)

(Continued)

TABLE 10.5   (*Continued*)
Various Reactor Designs Used along with Operational Conditions for Struvite Precipitation

Reactor Technology Involved	Operational Conditions	References
Fluidized bed reactor	**Mg:P:** 1.0 **Alkaline reagent:** 1 M NaOH **pH:** 9.0 **Mg source:** $MgCl_2 \cdot 6H_2O$	Su *et al.* (2014)
Fluidized bed reactor (Struvite precipitation & dewaterability technology)	**Mg:P:** 1.0 **Alkaline reagent:** 1 N NaOH **pH:** 8.13±0.05 **Mg source:** Brine **Temperature:** 25°C	Lahav *et al.* (2013)
Fluidized bed reactor	**Mg:P:** 1.0 **Alkaline reagent:** NaOH **pH:** 8.2±0.1 **Mg source:** $MgCl_2 \cdot 6H_2O$	Liu *et al.* (2013)
Fluidized bed reactor	**Mg:P:** 1.0 **Alkaline reagent:** NaOH **pH:** 8.2±0.1 **Mg source:** $MgCl_2 \cdot 6H_2O$	Ping *et al.* (2016)
Stirred tank reactor	**Mg:P:** 1 **Alkaline reagent:** 0.3 M NaOH **pH:** 8.8 **Mg source:** Seawater & $MgCl_2 \cdot 6H_2O$ **Stirring rate:** 1,400 rpm	Aguado *et al.* (2019)
Stirred tank reactor	**Mg:P:** 1.0–1.2 **Alkaline reagent:** 1 M NaOH **pH:** 7.0–10 **Mg source:** $MgCl_2 \cdot 6H_2O$	Rodrigues *et al.* (2019)
Microbial fuel cell	**pH:** 7.2–9.0 **Mg source:** $MgCl_2$ **Anode:** Mg; **Cathode:** Activated carbon with 20% polytetrafluoroethylene **Temperature:** 20±2°C	Merino-Jimenez *et al.* (2017)
Microbial reverse electrolysis cell (combined hydrogen production of 2.06 m^3 H_2/m^2 d)	**Voltage:** 0.4–0.12 V **Electrodes:** Titanium mesh cathode with platinum **Chamber:** Acetate fed single air chamber	Song et al. (2021)
Dual chamber microbial electrolysis cell (combined hydrogen production of 0.28 m^3 H_2/m^2 d)	**Mg:P:** 1.0 **pH:** 7.0 **Electrodes:** Carbon cloth connected by titanium **Membrane:** Nafion	Almatouq & Babatunde (2017)

(Continued)

TABLE 10.5   (Continued)
Various Reactor Designs Used along with Operational Conditions for Struvite Precipitation

Reactor Technology Involved	Operational Conditions	References
Ion exchange and isothermal supersaturation	**pH:** 8.0–11.0 **Mg source:** Marine **Temperature:** 25°C **Ion exchange resin:** Amberlite 1RC86 macroporous sulfonic resin, Lewatit CNP80 macroporous carboxylic resin, Lewatit SICOO microporous sulfonic resin	Ortueta *et al.* (2015)
Forward osmosis (nitrogen and phosphorus recovery as struvite)	**Alkaline reagent:** 0.1 M NaOH **pH:** 9.0 **Membrane:** Thin-film polyamide composite	Volpin *et al.* (2019)
Pre-Anaerobic membrane reactor MBR	*Continuous stirred tank reactor* **Mg:P:** 2.0 **pH:** 9.0–10.0 **Mg source:** $Mg(OH)_2$ *Batch stirred tank reactor* **Mg:P:** 1.5–2.0 **pH:** 9.0–9.5 **Mg source:** $Mg(OH)_2$	Haroon *et al.* (2020)
Stirred tank reactor (Struvite precipitation & dewaterability technology)	**Mg:P:** 1.5 **pH:** 7.5 **Mg Source:** $MgCl_2 \cdot 6H_2O$ **Stirring rate:** 800 rpm **Temperature:** 24–30°C	Bergmans et al. (2014)
Continuous in liquid phase and batch reactor in solid phase	**Mg:P:** 2.0 **Alkaline reagent:** 0.4 M NaOH & aeration **pH:** 8.5–8.7 **Mg Source:** $MgCl_2 \cdot 6H_2O$	Pastor *et al.* (2010)

10.4.7.1   Fluidized Bed Reactor (FBR)

FBR is the most common reactor type used for struvite precipitation. Many commercial installations around the world have utilized this design for the efficient production of marketable struvite. This type of reactor has two components: a fluidized bed component with increasing diameter from the point of wastewater inlet at the bottom and a settling zone at the top (Figure 10.2a). The difference of diameter on various regions ensures proper mixing of the reactants (Iqbal, 2008). The two components of the reactor are joined by a funnel of polyvinyl chloride to avoid settling precipitates in the settling zone.

The supernatant of the sludge is fed in to the reactor with the help of a peristaltic pump. It has a pH controller with automated input of NaOH in the reactor to maintain the pH. In addition, there are two other inlets for the supply of magnesium and recycled water fed at rates of 170–180 mL/min and 9–42 mL/min, respectively (Shim et al., 2020). There are two outlets, one at the bottom to remove the larger crystals from the reactor and the second one at the top to remove the liquid along with small crystals which are again fed back into the reactor as seeds.

The large crystals are promoted to form in the fluidized bed and small crystals along with the fluid that settles upwards in the settling zone by plug flow method. These crystals are removed by the second outlet and send to the clarifier for recycling back to the fluidized bed to form larger crystals in successive batches. In many reactors, there is provision for additional hydrocyclone units that facilitate proper mixing and pelletization (Lahav et al., 2013).

10.4.7.2   Stirred Tank Reactor (STR)

STR has a stirring system that helps in the homogenous mixing of reactants. These reactors possess several auxiliary devices for supplying reactants, wastewater, pH adjustment reagents, and struvite precipitate removal. In this reactor, the mixing of all the reactants occurs in the same region (Figure 10.2b). Stirring plays a major role in such reactors as it helps in mass transfer and hence the formation of crystals. The reactors are designed to perform both in batch and continuous modes. In the case of batch precipitation of struvite formation and precipitation occurs in the same reactor; whereas, in continuous precipitation, formation occurs at different regions. Many times, a mixed reactor is constructed where both the batch and continuous reactors can work side by side leading to the formation of struvite crystals. Stirred reactors are easy to build and manage and hence it is widely used for laboratory experiments (Guadie et al., 2014; Mehta & Batstone, 2013; Rodrigues et al., 2019). However, in the case of pilot-scale experiments, it requires high investment and the cost incurred by energy consumption is high making it impractical for large-scale applications.

10.4.7.3   Bioelectric System

Bioelectric systems for struvite precipitation are emerging as they help in simultaneous recovering of energy, biofuel, nutrients, or chemicals from wastewater. The process is carried out with the help of microorganisms that bring about redox reactions at the electrodes' vicinity. This system is cheap and easier to construct as compared to other fuel cells. The construction neither requires costly metal nor difficult working conditions. With the advancement of technology and the scope of science, these bioelectric systems are used to recover nutrients like struvite with instantaneous production of electricity (Kelly & He, 2014). There are two types of bioelectric systems generally being used for struvite crystallization namely MFCs and MECs.

10.4.7.3.1   Microbial Fuel Cells (MFCs)

MFCs are used to produce electricity with the help of a variety of microorganisms. In this system, the microorganisms act as a catalyst for the production

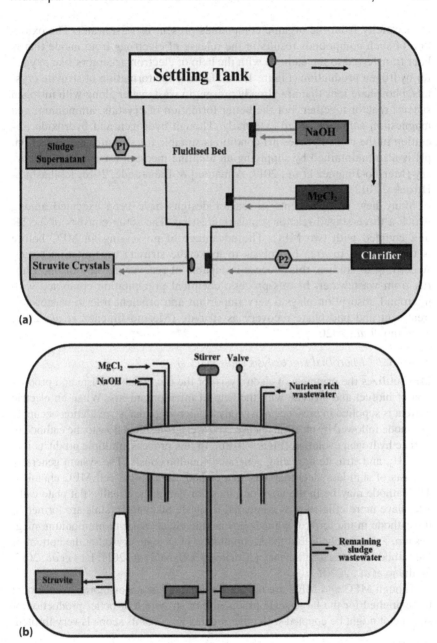

FIGURE 10.2 Reactor designs of (a) a fluidized bed reactor (FBR) and (b) a stirred tank reactor (STR) for struvite precipitation.

of electricity from the organic compounds present in wastewater. The oxidation of such compounds results in the release of electrons from anode that is later transferred to the cathode with the help of electron acceptors like oxygen for hydrogen production (Figure 10.3a). During the formation of struvite crystals, phosphate ions that are already present in wastewater along with nitrogen species reactor together. For the better formation of crystals, ammonium and magnesium salts are added externally. Though hydrogen and hydroxide liberation in the cell improves pH conditions suitable for struvite crystallization, pH is also maintained by supplying an alkaline medium for enhanced recovery (Merino-Jimenez et al., 2017; Almatouq & Babatunde, 2016; Ichihashi & Hirooka, 2012).

Many new advancements in reactor designs have been executed among which a three-staged reactor stands interesting. The setup consists of a STR unit coupled with two MFC. The advantage of possessing an MFC before is that it helps in urea hydrolysis to make the struvite precipitation easier. According to studies, this process is capable of recovering 82% of phosphorus from wastewater. In this process, chemical precipitation combined with microbial absorption plays a very important and efficient role in wastewater treatment and phosphate recovery as struvite (Merino-Jimenez et al., 2017; Siciliano et al., 2020).

10.4.7.3.2 Microbial Electrolysis Cells (MECs)

MEC utilizes the principle of electrolysis for the treatment of waste and production of biofuel and struvite with the help of microorganisms. When an electric current is supplied in presence of a catalyst (microorganisms), oxidation occurs at the anode followed by electron release. These electrons then flow to the cathode to initiate hydrogen evolution (Figure 10.3b). In this process, multiple products like H_2, CH_4, and struvite are being generated simultaneously. The system generally consists of stainless steel that acts as a cathode in a single-cell MEC chamber. The cathode may be in the form of a mesh or flat plate. Usually, flat plate cathodes have more efficiency as compared to mesh. Struvite crystals are formed at the cathode in the form of white solids having equal molar of ammonium, magnesium, and phosphate. But the accumulation of struvite crystal at the bottom of the cathode reduces its efficiency (Almatouq & Babatunde, 2017; Igos et al., 2017; Siciliano et al., 2020).

Though MFC and MEC are new technologies, it requires more advancement to be applied for the large-scale production of struvite. For better production of struvite, it might be coupled with other technologies but its scope is very limited.

10.4.7.4 Ion Exchange Technology (IET)

Ion exchange methods have been in use for several years for the process of nutrient recovery. This technology is very cheap, easy to use, recyclable, reusable, and regenerable. This technology is now being used for the production of struvite crystals with the help of certain resins. The resins used may be NH_4^+, PO_4^{3-}, or K^+ which are regenerated at the exchanger as ammonium or potassium

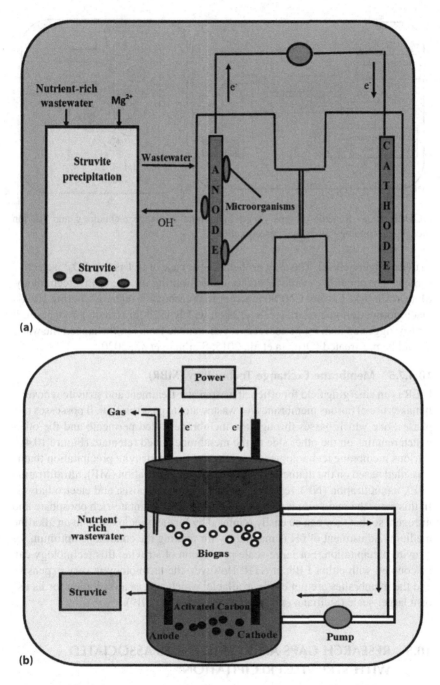

FIGURE 10.3   Reactor designs of bioelectric systems for struvite precipitation of (a) a microbial fuel cell and (b) a microbial electrolysis cell.

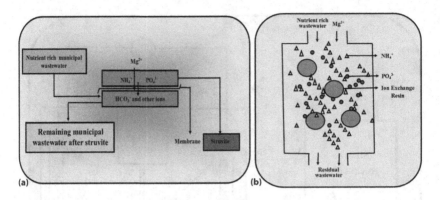

FIGURE 10.4 Reactor designs of (a) membrane exchange technology and (b) ion exchange technology for struvite precipitation.

struvite (Figure 10.4a). The only problem in the case of IET is that PO_4^{3-} selective sorbents are not easily available and reusable limiting the use of this technology. Many resins like Lewatit CNP80 macroporous carboxylic resin, Amberlite 1RC86 macroporous sulfonic resin, etc. have been widely used for struvite production. It is also suggested that, for better recovery of struvite from ion exchanger, the water should be pre-treated (Ortueta et al., 2015; Siciliano *et al.*, 2020).

10.4.7.5 Membrane Exchange Technology (MBR)

MBR is an emerging field for efficient wastewater treatment and struvite recovery. It makes use of porous membranes for wastewater to pass through. It possesses two phases, one which passes through the membrane called permeate and the other which remains on the other side of the membrane called retentate (Figure 10.4b). Various membrane technologies have been in use for struvite precipitation that is classified based on the diameter of the filter as microfiltration (MF), ultrafiltration (UF), nanofiltration (NF), reverse osmosis, forward osmosis, and electrodialysis. In this case, the main aim is to produce permeate or retentate-rich phosphate and nitrogen, so that they can be easily removed by further precipitation in an alkaline medium. Adjustment of pH is mandatory for making the conditions optimum for struvite precipitation. For large-scale production of struvite, this technology can be coupled with either FBR or STR. However, the technology is very expensive and the membranes are not easily available, which acts as a limitation for its use on a larger-scale (Siciliano *et al.*, 2020; Volpin *et al.*, 2019).

10.5 RESEARCH GAPS AND CHALLENGES ASSOCIATED WITH STRUVITE PRECIPITATION

Progress has been made in the field of struvite in the last two decades. However, there still exist some knowledge gaps and challenges that need to be addressed in further studies. Most of the work reported on struvite has been performed using synthetic formulations of wastewater (Gao *et al.*, 2020; Le Corre *et al.*, 2005; Tansel *et al.*, 2018). The results obtained from such experiments may not be suitable for real-time application. Hence, the focus should be shifted to real-time trails which also might help in the scale-up of the technology. Exploration of optimum physicochemical characteristics of wastewater in addition to operational parameters is necessary to gain better insights on reaction taking place and storage conditions. Most of the research deals only with the scope of struvite production. It has to be extended to plant growth studies for making it a commercial product. In addition, the mechanism behind struvite uptake by the plants is still unclear and more research on this facet has to be promoted. On the other hand, only the fertilizing applications of struvite have been widely considered (Oliveira *et al.*, 2020; Udert *et al.*, 2015). Other aspects and areas where struvite can be applied also have to be worked upon.

The external addition of magnesium is an important element of struvite crystallization. However, be it magnesium electrode or salt, it contributes to up to 75% of the total cost of the process. In order to address this issue, alternative low-cost sources of magnesium should be investigated. More attention to drying conditions of struvite should be given as it has a severe effect on sterilization and composition of mineral formed. The presence of harmful bacteria and heavy metal with struvite might be hazardous considering its application in the agricultural field (Bischel *et al.*, 2016; Decrey *et al.*, 2011). These are some of the areas that require more emphasis in further work for the development and establishment of the field on a larger scale.

10.6   CONCLUDING REMARKS AND FUTURE PERSPECTIVES

With the increase in population and urbanization, the generation of waste is increasing on an everyday basis. On the other hand, the infrastructure for effective treatment of sludge and reuse for drinking or irrigation purposes is still lacking. Struvite precipitation is an integrated field involving wastewater management, nutrient recycling, prevention of eutrophication, and production of organic fertilizer. In this view, struvite recovery from municipal wastewater can be economical due to the presence of high nutrient considered and ease of accessibility. This paper deals with various aspects of struvite crystallization from municipal wastewater with incorporated details of different precipitation technologies, significant operational conditions such as pH, temperature, source and dosage of magnesium, the concentration of co-present pollutants, and stirring rate. Several reactor designs that have been reported in the literature and commercially installed across the globe have been emphasized. In addition, research gaps and challenges associated with struvite production technologies have been highlighted for consideration in future studies. Some of the future perspectives of struvite research have been highlighted below:

- Source-separated collection of municipal wastewater can reduce the load of primary treatment, dilution, and improve assembly of concentrated nutrient-rich source for easy and effective recovery of struvite.

- Cost-effective pre-treatment techniques to improve solubilization of nutrients bound to organic compounds and heavy metals can enhance the amount of nutrient recovery.
- Seeding can improve the quality, size, and recovery rate of struvite crystals. Hence, future works on finding a lucrative seed material become essential.
- In addition to struvite formation, recovery of struvite analogs such as K-struvite and Na-struvite can improve retrieval of supplementary nutrients from wastewater.
- More research on emerging technologies such as MFC and MEC should be implemented to make the overall process economical as it involves simultaneous struvite, hydrogen, and electricity production.
- Incorporation of computational methods such as modeling, techno-economic, and life cycle assessments can provide better insights with respect to scale-up of struvite technologies.
- Biorefinery aspects such as combined algal growth prior to struvite precipitation can help in the assimilation of organic matter and heavy metal content for enhanced struvite recovery. This process can also benefit the utilization of algae for biodiesel or fertilizing applications.

DECLARATION OF INTEREST

The authors declare no conflict of interest to disclose.

ACKNOWLEDGMENTS

The authors thank the Department of Biotechnology and Medical Engineering of National Institute of Technology Rourkela for providing the necessary research facilities. The authors greatly acknowledge the Science and Engineering Research Board (SERB) of the Department of Science and Technology (DST), Government of India (GoI) for sponsoring the research through ASEAN-India Science Technology and Innovation Cooperation [File No. IMRC/AISTDF/CRD/2018/000082], and Technology Development Programme [File No. DST/TDT/Agro43/2020].

REFERENCES

Addagada, L. 2020. Enhanced phosphate recovery using crystal-seed-enhanced struvite precipitation: Process optimization with response surface methodology. *Journal of Hazardous, Toxic, and Radioactive Waste*, 24(4): 04020027.

Aguado, D., Barat, R., Bouzas, A., Seco, A., & Ferrer, J. 2019. P-recovery in a pilot-scale struvite crystallisation reactor for source separated urine systems using seawater and magnesium chloride as magnesium sources. *Science of the Total Environment*, 672: 88–96.

Ali, M. I., & Schneider, P. A. 2006. A fed-batch design approach of struvite system in controlled supersaturation. *Chemical Engineering Science*, 61(12): 3951–3961.

Almatouq, A., & Babatunde, A. O. 2016. Concurrent phosphorus recovery and energy generation in mediator-less dual chamber microbial fuel cells: Mechanisms and influencing factors. *International Journal of Environmental Research and Public Health*, *13*(4): 375.

Almatouq, A., & Babatunde, A. O. 2017. Concurrent hydrogen production and phosphorus recovery in dual chamber microbial electrolysis cell. *Bioresource Technology*, *237*: 193–203.

Antonini, S., Paris, S., Eichert, T., & Clemens, J. 2011. Nitrogen and phosphorus recovery from human urine by struvite precipitation and air stripping in Vietnam. *CLEAN – Soil, Air, Water*, *39*(12): 1099–1104.

Baierle, F., John, D. K., Souza, M. P., Bjerk, T. R., Moraes, M. S. A., Hoeltz, M., ... Schneider, R. C. S. 2015. Biomass from microalgae separation by electroflotation with iron and aluminum spiral electrodes. *Chemical Engineering Journal*, *267*: 274–281.

Barak, P., & Stafford, A. 2006. Struvite: A recovered and recycled phosphorus fertilizer. In *Proceedings of the 2006 Wisconsin Fertilizer, Aglime & Pest Management Conference*, vol. 45, p. 199.

Bayuseno, A. P., & Schmahl, W. W. 2020. Crystallization of struvite in a hydrothermal solution with and without calcium and carbonate ions. *Chemosphere*, *250*: 126245.

Bergmans, B. J. C., Veltman, A. M., Van Loosdrecht, M. C. M., Van Lier, J. B., & Rietveld, L. C. 2014. Struvite formation for enhanced dewaterability of digested wastewater sludge. *Environmental Technology*, *35*(5): 549–555.

Bischel, H. N., Schindelholz, S., Schoger, M., Decrey, L., Buckley, C. A., Udert, K. M., & Kohn, T. 2016. Bacteria inactivation during the drying of struvite fertilizers produced from stored urine. *Environmental Science and Technology*, *50*(23): 13013–13023.

Borojovich, E. J., Münster, M., Rafailov, G., & Porat, Z. 2010. Precipitation of ammonium from concentrated industrial wastes as struvite: A search for the optimal reagents. *Water Environment Research*, *82*(7): 586–591.

Campos, J. L., Crutchik, D., Franchi, Ó., & Pavissich, J. P. 2019. Nitrogen and phosphorus recovery from anaerobically pretreated agro-food wastes: A review. *Frontiers in Sustainable Food Systems*, *2*(1): 1–11.

Capdevielle, A., Sýkorová, E., Béline, F., & Daumer, M. 2014. Kinetics of struvite precipitation in synthetic biologically treated swine wastewaters. *Environmental Technology*, *35*(10): 1250–1262.

Capdevielle, A., Sýkorová, E., Béline, F., & Daumer, M. L. 2016. Effects of organic matter on crystallization of struvite in biologically treated swine wastewater. *Environmental Technology*, *37*(7): 880–892.

Cid, C. A., Jasper, J. T., & Hoffmann, M. R. 2018. Phosphate recovery from human waste via the formation of hydroxyapatite during electrochemical wastewater treatment. *ACS Sustainable Chemistry & Engineering*, *6*(3): 3135–3142.

Corona, F., Hidalgo, D., Martín-Marroquín, J. M., & Antolín, G. 2020. Study of the influence of the reaction parameters on nutrients recovering from digestate by struvite crystallisation. *Environmental Science and Pollution Research*, 1–13.

Crutchik, D., & Garrido, J. M. 2011. Struvite crystallization versus amorphous magnesium and calcium phosphate precipitation during the treatment of a saline industrial wastewater. *Water Science and Technology*, *64*(12): 2460–2467.

Crutchik, D., & Garrido, J. M. 2016. Kinetics of the reversible reaction of struvite crystallisation. *Chemosphere*, *154*: 567–572.

Crutchik, D., Morales, N., Vázquez-Padín, J. R., & Garrido, J. M. 2017. Enhancement of struvite pellets crystallization in a full-scale plant using an industrial grade magnesium product. *Water Science and Technology*, *75*(3): 609–618.

Decrey, L., Udert, K. M., Tilley, E., Pecson, B. M., & Kohn, T. 2011. Fate of the pathogen indicators phage ΦX174 and *Ascaris suum* eggs during the production of struvite fertilizer from source-separated urine. *Water Research*, *45*(16): 4960–4972.

Desmidt, E., Ghyselbrecht, K., Monballiu, A., Verstraete, W., & Meesschaert, B. D. 2012. Evaluation and thermodynamic calculation of ureolytic magnesium ammonium phosphate precipitation from UASB effluent at pilot scale. *Water Science and Technology*, *65*(11): 1954–1962.

Doino, V., Mozet, K., Muhr, H., & Plasari, E. 2011. Study on struvite precipitation in a mechanically stirring fluidized bed reactor. *Chemical Engineering Transactions*, 24: 679–684.

Esakku, S., Selvam, A., Joseph, K., & Palanivelu, K. 2005. Assessment of heavy metal species in decomposed municipal solid waste. *Chemical Speciation & Bioavailability*, *17*(3): 95–102.

Evans, A. E., Hanjra, M. A., Jiang, Y., Qadir, M., & Drechsel, P. 2012. Water quality: Assessment of the current situation in Asia. *International Journal of Water Resources Development*, *28*(2): 195–216.

Fang, C., Zhang, T., Jiang, R., & Ohtake, H. 2016. Phosphate enhance recovery from wastewater by mechanism analysis and optimization of struvite settleability in fluidized bed reactor. *Scientific Reports*, *6*(1): 1–10.

Gao, F., Wang, L., Wang, J., Zhang, H., & Lin, S. 2020. Nutrient recovery from treated wastewater by a hybrid electrochemical sequence integrating bipolar membrane electrodialysis and membrane capacitive deionization. *Environmental Science: Water Research and Technology*, *6*(2): 383–391.

Gilbert, N. 2009. The disappearing nutrient. *Nature*, *461*(7265): 716–718.

Guadie, A., Xia, S., Jiang, W., Zhou, L., Zhang, Z., Hermanowicz, S. W., ... Shen, S. 2014. Enhanced struvite recovery from wastewater using a novel cone-inserted fluidized bed reactor. *Journal of Environmental Sciences*, *26*(4): 765–774.

Gurr, T. M. 2012. Phosphate rock. *Mining Engineering*, *64*(6): 81–84.

Hallas, J. F., Mackowiak, C. L., Wilkie, A. C., & Harris, W. G. 2019. Struvite phosphorus recovery from aerobically digested municipal wastewater. *Sustainability*, *11*(2): 376.

Hanhoun, M., Montastruc, L., Azzaro-pantel, C., Biscans, B., Frèche, M., & Pibouleau, L. 2011. Temperature impact assessment on struvite solubility product: A thermodynamic modeling approach. *Chemical Engineering Journal*, *167*(1): 50–58.

Hanhoun, M., Montastruc, L., Azzaro-Pantel, C., Biscans, B., Frèche, M., & Pibouleau, L. 2013. Simultaneous determination of nucleation and crystal growth kinetics of struvite using a thermodynamic modeling approach. *Chemical Engineering Journal*, *215–216*: 903–912.

Haroon, M., Jegatheesan, V., & Navaratna, D. 2020. The potential of adopting struvite precipitation as a strategy for the removal of nutrients from pre-AnMBR treated abattoir wastewater. *Journal of Environmental Management*, *259*: 109783. 83

Heinonen-Tanski, H., & Van Wijk-Sijbesma, C. 2005. Human excreta for plant production. *Bioresource Technology*, *96*(4): 403–411.

Hester, R. E.; Harrison, R. M. (Eds.). 2013. *Waste as a Resource*. Cambridge, UK: RSC Publishing.

Hu, L., Yu, J., Luo, H., Wang, H., Xu, P., & Zhang, Y. 2020. Simultaneous recovery of ammonium, potassium and magnesium from produced water by struvite precipitation. *Chemical Engineering Journal*, *382*: 123001.

Huang, H., Li, J., Li, B., Zhang, D., Zhao, N., & Tang, S. 2019. Comparison of different K-struvite crystallization processes for simultaneous potassium and phosphate recovery from source-separated urine. *Science of the Total Environment, 651*: 787–795.

Huang, H., Zhang, D., Wang, W., Li, B., Zhao, N., Li, J., & Dai, J. 2019. Alleviating Na+ effect on phosphate and potassium recovery from synthetic urine by K-struvite crystallization using different magnesium sources. *Science of the Total Environment, 655*: 211–219.

Hug, A., & Udert, K. M. 2013. Struvite precipitation from urine with electrochemical magnesium dosage. *Water Research, 47*(1): 289–299.

Huygens, D., Saveyn, H., Tonini, D., Eder, P., & Delgado Sancho, L. 2019. Technical proposals for selected new fertilising materials under the Fertilising Products Regulation (Regulation (EU) 2019/1009): Process and quality criteria, and assessment of environmental and market impacts for precipitated phosphate salts & derivates, thermal oxidation materials & derivates and pyrolysis & gasification materials. *Publications Office of the European Union*.

Ichihashi, O., & Hirooka, K. 2012. Removal and recovery of phosphorus as struvite from swine wastewater using microbial fuel cell. *Bioresource Technology, 114*: 303–307.

Igos, E., Besson, M., Navarrete Gutiérrez, T., Bisinella de Faria, A. B., Benetto, E., Barna, L., … Spérandio, M. 2017. Assessment of environmental impacts and operational costs of the implementation of an innovative source-separated urine treatment. *Water Research, 126*: 50–59.

Iqbal, M., Bhuiyan, H., & Mavinic, D. S. 2008. Assessing struvite precipitation in a pilot-scale fluidized bed crystallizer. *Environmental Technology, 29*(11): 1157–1167.

Ishii, S. K. L., & Boyer, T. H. 2015. Life cycle comparison of centralized wastewater treatment and urine source separation with struvite precipitation: Focus on urine nutrient management. *Water Research, 79*: 88–103.

Jardin, N., & Pöpel, H. J. 2001. Refixation of phosphates released during bio-p sludge handling as struvite or aluminium phosphate. *Environmental Technology, 22*(11): 1253–1262.

Jasinski, S. 2014. *Phosphate rock USGS 2013*, vol. 703, pp. 2013–2014.

Kataki, S., West, H., Clarke, M., & Baruah, D. C. 2016. Phosphorus recovery as struvite from farm, municipal and industrial waste: Feedstock suitability, methods and pre-treatments. *Waste Management, 49*: 437–454.

Kelly, P. T., & He, Z. 2014. Bioresource Technology Nutrients removal and recovery in bioelectrochemical systems: A review. *Bioresource Technology, 153*: 351–360.

Kinuthia, G. K., Ngure, V., Beti, D., Lugalia, R., Wangila, A., & Kamau, L. 2020. Levels of heavy metals in wastewater and soil samples from open drainage channels in Nairobi, Kenya: Community health implication. *Scientific Reports*, 10(1): 1–13.

Krishnamoorthy, N., Dey, B., Arunachalam, T., & Paramasivan, B. 2020. Effect of storage on physicochemical characteristics of urine for phosphate and ammonium recovery as struvite. *International Biodeterioration and Biodegradation, 153*: 105053.

Krishnamoorthy, N., Dey, B., Unpaprom, Y., Ramaraj, R., Maniam, G. P., Govindan, N., … Paramasivan, B. 2021a. Engineering principles and process designs for phosphorus recovery as struvite: A comprehensive review. *Journal of Environmental Chemical Engineering*, 105579.

Krishnamoorthy, N., Arunachalam, T., & Paramasivan, B. 2021b. A comparative study of phosphorus recovery as struvite from cow and human urine. *Materials Today: Proceedings*. https://doi.org/10.1016/j.matpr.2021.04.587

Kruk, D. J., Elektorowicz, M., & Oleszkiewicz, J. A. 2014. Struvite precipitation and phosphorus removal using magnesium sacrificial anode. *Chemosphere*, *101*: 28–33.

Lahav, O., Telzhensky, M., Zewuhn, A., Gendel, Y., Gerth, J., & Calmano, W. 2013. Struvite recovery from municipal-wastewater sludge centrifuge supernatant using seawater NF concentrate as a cheap Mg (II) source. *Separation and Purification Technology*, *108*: 103–110.

Latifian, M., Liu, J., & Mattiasson, B. 2012. Struvite-based fertilizer and its physical and chemical properties. *Environmental Technology*, *33*(24): 2691–2697.

Law, K. P., & Pagilla, K. R. 2018. Phosphorus recovery by methods beyond struvite precipitation. *Water Environment Research*, *90*(9): 840–850.

Le Corre, K. S., Valsami-Jones, E., Hobbs, P., & Parsons, S. A. 2005. Impact of calcium on struvite crystal size, shape and purity. *Journal of Crystal Growth*, *283*(3–4): 514–522.

Lee, E. Y., Oh, M. H., Yang, S. H., & Yoon, T. H. 2015. Struvite crystallization of anaerobic digestive fluid of swine manure containing highly concentrated nitrogen. *Asian-Australasian Journal of Animal Sciences*, *28*(7): 1053.

Li, B., Boiarkina, I., Yu, W., Ming, H., & Munir, T. 2018. Phosphorous recovery through struvite crystallization: Challenges for future design science of the total environment phosphorous recovery through struvite crystallization: Challenges for future design. *Science of the Total Environment*, *648*: 1244–1256.

Li, B., Huang, H. M., Boiarkina, I., Yu, W., Huang, Y. F., Wang, G. Q., & Young, B. R. 2019. Phosphorus recovery through struvite crystallisation: Recent developments in the understanding of operational factors. *Journal of Environmental Management*, *248*: 109254.

Li, B., Huang, H., Sun, Z., Zhao, N., Munir, T., Yu, W., & Young, B. 2020. Minimizing heavy metals in recovered struvite from swine wastewater after anaerobic biochemical treatment: Reaction mechanisms and pilot test. *Journal of Cleaner Production*, *272*: 122649.

Li, H., Yao, Q., Wang, Y., Li, Y., & Zhou, G. 2015. Biomimetic synthesis of struvite with biogenic morphology and implication for pathological biomineralization, *Scientific Reports*, *5*(1): 1–8.

Liu, J., Zheng, M., Wang, C., Liang, C., Shen, Z., & Xu, K. 2020. A green method for the simultaneous recovery of phosphate and potassium from hydrolyzed urine as value-added fertilizer using wood waste. *Resources, Conservation and Recycling*, *157*: 104793.

Liu, X., Hu, Z., Zhu, C., Wen, G., Meng, X., & Lu, J. 2013. Influence of process parameters on phosphorus recovery by struvite formation from urine. *Water Science and Technology*, *68*(11): 2434–2440.

Liu, Z., Zhao, Q., Wang, K., Lee, D., Qiu, W., & Wang, J. 2008. Urea hydrolysis and recovery of nitrogen and phosphorous as MAP from stale human urine. *Journal of Environmental Sciences*, *20*(8): 1018–1024.

Liu, Z., Zhao, Q., Wei, L., Wu, D., & Ma, L. 2011. Effect of struvite seed crystal on MAP crystallization. *Journal of Chemical Technology and Biotechnology*, *86*(11): 1394–1398.

Luo, Y., Guo, W., Hao, H., Duc, L., Ibney, F., Zhang, J., ... Wang, X. C. 2014. A review on the occurrence of micropollutants in the aquatic environment and their fate and removal during wastewater treatment. *Science of the Total Environment*, *473–474*: 619–641.

Maaß, O., Grundmann, P., & Von Bock Und Polach, C. 2014. Added-value from innovative value chains by establishing nutrient cycles via struvite. *Resources, Conservation and Recycling*, *87*: 126–136.

Mahy, J. G., Wolfs, C., Mertes, A., Vreuls, C., Drot, S., Smeets, S., ... Lambert, S. D. 2019. Advanced photocatalytic oxidation processes for micropollutant elimination from municipal and industrial water. *Journal of Environmental Management, 250*: 109561.

Matassa, S., Batstone, D. J., Schnoor, J., & Verstraete, W. 2015. Can direct conversion of used nitrogen to new feed and protein help feed the world? *ACS Publications, 49*(9): 5247–5254.

Mehta, C. M., & Batstone, D. J. 2013. Nucleation and growth kinetics of struvite crystallization. *Water Research, 47*(8): 2890–2900.

Merino-Jimenez, I., Celorrio, V., Fermin, D. J., Greenman, J., & Ieropoulos, I. 2017. Enhanced MFC power production and struvite recovery by the addition of sea salts to urine. *Water Research, 109*: 46–53.

Morales, N., Boehler, M. A., Buettner, S., Liebi, C., & Siegrist, H. 2013. Recovery of N and P from urine by struvite precipitation followed by combined stripping with digester sludge liquid at full scale. *Water, 5*(3):1262–1278.

Oliveira, V., Dias-Ferreira, C., Labrincha, J., Rocha, J. L., & Kirkelund, G. M. 2020. Testing new strategies to improve the recovery of phosphorus from anaerobically digested organic fraction of municipal solid waste. *Journal of Chemical Technology and Biotechnology, 95*(2): 439–449.

Ortueta, M., Celaya, A., Mijangos, F., & Muraviev, D. J. S. E. 2015. Ion exchange synthesis of struvite accompanied by isothermal supersaturation: Influence of polymer matrix and functional groups Type. *Solvent Extraction and Ion Exchange, 33*(1): 65–74.

Pastor, L., Mangin, D., Ferrer, J., & Seco, A. 2010. Struvite formation from the supernatants of an anaerobic digestion pilot plant. *Bioresource Technology, 101*(1): 118–125.

Peng, L., Dai, H., Wu, Y., Peng, Y., & Lu, X. 2018. A comprehensive review of phosphorus recovery from wastewater by crystallization processes. *Chemosphere, 197*: 768–781.

Ping, Q., Li, Y., Wu, X., Yang, L., & Wang, L. 2016. Characterization of morphology and component of struvite pellets crystallized from sludge dewatering liquor: Effects of total suspended solid and phosphate concentrations. *Journal of Hazardous Materials, 310*: 261–269.

Prywer, J., & Torzewska, A. (2010). Biomineralization of struvite crystals by *Proteus mirabilis* from artificial urine and their mesoscopic structure. *Crystal Research and Technology*, 45(12): 1283–1289.

Quintana, M., Colmenarejo, M. F., Barrera, J., García, G., García, E., & Bustos, A. 2004. Use of a byproduct of magnesium oxide production to precipitate phosphorus and nitrogen as struvite from wastewater treatment liquors. *Journal of Agricultural and Food Chemistry, 52*(2): 294–299.

Rahaman, M. S., Ellis, N., & Mavinic, D. S. 2008. Effects of various process parameters on struvite precipitation kinetics and subsequent determination of rate constants. *Water Science and Technology, 57*(5): 647–654.

Rodrigues, D. M., do Amaral Fragoso, R., Carvalho, A. P., Hein, T., & Guerreiro de Brito, A. 2019. Recovery of phosphates as struvite from urine-diverting toilets: Optimization of pH, Mg: PO_4 ratio and contact time to improve precipitation yield and crystal morphology. *Water Science and Technology, 80*(7): 1276–1286.

Ryu, H. D., Lim, C. S., Kim, Y. K., Kim, K. Y., & Lee, S. I. 2012. Recovery of struvite obtained from semiconductor wastewater and reuse as a slow-release fertilizer. *Environmental Engineering Science, 29*(6): 540–548.

Saerens, B., Geerts, S., & Weemaes, M. 2021. Phosphorus recovery as struvite from digested sludge – experience from the full scale. *Journal of Environmental Management, 280*: 111743.

Saidou, H., Korchef, A., Moussa, S. B., & Amor, M. B. 2015. Study of Cd^{2+}, Al^{3+}, and SO_4 ions influence on struvite precipitation from synthetic water by dissolved CO_2 degasification technique. *Open Journal of Inorganic Chemistry*, 5(03): 41.

Sakthivel, S. R., Tilley, E., & Udert, K. M. 2012. Wood ash as a magnesium source for phosphorus recovery from source-separated urine. *Science of the Total Environment*, 419: 68–75.

Salsabili, A., Salleh, M. A. M., & Zohoori, M. 2014. Evaluating the effects of operational parameters and conditions on struvite crystallization and precipitation with the focus on temperature. *Advances in Natural and Applied Sciences*, 8(4): 175–180.

Sathiasivan, K., Ramaswamy, J., & Rajesh, M. 2021. Struvite recovery from human urine in inverse fluidized bed reactor and evaluation of its fertilizing potential on the growth of *Arachis hypogaea*. *Journal of Environmental Chemical Engineering*, 9(1): 104965.

Sengupta, S., & Pandit, A. 2011. Selective removal of phosphorus from wastewater combined with its recovery as a solid-phase fertilizer. *Water Research*, 45(11): 3318–3330.

Shim, S., Won, S., Reza, A., Kim, S., Ahmed, N., & Ra, C. 2020. Design and optimization of fluidized bed reactor operating conditions for struvite recovery process from swine wastewater. *Processes*, 8(4): 422.

Siciliano, A., & De Rosa, S. 2014. Recovery of ammonia in digestates of calf manure through a struvite precipitation process using unconventional reagents. *Environmental Technology*, 35(7): 841–850.

Siciliano, A., Limonti, C., Curcio, G. M., & Molinari, R. 2020. Advances in struvite precipitation technologies for nutrients removal and recovery from aqueous waste and wastewater. *Sustainability*, 12(18): 7538.

Song, Y. H., Qiu, G. L., Yuan, P., Cui, X. Y., Peng, J. F., Zeng, P., ... Qian, F. 2011. Nutrients removal and recovery from anaerobically digested swine wastewater by struvite crystallization without chemical additions. *Journal of Hazardous Materials*, 190(1–3): 140–149.

Song, Y. H., Hidayat, S., Effendi, A. J., & Park, J. Y. 2021. Simultaneous hydrogen production and struvite recovery within a microbial reverse-electrodialysis electrolysis cell. *Journal of Industrial and Engineering Chemistry*, 94: 302–308.

Srivastava, V., Gupta, S. K., Singh, P., Sharma, B., & Singh, R. P. 2018. Biochemical, physiological, and yield responses of lady's finger (*Abelmoschus esculentus L.*) grown on varying ratios of municipal solid waste vermicompost. *International Journal of Recycling of Organic Waste in Agriculture*, 7(3): 241–250.

Su, C., Ruffel, R., Abarca, M., Daniel, M., Luna, G. De, & Lu, M. 2014. Phosphate recovery from fluidized-bed wastewater by struvite crystallization technology. *Journal of the Taiwan Institute of Chemical Engineers*, 45(5): 2395–2402.

Tansel, B., Lunn, G., & Monje, O. 2018. Struvite formation and decomposition characteristics for ammonia and phosphorus recovery: A review of magnesium-ammonia-phosphate interactions. *Chemosphere*, 194: 504–514.

Theregowda, R. B., González-Mejía, A. M., Ma, X., & Garland, J. 2019. Nutrient recovery from municipal wastewater for sustainable food production systems: An alternative to traditional fertilizers. *Environmental engineering science*, 36(7): 833–842.

Tilley, E., Atwater, J., & Mavinic, D. 2008. Effects of storage on phosphorus recovery from urine. *Environmental Technology*, 29(7): 807–816.

Udert, K. M., Buckley, C. A., Wächter, M., McArdell, C. S., Kohn, T., Strande, L., ... Etter, B. 2015. Technologies for the treatment of source-separated urine in the eThekwini Municipality. *Water SA*, 41(2): 212–221.

Van Kauwenbergh, S. J., Stewart, M., & Mikkelsen, R. 2013. World reserves of phosphate rock. A dynamic and unfolding story. *Better Crops, 97*: 18–20.

Vodyanitskii, Y. N. 2016. Standards for the contents of heavy metals in soils of some states. *Annals of Agrarian Sciences, 14*(3): 257–263.

Volpin, F., Heo, H., Hasan, A., Cho, J., Phuntsho, S., & Kyong, H. 2019. Techno-economic feasibility of recovering phosphorus, nitrogen and water from dilute human urine via forward osmosis. *Water Research, 150*: 47–55.

Werle, S., & Dudziak, M. 2012. Analysis of organic and inorganic contaminants in dried sewage sludge and by-products of dried sewage sludge gasification. *Energies, 19*: 137–144.

Williams, A. T., Zitomer, D. H., & Mayer, B. K. 2015. Ion exchange-precipitation for nutrient recovery from dilute wastewater. *Environmental Science: Water Research & Technology, 1*(6): 832–838.

Wilsenach, J. A., & Van Loosdrecht, M. C. M. 2004. Effects of separate urine collection on advanced nutrient removal processes. *Environmental Science and Technology, 38*(4): 1208–1215.

Wu, S., Zou, S., Liang, G., Qian, G., & He, Z. 2018. Enhancing recovery of magnesium as struvite from landfill leachate by pretreatment of calcium with simultaneous reduction of liquid volume via forward osmosis. *Science of the Total Environment, 610–611*: 137–146.

Ye, X., Gao, Y., Cheng, J., Chu, D., Ye, Z., & Chen, S. 2018. Numerical simulation of struvite crystallization in fluidized bed reactor. *Chemical Engineering Science, 176*: 242–253.

Yetilmezsoy, K., & Sapci-Zengin, Z. 2009. Recovery of ammonium nitrogen from the effluent of UASB treating poultry manure wastewater by MAP precipitation as a slow release fertilizer. *Journal of Hazardous materials, 166*(1): 260–269.

Zamora, P., Georgieva, T., Salcedo, I., Elzinga, N., Kuntke, P., & Buisman, C. J. 2017. Long-term operation of a pilot-scale reactor for phosphorus recovery as struvite from source-separated urine. *Journal of Chemical Technology & Biotechnology, 92*(5): 1035–1045.

Zhang, D. M., Chen, Y. X., Jilani, G., Wu, W. X., Liu, W. L., & Han, Z. Y. 2012. Optimization of struvite crystallization protocol for pretreating the swine wastewater and its impact on subsequent anaerobic biodegradation of pollutants. *Bioresource Technology, 116*: 386–395.

Zhang, Y., Desmidt, E., Van Looveren, A., Pinoy, L., Meesschaert, B., & Van Der Bruggen, B. 2013. Phosphate separation and recovery from wastewater by novel electrodialysis. *Environmental Science and Technology, 47*(11): 5888–5895.

Zou, H., & Wang, Y. 2016. Phosphorus removal and recovery from domestic wastewater in a novel process of enhanced biological phosphorus removal coupled with crystallization. *Bioresource Technology, 211*, 87–92.

11 Algae-Based Industrial Wastewater Treatment Methods and Applications

Raghunath Satpathy
School of Biotechnology, Gangadhar Meher University,
Amruta Vihar, Sambalpur, Odisha, India

CONTENTS

DOI: 10.1201/9781003201076-11

11.1 INTRODUCTION

Due to rapid industrialization process throughout the globe in the last 50 years, the entire world produces larger volume of the industrial effluent. Majority of these industries uses the water, hence a significant volume of wastewater is produced from them. These industrial effluents are generally discharged into other water sources by either untreated or inadequately treated manner. Therefore, this creates the problem of water pollution both in surface and sub-soil region. The wastewater from industry contains several components, some of which are nutrients and some others toxic, in addition to the high level of total dissolved solids (TDS) (Cai et al., 2013; Renuka et al., 2013; Sonune & Ghate, 2004). To characterize industrial wastewater, several parameters have to be considered and these are divided into three main categories as given below.

11.1.1 PHYSICAL CHARACTERISTICS

Typically, different physical characteristics are considered for quantification of industrial wastewater, such as the color, odor, temperature, and total solid content (floating matter and its contents in suspension and colloidal contents).

11.1.2 CHEMICAL CHARACTERISTICS

Most of the chemical characteristics shown by the pollutants are produced by dissolved materials which may include organic matter, nutrients, metals, and agrochemicals.

11.1.3 BIOLOGICAL CHARACTERISTICS

Microorganisms are mainly responsible for defining biological characteristics. The principal groups of organisms found in, surface and wastewater are classified as Protista, plants, and animals. However with the exceptions for bacteria and fungi, the other microorganisms are usually absent in the industrial effluents.

11.2 TRADITIONAL METHODS OF WASTEWATER TREATMENT AND CHALLENGES

Industrial wastewater treatment requires to be handled in a systematic and planned way. A treatment program typically includes identification of pollutant(s), determination of total pollution load with the assistance of flow rate and measurement of the quantity of pollutant(s). The treatment programs are separately planned for concentrated as well as dilute effluents. The treatment methods can be classified into three main categories: such as physical treatment, chemical treatment, and biological methods (Li et al., 2018; Peters et al., 1985; Segneanu et al., 2013).

11.2.1 PHYSICAL TREATMENT

Physical method is used for removal of bricks, wood pieces, paper, etc. For the breaking of the solid material, commenting devices like grinders, cutters, or shredders are also used. Grit chambers are used to arrest sand, dust, stone, cinders, and other heavy inorganic settleable materials. Oil and grease from the effluent are removed by grease traps. Basically, the sedimentation tanks are used to remove the suspended organic solids from the wastewater, as a part of the preliminary treatment process.

11.2.2 CHEMICAL TREATMENT

This process of treatment includes methods like neutralization, coagulation, and flocculation and other chemical methods for removal of a variety of toxic substances and nutrients. Neutralization is a method when an alkaline or acidic substances is added to a solution, the medium correspondingly changes due to neutralization reactions. In the flocculation process, due to the addition of a coagulating chemical precipitation occurs and can be observed as foaming. Subsequent movement of the wastewater to the settling tank can remove other suspended solids by absorption or mechanical agglomeration. The chemical treatment methods are usually cost-prohibitive and not as efficient as the biological treatment methods.

11.2.3 BIOLOGICAL TREATMENT

Basically, in most cases one or more essential nutrient is missing in the industrial effluents, therefore it is always desirable to treat them biologically along with the sewage. Because the sewage will supply the nutrient, hence enable the growth of bacteria lead to breakdown of the pollutants. Biological treatment process utilizes both the resources of natural flora and fauna for the treatment of wastewater. Low-cost biological treatment methods usually give emphasis on land treatment of wastewater and its use in crop irrigation for the removal of pollution load. There is now, more than ever before, an imperative need to develop efficient biological treatment system to suit the types of industrial effluents as well as specific needs. Basically, the aerobic treatment methods remove organic materials. If dissolved O_2 and proper environmental conditions are present, certain microorganisms utilize organic waste as food, converting it into simpler non-polluting compounds such as CO_2, nitrates, sulfates, and water. Aerobic biological treatment of industrial wastes requires for its successful performance a community of proper biological micro-organisms, food material, appropriate environment, and supply of dissolved O_2. In addition, the temperature also affects the metabolic activities of the microbes and plays a crucial role in determining the efficiency of the process.

Especially, the application of the microalgae for the bioremediation purpose have proved their potential as they are flexible and are capable in adapting themselves in several adverse toxic environments caused due to exposure to the industrial wastewater. Recent studies have successfully demonstrated the success of using algal cultures to remove some of the components from wastewater and use them as nutrients. Different types of physiological adaptation such as growth pattern of algae, rate of photosynthesis as well as other metabolic parameters can be used to evaluate the bioremediation potential of the algae after their treatment to industrial wastewater (Gupta & Bux, 2019).

11.3 ALGAL BIODIVERSITY AND ITS ROLE IN WASTEWATER TREATMENT

Microalgae are one group of photosynthetically active microorganisms which are abundant in nature, also can grow in both autotrophic and heterotrophic mode. The cultivation of the microalgae has two major goals such as: to generate platform for bioremediation of environmental pollutant as well as to generate renewal energy. Microalgae are now frequently used in wastewater treatment as they possess enormous potential for removal of N (nitrogen) and P (phosphorus) compounds from wastewater (Abdel-Raouf et al., 2012; Hammouda et al., 1995; Li et al., 2019) (Table 11.1). The algae possess unique characteristics such as high growth rate and productivity, high biomass content, and higher tolerance to toxic pollutants hence make it suitable for wastewater treatment as well as biofuel production. Selection of a specific microalga species to perform wastewater treatment are based on the knowledge about the presence of the algal species in wastewater (Aravantinou et al., 2013; Beuckels et al., 2015; Christenson & Sims, 2011). For example, wastewater from the agricultural industry consists of urea, ammonium, organic acids, phenolic compounds, and pesticides so this may be the constraint to algal growth when treating the wastewater. Travieso et al. (2008) designed an anaerobic fixed bed reactor in the algal pond system for the effective treatment of distillery wastewater, capable to remove about 80% of organic nitrogen, ammonia, and total phosphorus content. Hodaifa et al. (2008) treated industrial wastewater with the microalgae *Scenedesmus obliquus* to remove potassium salts and other minerals. Metals from the wastewater are basically taken up by algae through a physical adsorption method and subsequently slowly transported into the cytoplasm by the process known as chemisorption. Shehata et al. tested different concentrations of toxic substances such as copper, cadmium, nickel, zinc, and lead in the cultured *Scenedesmus* and observed the less toxic effect of nickel in comparison to copper for the growth of algae. Microalgae are capable of growing on a number of carbon compounds. Industrial wastewater also rich in the several types of the organic compounds are utilized in the growth process of the algae and

TABLE 11.1
Algal Cultures Used in Industrial Wastewater Treatment Processes

S. No.	Name of Algae	Type of Wastewater Contamination	Bioremediation Strategy	References
1	Chlamydomonas mexicana	Carbamazepine	Degradation	Xiong et al. (2016)
2	Chlorella vulgaris	Remazol dye removal (black, red, golden yellow)	Adsorption	Aksu and Tezer (2005)
3	Scenedesmus quadricauda	Pharmaceutical wastewater	Absorption	Vanerkar et al. (2015)
4	Lyngbya lagerlerimi	Dye: Orange II	Degradation	Vanerkar et al. (2015)
5	Chlorella vulgaris	Dye: G-Red	Degradation	El-Sheekh et al. (2009)
6	Chlorella vulgaris	Synthetic wastewater	Degradation	Xu et al. (2016)
7	Microalgae consortium	Wastewater containing emerging contaminants: 4-octylphenol, galaxolide, and tributyl phosphate concentrations	Degradation	Matamoros et al. (2016)
8	Oscillatoria sp. OSC Oscillatoria sp.	Oil compounds in wastewater	Degradation	Abed and Köster (2005)
9	Chlorella sorokiniana + Ralstonia basilensis	Toxic compounds	Degradation	Guieysse et al. (2002)
10	Consortium of microalgae and cyanobacteria	Industrial wastewater	Degradation	Van Den Hende et al. (2014)
11	Chlorella vulgaris + Lemna minuscula	Recalcitrant effluent	Degradation	Valderrama et al. (2002)

contribute to bioremediation process (Chen & Johns, 1991; Omar, 2002; Shehata & Badr, 1980; Yu & Wang, 2004).

11.3.1 CHALLENGES IN THE APPLICATION OF MICROALGAE WITH WASTEWATER TREATMENT

Although research for utilizing the microalgae with wastewater treatment process having many advantages, also some challenges are exists as described below (Amenorfenyo et al., 2019; Udaiyappan et al., 2017):

- The growth of the microalgae requires the utilization of certain micronutrients and supply of which may raise the overall cost of the process.
- The presence of toxic substances in most industrial wastewater may interfere with the photosynthetic process of the microalgae.

- In addition, the microalgae-based pond systems require larger land areas, therefore, it affects the design of the system especially in the urban areas, where obtaining the suitable land is limited and also land prices are high.

11.4 EXPERIMENTAL METHODS FOR ALGAL TREATMENT OF INDUSTRIAL WASTEWATER

The wastewater samples are collected preferably at an interval of 30 days during the year from inlet and outlet sources of industrial wastewater. Further several properties such as conductivity, temperature of the sample water, salinity and dissolved oxygen (DO), and the collected wastewater are calculated on-site using different probe systems/sensors. In addition, calculating other chemical properties like biochemical oxygen demand (BOD), chemical oxygen demand (COD), total solids, total suspended solids, and TDS are shown in Table 11.2.

TABLE 11.2
Experimental Methods for the Measurement of Different Properties of Industrial Wastewater

S. No.	Properties	Mode of Measurement
1	pH	Measured by using pH meter
2	Electrical conductivity and salinity	Measured by using electrical conductivity and salinity sensors
3	Total solids	Total solid (mg/mL) = (Final weight of the beaker containing residues in mg − initial weight of the beaker)/sample volume in mL
4	Total dissolved solids	Total dissolved solid (mg/mL) = (Final weight of the beaker containing residues in mg − initial weight of the beaker)/sample volume in mL
5	Total suspended solids	Total suspended solids (mg/ L) = Total solids (mg/L) − Total dissolved solids (mg/L)
6	Dissolved oxygen (DO)	Titration methods given by the formulae $DO (mg/L) = (8 \times 1,000 \times N) \times vV$ where V = sample volume taken (mL) v = titrant volume used N = normality of the titrant
7	Biological oxygen demand (BOD)	BOD (mg/L) = (dissolved oxygen of freshly prepared sample − dissolved oxygen of wastewater after 5 days incubation at 20°C)/ fraction of wastewater sample used for total combined volume
8	Chemical oxygen demand (COD)	$COD (mg/L) = 8 \times M$ (molarity of sodium thiosulfate) $\times (B − A) \times 1,000 \times S$ where M = concentration of the titrant A = titrant volume used for blank (mL) B = titrant volume used for sample (mL) S = water sample volume taken (mL)

11.4.1   MEDIA PREPARATION FOR MICROALGAE CULTIVATION FOR INDUSTRIAL WASTEWATER TREATMENT

Basically, the preparation of the media plays a major role in the growth of the desired algae to be used for the bioremediation of the industrial wastewater. Carbon, nitrogen, and phosphorous are three basic nutrients contained as the components of the media. Along with this, a small amount of the micronutrient is added in the media. The algae are photosynthetically active organism and utilize CO_2 (carbon dioxide) from the atmosphere to do photosynthesis, and subsequently release oxygen to the atmosphere. As it is related to growth of the algae, hence provision for supply of CO_2 is essential. In addition, nitrogen is the second most important nutrient essential for the growth of the algae besides the carbon. Usually, the nitrogen demand for the algal growth is variable in nature and depends upon type of the algal strain. The limitation of the nitrogen in the algal culture limitation is typically reflected by discoloration of the algae in the culture and also quantifying the accumulation of organic compounds. Another important compound is phosphorous act as essential component utilized for the growth and metabolic process of the microalgae. Phosphorous is usually externally supplied as orthophosphate (PO_4^{2-}). In addition to the above compounds the algal culture also requires small amounts of other substances known as micronutrients. Several important categories of compounds are treated as micronutrients include sulfur (S), sodium (Na), potassium (K), iron (Fe), magnesium (Mg), and calcium (Ca). Also many types of the trace elements are required for the culture including boron (B), copper (Cu), manganese (Mn), zinc (Zn), molybdenum (Mo), cobalt (Co), vanadium (V), and selenium (Se). These trace compounds play an important role in the catalyzation activity of the enzyme reactions and are responsible for biosynthesis of many compounds. After formulation and preparation of the medium for the growth of specific algae it, can be used further for effective removal of the nutrients from the wastewater by producing algal biomass. Several media have been formulated for the growth of the microalgae as described in Tables 11.3–11.5 (Dębowski et al., 2012; Grobbelaar, 2013; Şirin & Sillanpää, 2015; Sriram & Seenivasan, 2012).

TABLE 11.3
Composition of BG 11 Media

Compound	Quantity
$NaNO_3$	1 g
K_2HPO_4	0.04 g
$MgSO_4 \cdot 7H_2O$	0.075 g
$CaCl_2 \cdot 2H_2O$	0.036 g
Citric acid	0.006 g
Ferric ammonium citrate	0.006 g
EDTA (disodium salt)	0.001 g
Glucose	0.51g
Distilled water	1.0 L

TABLE 11.4

Composition of Fog's Medium

Compound	Quantity
Magnesium sulfate hepta hydrate ($MgSO_4 \cdot 7H_2O$)	0.2 g
Di-potassium hydrogen phosphate (K_2HPO_4)	0.2 g
Micronutrients solution	1 mL
Calcium chloride hydrated ($CaCl_2 \cdot H_2O$)	0.1 g
Fe-EDTA solution: In hot water 745 mg of Na_2 EDTA is dissolved and 557 mg of $FeSO_4 \cdot 7H_2O$ is added. Then the solution is boiled and the volume is made to 100 mL	5 mL
Distilled water	1 L
Agar (Difco)	12 g

TABLE 11.5

Composition of Micronutrient Solutions

Compound	Quantity
Hydrated manganese chloride ($MnCl_2 \cdot 4H_2O$)	181 mg
Boric acid	286 mg
Zinc sulfate hepta hydrate ($ZnSO_4 \cdot 7H_2O$)	22 mg
Sodium molybdate ($Na_2MoO_4 \cdot 2H_2O$)	39 mg
Copper sulfate pentahydrate ($CuSO_4 \cdot 5H_2O$)	8 mg
Distilled water	100 mL

11.4.2 CONSIDERATION OF FACTORS AFFECTING ALGAL TREATMENT TO WASTEWATER

Successful treatment of industrial wastewater with microalgae requires the understanding about the essential environmental factors for their growth. The enhanced growth rate of microalgae is influenced by several types of physical, chemical, and biological factors (Figure 11.1). Two types of systems such as open and closed one are followed for the cultivation of microalgae. The closed system is associated with the controlled growth and the open system is more dependent upon the external environmental factors such light, air, etc. Also, considering the economic point of view, the open systems of microalgae culture are having advantageous and viable (Larsdotter, 2006; Mohsenpour et al., 2020).

11.4.3 BIOREACTOR DESIGN AND COMPARISON WITH OTHER STUDIES

In general, efforts have been given in the selection process of specific algal strain and study about the development of its cultivation systems to make the process cost-effective as well as production efficient. Currently, the microalgae cultivation technology is conducted both at laboratory as well as commercial scales.

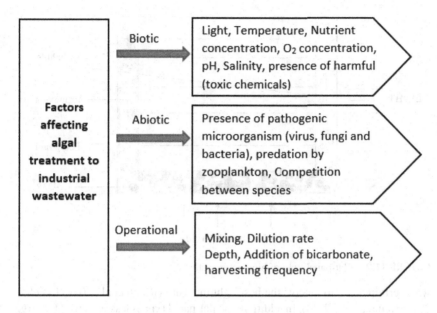

FIGURE 11.1   The factors considered for the algal treatment to wastewater.

However, the cultivation systems of microalgae are basically depends on the capital cost, purpose for which it is produced, cost for nutrient supply, and so on. Also, the microalgae cultivation can be done either in the open systems (ponds) or in the closed control systems called as photobioreactors (PBRs), designed as per the requirement. As far as of the algae-based wastewater treatment is concerned, mostly the open systems are preferred as they provide easy in the operation system and require low investment (Acién et al., 2012; Chisti, 2007; Klinthong et al., 2015; Xu et al., 2009).

During the treatment process of industrial wastewater the microalgae absorbs essential nutrients and release the free oxygen by their photosynthetic activities that are subsequently utilized by the other aerobic microorganisms. Commonly, the ponds and PBR systems are being utilized in microalgae-based wastewater treatment purpose. The different types of reactor systems that are available for the treatment of wastewater are given below (Ting et al., 2017).

11.4.3.1   Flat-Panel System for Industrial and Municipal Wastewater Treatment

A flat-panel system is a common form of PBR that appears as a rectangular box used for the cultivation of algae in wastewater. The composition of flat panels are made up of either transparent or semi-transparent materials such as glass, polycarbonate sheet, etc., that allows the light to easily penetrate to the algal culture (Figure 11.2).

Mixing in this type of bioreactor is done by the air bubbles that are generated from the air sparger. A pump is also used to facilitate the movement of air bubbles through the algal cell suspension. The reactor may be kept at an inclined position

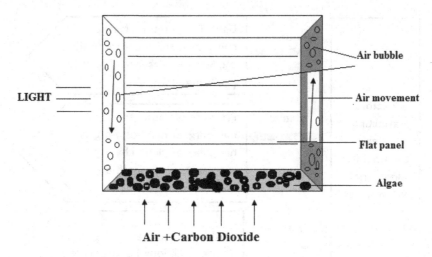

FIGURE 11.2   Flat-panel type of bioreactor systems.

with certain angle to receive the best light intensity (Riaño et al., 2011; Tan et al., 2014; Tuantet et al., 2014). In addition, the flat-panel type is less preferred for large-scale purposes due to the large area required to maintain the compartments in the reactor and also difficulty in addressing the problem arises due to biofouling, sterilization, controlling of temperature and aeration, etc. (Gupta et al., 2015; Ugwu, et al., 2008; Yu et al., 2009). Besides this, the flat-panel types of bioreactors are used for microalgal cultivation because of their advantages of large surface/volume ratio, short light/dark cycles, simple structure, and easy scale-up.

11.4.3.2   Tubular Reactor System for Industrial and Agricultural Wastewater Treatment

The tubular reactor is a design of horizontal tube PBRs and contains the largest surface to volume ratio for the maximum exposure of the microalgae to light. The construction of the tubular reactor systems is horizontal, inclined, spiral, helicoidal in shape with many variations (Figure 11.3). The range for the diameter of the

FIGURE 11.3   Tubular type of bioreactor systems.

tubes used in this type of bioreactors is from 10 mm to 60 mm and having length comprises of several hundred meters. The typical parameter of these types of the bioreactor is flow velocity of the culture and mixing efficiency.

The pump is used for continuous flowing of the culture liquid and finally the algal biomass can be harvested in the continuous mode of operation (Menke et al., 2012; Michels et al., 2014; Yang et al., 2008).

This reactor system can be designed in such a way that, it is able to use both natural and artificial light provided externally. Other features of the bioreactor are the variation of the hydrodynamic stress on the algae as well as the gas transfer to the culture as they depend on the flow characteristics and the air supply technique adopted. The coil types of the architecture in the bioreactor can be designed to enhance the efficiency of the growth of the microalgae due to its larger ratio of surface area to culture volume to receive maximum light intensity (Briassoulis et al., 2010; Wang et al., 2017; Yen et al., 2019).

114.3.3 Plastic Bag Reactors for Industrial and Agricultural Wastewater Treatment

The plastic bag bioreactors are a type of closed system bioreactor and having three basic components: plastic bags, frame for supporting the plastic, and an aeration systems. In wastewater treatment methods, these reactors are considered to be the cheapest as compared to other PBRs. One advantage of plastic bag bioreactors is the microalgae used are always pure without bacteria. Several types of plastic bag bioreactors have been developed for the microalgae cultivation purpose (Chinnasamy et al., 2010; Cicci et al., 2013; Zhu et al., 2014). The microalgal cultivation in this type of bioreactor depends upon temperature, whether indoor or outdoor, seasonal variations, location, bag dimensions, etc., and types of the algal strain. However, the algal production in plastic bag reactors is still under development stage. As importance of many other parameters such as implementation cost, monitoring the real productivity of the algae, nutrient supply cost as well as plastic bag replacing costs associated with the reactor are to addressed (Ting et al., 2017; Zittelli et al., 2013).

11.4.3.4 Cylinder Type Bioreactor System for Industrial Wastewater Treatment

In the cylinder type bioreactor systems, the microalgae remain in the suspended stage. This type of bioreactors is made up of transparent resins and borosilicate glass. The cylinder types of PBR contain two types of setup such as bubbling type and airlift style to facilitate the microalgae to grow. These bioreactor are classified into two main types as per the architecture is concerned. One contains internal loop type vessel design in which the air is sparged inside through the tubular structure and further exit to the outer zone. The other types contain an external loop that facilitates the air circulation through the separate tube. One of the important features of this bioreactor is effective mass transfer between the gas and liquid and having circular mixing nature. This facilitates the algae to utilize the sunlight and air by the microalgae thereby enhancing their photosynthetic efficiency. However, major drawback is complexity in operation as well as maintenance cost (Barbosa et al., 2003; Bharathiraja et al., 2015; Janssen et al., 2003).

11.5 ADVANTAGES OF MICROALGAE WASTEWATER TREATMENT OVER CONVENTIONAL TREATMENT

In comparison to the other conventional type of wastewater treatment methods, the microalgae-based methods offer many advantages hence implemented widely in the process.

11.5.1 Nutrient Removal

Several forms of nutrient such as ammonia (NH_4), nitrite nitrate, and orthophosphate are removed as it is utilized as the algal food sources, thereby helps to remediate the nitrogen from the industrial wastewater and avoids the nitrogen pollution.

11.5.2 Pathogen Removal

Dissolving of the carbon dioxide, as well as the H^+ ions by the wastewater, enhances the formation of bicarbonate. This ultimately increases the pH (up to 9.4) of the wastewater and this change of pH is able to eradicate the several pathogenic strains of *E. coli* and other pathogenic species.

11.5.3 Aeration by Photosynthesis

The photosynthetic process of the algae helps in the decrease of BOD and COD of the wastewater also consumes the CO_2.

11.5.4 Heavy Metal Removal

Several strains of algae are efficient for absorption of heavy metals.

11.5.5 Decreased Energy Requirements and Cost

The cultivated microalgae with wastewater treatment has been proved as an efficient process and suitably used for enhancing the overall efficiency and lowering cost.

11.5.6 Useful Biomass Production

The microalgae culture is responsible for diversity of products and that can be used in medical applications and production of biofuels and other substances.

11.6 CONCLUSIONS

Microalgae-based technology have a great potential to address the bioremediation of industry-based wastewater to combat the water-based pollution problem that occurs globally. Recently, implementation of several modes of cultivation strategy of the microalgae and their treatment strategy to industrial wastewater are widely studied by researchers. Being a photosynthetic microorganism, the microalgae have shown potential in remediation of wastewater by using them in different cultivation methods. Microalgae-based industrial wastewater treatment is considered as one of the efficient methods for bioremediation in comparison to the other conventional method. This chapter deals with the methods for the quality measurement of the industrial wastewater and algal treatment methods. In addition, the experimental methods involved in industrial wastewater quality measurement, microalgae diversity in the treatment methods, environmental factors to be considered during treatment methods, and important bioreactor systems with their advantage and disadvantage have been discussed. Four types of bioreactors such as flat panel, tubular, plastic bag PBRs, and cylindrical have been discussed for their large-scale utilization for algal growth and their unique features. Since each of these reactors have its own *pros* and *cons*, so new hybrid type of bioreactor development is desirable in order to improve the bioremediation performance of wastewater. Although the microalgae-based bioremediation strategies of industrial wastewater method are capable of removal of toxic pollutants but technologically the process is limited in terms of simultaneous removal of odor and color. Hence other technologies should be combined in order to enhance the efficiency of the process.

REFERENCES

Abdel-Raouf, N., Al-Homaidan, A. A., & Ibraheem, I. B. M. 2012. Microalgae and wastewater treatment. *Saudi Journal of Biological Sciences*, 19:257–275.

Abed, R. M., & Köster, J. 2005. The direct role of aerobic heterotrophic bacteria associated with cyanobacteria in the degradation of oil compounds. *International Biodeterioration & Biodegradation*, 55:29–37.

Acién, F. G., Fernández, J. M., Magán, J. J., & Molina, E. 2012. Production cost of a real microalgae production plant and strategies to reduce it. *Biotechnology Advances*, 30:1344–1353.

Aksu, Z., & Tezer, S. 2005. Biosorption of reactive dyes on the green alga *Chlorella vulgaris. Process Biochemistry*, 40:1347–1361.

Amenorfenyo, D. K., Huang, X., Zhang, Y., Zeng, Q., Zhang, N., Ren, J., & Huang, Q. 2019. Microalgae brewery wastewater treatment: Potentials, benefits and the challenges. *International Journal of Environmental Research and Public Health*, 16: 1910.

Aravantinou, A. F., Theodorakopoulos, M. A., & Manariotis, I. D. 2013. Selection of microalgae for wastewater treatment and potential lipids production, *Bioresource Technology*, 147:130–134.

Barbosa, M. J., Janssen, M., Ham, N., Tramper, J., & Wijffels, R. H. 2003. Microalgae cultivation in air-lift reactors: Modeling biomass yield and growth rate as a function of mixing frequency. *Biotechnology and Bioengineering*, 82:170–179.

Beuckels, A., Smolders, E., & Muylaert, K. 2015. Nitrogen availability influences phosphorus removal in microalgae-based wastewater treatment. *Water Research*, 77: 98–106.

Bharathiraja, B., Chakravarthy, M., Kumar, R. R., Yogendran, D., Yuvaraj, D., Jayamuthunagai, J., ... & Palani, S. 2015. Aquatic biomass (algae) as a future feed stock for bio-refineries: A review on cultivation, processing and products. *Renewable and Sustainable Energy Reviews*, 47:634–653.

Briassoulis, D., Panagakis, P., Chionidis, M., Tzenos, D., Lalos, A., Tsinos, C., ... & Jacobsen, A. 2010. An experimental helical-tubular photobioreactor for continuous production of *Nannochloropsis* sp. *Bioresource Technology*, 101:6768–6777.

Cai, T., Park, S. Y., & Li, Y. 2013. Nutrient recovery from wastewater streams by microalgae: Status and prospects. *Renewable and Sustainable Energy Reviews*, 19: 360–369.

Chen, F., & Johns, M. R. 1991. Effect of C/N ratio and aeration on the fatty acid composition of heterotrophic Chlorella sorokiniana. *Journal of Applied Phycology*, 3: 203–209.

Chinnasamy, S., Bhatnagar, A., Claxton, R., & Das, K. C. 2010. Biomass and bioenergy production potential of microalgae consortium in open and closed bioreactors using untreated carpet industry effluent as growth medium. *Bioresource Technology*, 101: 6751–6760.

Chisti, Y. 2007. Biodiesel from microalgae. *Biotechnology Advances*, 25: 294–306.

Christenson, L., & Sims, R. 2011. Production and harvesting of microalgae for wastewater treatment, biofuels, and bioproducts. *Biotechnology Advances*, 29: 686–702.

Cicci, A., Stoller, M., & Bravi, M. 2013. Microalgal biomass production by using ultra- and nanofiltration membrane fractions of olive mill wastewater. *Water Research*, 47: 4710–4718.

Dębowski, M., Zieliński, M., Krzemieniewski, M., Dudek, M., & Grala, A. 2012. Microalgae – cultivation methods. *Polish Journal of Natural Sciences*, 27: 151–164.

El-Sheekh, M. M., Gharieb, M. M., & Abou-El-Souod, G. W. 2009. Biodegradation of dyes by some green algae and cyanobacteria. *International Biodeterioration & Biodegradation*, 63: 699–704.

Grobbelaar, J. U. 2013. Mass production of microalgae at optimal photosynthetic rates. In *Photosynthesis* (pp. 357–371). InTech.

Guieysse, B., Borde, X., Muñoz, R., Hatti-Kaul, R., Nugier-Chauvin, C., Patin, H., & Mattiasson, B. 2002. Influence of the initial composition of algal-bacterial microcosms on the degradation of salicylate in a fed-batch culture. *Biotechnology Letters*, 24: 531–538.

Gupta, P. L., Lee, S. M., & Choi, H. J. 2015. A mini review: Photobioreactors for large scale algal cultivation. *World Journal of Microbiology and Biotechnology*, 31:1409–1417.

Gupta, S., & Bux, F. 2019. *Application of Microalgae in Wastewater Treatment*. Springer, Switzerland.

Hammouda, O., Gaber, A., & Abdelraouf, N. 1995. Microalgae and wastewater treatment. *Ecotoxicology and Environmental safety*, 31: 205–210.

Hodaifa, G., Martínez, M. E., & Sánchez, S. 2008. Use of industrial wastewater from olive-oil extraction for biomass production of *Scenedesmus obliquus*. *Bioresource Technology*, 99:1111–1117.

Janssen, M., Tramper, J., Mur, L. R., & Wijffels, R. H. 2003. Enclosed outdoor photobioreactors: Light regime, photosynthetic efficiency, scale-up, and future prospects. *Biotechnology and Bioengineering*, 81:193–210.

Klinthong, W., Yang, Y. H., Huang, C. H., & Tan, C. S. 2015. A review: Microalgae and their applications in CO_2 capture and renewable energy. *Aerosol and Air Quality Research*, 15: 712–742.

Larsdotter, K. 2006.Wastewater treatment with microalgae – A literature review. *Vatten*, 62:, 31.

Li, J., Li, J., Gao, R., Wang, M., Yang, L., Wang, X., ... & Peng, Y. 2018. A critical review of one-stage anammox processes for treating industrial wastewater: Optimization strategies based on key functional microorganisms. *Bioresource Technology*, 265:498–505.

Li, K., Liu, Q., Fang, F., Luo, R., Lu, Q., Zhou, W., ... & Ruan, R. 2019. Microalgae-based wastewater treatment for nutrients recovery: A review. *Bioresource Technology*, 291:121934.

Matamoros, V., Uggetti, E., García, J., & Bayona, J. M. 2016. Assessment of the mechanisms involved in the removal of emerging contaminants by microalgae from wastewater: A laboratory scale study. *Journal of Hazardous Materials*, 301: 197–205.

Menke, S., Sennhenn, A., Sachse, J. H., Majewski, E., Huchzermeyer, B., & Rath, T. 2012. Screening of microalgae for feasible mass production in industrial hypersaline wastewater using disposable bioreactors. *CLEAN – Soil, Air, Water*, 40:1401–1407.

Michels, M. H., Vaskoska, M., Vermuë, M. H., & Wijffels, R. H. 2014. Growth of *Tetraselmis suecica* in a tubular photobioreactor on wastewater from a fish farm. *Water Research*, 65: 290–296.

Mohsenpour, S. F., Hennige, S., Willoughby, N., Adeloye, A., & Gutierrez, T. 2020. Integrating micro-algae into wastewater treatment: A review. *Science of the Total Environment*, 142168.

Omar, H. H. 2002. Bioremoval of zinc ions by *Scenedesmus obliquus* and *Scenedesmus quadricauda* and its effect on growth and metabolism. *International Biodeterioration & Biodegradation*, 50:95–100.

Peters, R. W., Walker, T. J., Eriksen, E., Peterson, J. E., Chang, T. K., Ku, Y., & Lee, W. M. 1985. Wastewater treatment: Physical and chemical methods. *Journal (Water Pollution Control Federation)*, 57: 503–517.

Renuka, N., Sood, A., Ratha, S. K., Prasanna, R., & Ahluwalia, A. S. 2013. Evaluation of microalgal consortia for treatment of primary treated sewage effluent and biomass production. *Journal of Applied Phycology*, 25:1529–1537.

Riaño, B., Molinuevo, B., & García-González, M. C. 2011. Treatment of fish processing wastewater with microalgae-containing microbiota. *Bioresource Technology*, 102: 10829–10833.

Segneanu, A. E., Orbeci, C., Lazau, C., Sfirloaga, P., Vlazan, P., Bandas, C., & Grozescu, I. 2013. Waste water treatment methods. In *Water Treatment* (pp. 53–80). InTech.

Shehata, S. A., & Badr, S. A. 1980. Growth response of *Scenedesmus* to different concentrations of copper, cadmium, nickel, zinc, and lead. *Environment International*, 4: 431–434.

Şirin, S., & Sillanpää, M. 2015. Cultivating and harvesting of marine alga *Nannochloropsis oculata* in local municipal wastewater for biodiesel. *Bioresource Technology*, 191: 79–87.

Sonune, A., & Ghate, R. 2004. Developments in wastewater treatment methods. *Desalination*, 167:55–63.

Sriram, S., & Seenivasan, R. 2012. Microalgae cultivation in wastewater for nutrient removal. *Algal Biomass Utln*, 3:9–13.

Tan, X., Chu, H., Zhang, Y., Yang, L., Zhao, F., & Zhou, X. 2014. *Chlorella pyrenoidosa* cultivation using anaerobic digested starch processing wastewater in an airlift circulation photobioreactor. *Bioresource Technology*, 170: 538–548.

Ting, H., Haifeng, L., Shanshan, M., Zhang, Y., Zhidan, L., & Na, D. 2017. Progress in microalgae cultivation photobioreactors and applications in wastewater treatment: A review. *International Journal of Agricultural and Biological Engineering*, 10: 1–29.

Travieso, L., Benítez, F., Sánchez, E., Borja, R., León, M., Raposo, F., & Rincón, B. 2008. Assessment of a microalgae pond for post-treatment of the effluent from an anaerobic fixed bed reactor treating distillery wastewater. *Environmental Technology*, 29: 985–992.

Tuantet, K., Temmink, H., Zeeman, G., Janssen, M., Wijffels, R. H., & Buisman, C. J. 2014. Nutrient removal and microalgal biomass production on urine in a short light-path photobioreactor. *Water Research*, 55: 162–174.

Udaiyappan, A. F. M., Hasan, H. A., Takriff, M. S., & Abdullah, S. R. S. 2017. A review of the potentials, challenges and current status of microalgae biomass applications in industrial wastewater treatment. *Journal of Water Process Engineering*, 20:8–21.

Ugwu, C. U., Aoyagi, H., & Uchiyama, H. 2008. Photobioreactors for mass cultivation of algae. *Bioresource Technology*, 99:4021–4028.

Valderrama, L. T., Del Campo, C. M., Rodriguez, C. M., de-Bashan, L. E., & Bashan, Y. 2002. Treatment of recalcitrant wastewater from ethanol and citric acid production using the microalga *Chlorella vulgaris* and the macrophyte *Lemna minuscula*. *Water Research*, 36: 4185–4192.

Van Den Hende, S., Carré, E., Cocaud, E., Beelen, V., Boon, N., & Vervaeren, H. 2014. Treatment of industrial wastewaters by microalgal bacterial flocs in sequencing batch reactors. *Bioresource Technology*, 161: 245–254.

Vanerkar, A. P., Fulke, A. B., Lokhande, S. K., Giripunje, M. D., & Satyanarayan, S. 2015. Recycling and treatment of herbal pharmaceutical wastewater using *Scenedesmus quadricauda*. *Current Science*, 979–983.

Wang, Y., Ho, S. H., Cheng, C. L., Nagarajan, D., Guo, W. Q., Lin, C., ... & Chang, J. S. 2017. Nutrients and COD removal of swine wastewater with an isolated microalgal strain *Neochloris aquatica* CL-M1 accumulating high carbohydrate content used for biobutanol production. *Bioresource Technology*, 242:7–14.

Xiong, J. Q., Kurade, M. B., Abou-Shanab, R. A., Ji, M. K., Choi, J., Kim, J. O., & Jeon, B. H. 2016. Biodegradation of carbamazepine using freshwater microalgae Chlamydomonas mexicana and *Scenedesmus obliquus* and the determination of its metabolic fate. *Bioresource Technology*, 205: 183–190.

Xu, L., Weathers, P. J., Xiong, X. R., & Liu, C. Z. 2009. Microalgal bioreactors: Challenges and opportunities. *Engineering in Life Sciences*, 9:178–189.

Xu, Y., Wang, Y., Yang, Y., & Zhou, D. 2016. The role of starvation in biomass harvesting and lipid accumulation: Co-culture of microalgae–bacteria in synthetic wastewater. *Environmental Progress & Sustainable Energy*, 35: 103–109.

Yang, C. F., Ding, Z. Y., & Zhang, K. C. 2008. Growth of Chlorella pyrenoidosa in wastewater from cassava ethanol fermentation. *World Journal of Microbiology and Biotechnology*, 24:2919–2925.

Yen, H. W., Hu, I. C., Chen, C. Y., Nagarajan, D., & Chang, J. S. 2019. Design of photobioreactors for algal cultivation. In *Biofuels from Algae* (pp. 225–256). Elsevier.

Yu, G., Li, Y., Shen, G., Wang, W., Lin, C., Wu, H., & Chen, Z. 2009. A novel method using CFD to optimize the inner structure parameters of flat photobioreactors. *Journal of Applied Phycology*, 21: 719–727.

Yu, R. Q., & Wang, W. X. 2004. Biokinetics of cadmium, selenium, and zinc in freshwater alga *Scenedesmus obliquus* under different phosphorus and nitrogen conditions and metal transfer to *Daphnia magna*. *Environmental Pollution*, 129: 443–456.

Zhu, L., Hiltunen, E., Shu, Q., Zhou, W., Li, Z., & Wang, Z. 2014. Biodiesel production from algae cultivated in winter with artificial wastewater through pH regulation by acetic acid. *Applied Energy*, 128:103–110.

Zittelli, G. C., Biondi, N., Rodolfi, L., & Tredici, M. R. 2013. Photobioreactors for mass production of microalgae. *Handbook of Microalgal Culture: Applied Phycology and Biotechnology*, 2: 225–266.

12 The Enzymatic Treatment of Animal Wastewater and Manure

Ayesha Kashif,[1] Ayesha Batool,[2]
Ashfaq Ahmad Khan,[3] and
Muhammad Kashif Shahid[4]
[1]Department of Senior Health Care, Eulji
University, Daejeon, Republic of Korea
[2]Department of Chemistry, Quaid e Azam
University, Islamabad, Pakistan
[3]Department of Chemistry, Govt. Postgraduate
College, Haripur, KPK, Pakistan
[4]Research Institute of Environment &
Biosystem, Chungnam National University,
Daejeon, Republic of Korea

CONTENTS

12.1 INTRODUCTION

Since ancient times, animals have been played an essential role in human life either as a source of food or facilitator in traveling, trade, sports, festivals, and other aspects of life. The food and non-food utilization of animals is increasing day by day due to the growing population in every part of the world. Therefore, the animal farming is also expanded to meet the needs of time (Ramankutty et al. 2018). The rising number of animal farms and subsequently, higher production of chickens, goats, cattle, pigs, and aquaculture are responsible for several environmental challenges. These challenges include the management and treatment of animal wastewater and manure generated from animal farms, slaughterhouses, and meat processing facilities (Kunz, Miele, and Steinmetz 2009; Liu et al. 2021). The major origin of animal waste includes the

DOI: 10.1201/9781003201076-12

effluent of dairy and farm sheds (consisting of waste feed, residual milk, dung, urine, and cleaning water), renderings, poultry litter (a mixture of water, manure, spilled feed, bedding stuff, and feathers), daily manure, and other wastes from end processing of livestock. The poor management or uncontrolled movement of animal wastewater and manure in the environment can cause several infections and diseases in animals and humans since they carry numerous pathogens (Manyi-Loh et al. 2016).

Animal wastewater and manure carry essential nutrients, organic matter and other trace elements and used as fertilizer and irrigation to achieve several advantages such as carbon in rich soil, improved water holding capacity, higher infiltration rates, and greater cation-exchange capacity (Maillard and Angers 2014; Leno et al. 2021). However, it can compromise the security of environment due to the nutrients accumulation in soils causing possible groundwater and surface water contamination (Ndambi et al. 2019). Hence, beside the agricultural benefits of animal wastewater and manure, it is highly necessary to address their potential role in spread of pollution and pathogenic diseases. Animal manure harbors a range of pathogenic microbes up to different level and kinds depending on the animal species, feed, age, and health of the animal. The most common forms of pathogenic microbes found in animal manure include *Campylobacter* spp., *Salmonella* spp., *Yersinia enterocolitica*, *Listeria monocytogenes*, *Escherichia coli*, *Giardia lamblia*, and *Cryptosporidium parvum* (Manyi-Loh et al. 2016). These pathogens may spread into the environment through uncontrolled and unplanned discharge by runoff either from land application of manure or livestock premises. Another way of pathogens dissemination is their infiltration into the groundwater and soils.

Several physical, chemical, and biological processes are known for the treatment of animal wastewater and manure and subsequently, their conversion into worthwhile products such as organic fertilizers and biogas (Neshat et al. 2017). Generally, the biological processes are favored due to simple and cost effective process. The main challenge in the smooth operation of biological process is the presence of greases and fats in animal wastewater, which come from the slaughterhouses, dairy, and meat processing facilities (Bustillo-Lecompte and Mehrvar 2015). Therefore, pretreatment for greases and fats is essential prior to biological treatment of animal wastewater. Anaerobic digestion is widely known process for treatment of manure and its conversion into biogas (Neshat et al. 2017). However, the slowly degradable lignocellulose present in animal manure influences the production rate of biogas (Triolo et al. 2011).

The enzymatic treatment of animal wastewater and manure is a promising process to control the lignocelluloses and fats biotransformation in the ensuing bioprocesses to improve the productivity. Enzymes are known as biological molecules (usually proteins), which serve in accelerating the reaction rates for all the reactions happening within the cells (Gaytán, Burelo, and Loza-Tavera 2021). Nowadays, enzymes are also applied for catalyzing reactions outside the cells as well as for the industrial perspectives (Shahid et al. 2020). These molecules are essential for life and assist a numerous essential functions in the body (e.g., metabolism and digestion). Several studies reported the application of enzymatic bioprocesses in treatment of animal wastewater and manure. This chapter details a brief overview of enzymatic bioprocesses of enzymatic bioprocesses emphasizing the mechanisms and process conditions.

12.2   ENZYMATIC TREATMENT OF ANIMAL WASTEWATER

Animal wastewater carries greases and fats from slaughterhouses, dairies, and meat processing units that limit the efficiency of biological wastewater treatment. These fats and greases can alter the biological processes via accumulation onto the surface of sludge, lowering the transfer rate of oxygen to aerobic microbes, and substrate solutions to biomass. Furthermore, fat and grease matter can mitigate the activity of sludge and result in the growth of filamentous microbes that in turn affect sludge sedimentation. Hence, the hydrolysis of fat and grease matter in pretreatment is essential to enhance the efficiency of biological wastewater treatment.

In general, the wastewater carrying grease and fat matter is pretreated by several conventional physiochemical processes such as adsorption, filtration, membrane separation, coagulation, and flotation. However, the efficiency of these processes is conditioned with the composition of water such as emulsified and colloidal particles which are difficult to remove through these processes and also affect operational costs. Recently, an enzymatic process for the hydrolysis of grease and fat content achieved much attention (Rafiee and Rezaee 2021). Lipase is a widely applied enzyme in the degradation of fats and greases in animal wastewater. It is helpful for structure alteration of fats through catalyzing the discharge of simple fatty acids from complex triglycerides (Cheng et al. 2021). Lipases can be produced by solid-state fermentation or submerged fermentation. Solid-state fermentation is often favored due to the simple operation and excessive availability of low-cost substrates derived from agro-waste (e.g., soybean meal, coconut, wheat bran, etc.).

Fungal lipases obtained from solid-state fermentation have been effectively applied in treatment of wastewater obtained from slaughterhouses, poultry, dairy, and fish processing units. A study examined the efficiency of lipase enzyme in pretreatment of fish-processing effluent and found significant increase in removal efficiency (79.9–90.9%) for chemical oxygen demand (COD) when 0.2% w/v dosage of enzymes is applied to hydrolyze 1.5 g/L of oil and grease matter (Alexandre et al. 2011). Another study used *Penicillium* sp. lipase and achieved higher COD removal efficiency (32–90%) during treatment of dairy wastewater (1.2 g/L fat and grease content) (Rosa et al. 2009). Table 12.1 details the summary of several studies reported on the enzymatic degradation of grease and fat matter during treatment of animal wastewater.

Lipase-derived hydrolysis of triglycerides is accomplished via a ping-pong bi-bi mechanism that is a sequence of reactions, which produces intermediate monoglycerides and diglycerides (Chang, Chan, and Song 2021). The triglyceride and lipase binds together to make a lipase–triglyceride complex. Later, this complex transforms into an intermediate complex and glycerol through isomerization. Moreover, an intermediate complex interacts with three molecules of H_2O and makes a binary complex. Finally, the binary complex goes through unimolecular isomerization, where enzyme is regenerated and fatty acids are released (Liew et al. 2020). The activity of lipase significantly depends on the operation

TABLE 12.1

A Summary of Different Studies Reported on the Application of Lipases in Animal Wastewater Treatment

Source of Animal Wastewater	Grease and Fat Conc. (g/L)	Lipase Source	Lipase Dosage (w/v %)	Operational Conditions (pH; Temp.; Time)	Advantages	Reference
Poultry industry wastewater	1.2	*Penicillium restrictum*	0.1	7; 35°C; 22 h	An increased COD removal (53–85%) and biogas production (37–175 mL) after 4 d	(Valladão, Freire, and Cammarota 2007)
Swine meat processing wastewater	10	*Penicillium restrictum*	5	6; 45°C; 9–15 h	100.1 µmol of free acid/mL	(Rigo et al. 2008)
	10	Commercial Lipolase 100T	5	6; 37.5°C; 9–15 h	52.1 µmol of free acid/mL	(Rigo et al. 2008)
Dairy wastewater	1.2	*Pseudomonas aeruginosa* KM110	10	7; 30°C; 24 h	An increased COD removal (66–99%) and biogas production (2.33–4.71 L)	(Mobarak-Qamsari et al. 2012)
	1.2	*Penicillium* sp.	0.1	7; 30°C; 24 h	An increased COD removal (32–90%) and 8 fold increased free acid release	(Rosa et al. 2009)
Fish-processing unit	1.5	*Penicillium simplicissimum*	0.2	–; 30°C; 8 h	90.9% COD removal efficiency	(Alexandre et al. 2011)
wastewater	1.5	*Penicillium simplicissimum*	0.5	–; 30°C; 8 h	85% COD removal efficiency	(Alexandre et al. 2011)

conditions such as pH, temperature, concentration of the enzyme, and the hydrolysis. The optimum temperature is found between 30 and 50°C for the hydrolysis of grease and fat via lipase activity (Meng et al. 2017). The appropriate pH varies between 6 and 8, depending on the produces microbes. The hydrolysis time is another influential constraint in enzymatic degradation of fats and greases. A study reported sufficient hydrolysis of lipids in 24 h, whereas no noticeable change was observed in the hydrolysis rates by exceeding the hydrolysis time up to 36 h (Meng et al. 2017).

12.3 ENZYMATIC TREATMENT OF ANIMAL MANURE

The production of biogas from animal manure by anaerobic digestion can substitute fossil fuels in power generation, whereas the remaining digestate can be applied as a fertilizer to agricultural fields. However, the major limitation in production of biogas is the lignocellulose proportion in animal manure i.e., 30–80% (Triolo et al. 2011). A study reported 32% and 69% biodegradability of cattle and pig manure, respectively. The higher lignocellulose matter in cattle manure results in lower biodegradability as compared with pig manure.

Generally, anaerobic digestion accomplishes in four stages to generate the biogas from animal manure, which are hydrolysis, acidogenesis, acetogenesis, and methanogenesis. The product of one step functions as the substrates for the succeeding step, causing transformation of organic substance into biogas (Parawira 2012). The hydrolysis of lignin, cellulose, and hemicellulose in animal manure is generally a rate-restrictive stage that needs an active technique to improve the manure biodegradation and biogas generation. Hence, the pretreatment of animal manure is necessary to confirm the lignin, cellulose, and hemicellulose are easily available to microbes and increase hydrolysis and production of biogas.

In comparison with other physicochemical, chemical, microbial, and mechanical methods, enzymatic hydrolysis is favored due to its possessions of great adaptability and selectivity, ecofriendly, and low-cost operation. Table 12.2 summarizes the results of several studies reported on biogas generation by applying enzymes in the anaerobic digestion of animal manure. The enzyme activity and the proficiency of enzyme hydrolysis can be influenced by several constraints such as environmental conditions, system configuration, enzyme formulation, and adding techniques (Parawira 2012).

There are two ways of applying enzymes in treatment of animal manure including direct dosage of enzyme in digester, or application in pretreatment before the anaerobic digestion process. Several studies reported the higher production of biogas during application of enzymes in pretreatment (Sutaryo, Ward, and Møller 2014). In case of direct addition in the digester, the microbes can degrade the enzymes. The optimum temperature for the enzymatic treatment of animal manure is 35–50°C that is appropriate for the microbial growth. Moreover, an increase in temperature within optimal range, results in higher rate of hydrolysis and biogas production. A study highlighted that the reduction in hydrolysis rate by rising the temperature >60°C due to the lower activity of enzyme in result of enzyme denaturation (Quiñones et al. 2012).

12.4 CONCLUSION

Anaerobic processes are usually applied for the treatment of animal wastewater and manure and generate biogas. However, the productivity of the processes can be declined due to the grease and fat contents in animal wastewater and lignocellulose matter in animal manure. The application of enzymes can assist in mitigation of these limitations via hydrolysis of grease, fats, and lignocellulose contents

TABLE 12.2
A Summary of Different Studies Reported on the Application of Lipases in Animal Wastewater Treatment

Source of Animal Manure	Enzymes	Enzyme Addition	Temperature (°C)	Biogas Yields Increase (%)	Reference
Chicken manure	*Cellulase*	Direct addition in anaerobic digester	40	11.2 (maximum); 9.4 (after 60 d)	(Weide et al. 2020)
Solid cattle manure	*Mixture of protease, pectin esterase, xylanase, cellulase, hemicellulase, lipase, xylan esterase, pectinase, amylase, and glucosidase*	Pretreatment before anaerobic digestion	40	106.06	(Quiñones et al. 2012)
Horse manure	*Cellulase*	Direct addition in anaerobic digester	40	4.6 (maximum); −2.3 (after 60 d)	(Weide et al. 2020)
Cow manure and corn straw	*Protease*	Direct addition in anaerobic digester	37	1.47	(Wang et al. 2016)
	Cellulase	Pretreatment before anaerobic digestion	55	103.2	(Wang et al. 2016)
	Amylase	Direct addition in anaerobic digester	37	110.79	(Wang et al. 2016)
Dairy cattle manure	*Mixture of protease, cellulase, and pectate lyase*	Pretreatment before anaerobic digestion	50	4.44	(Sutaryo, Ward, and Møller 2014)
		Direct addition in anaerobic digester	35	4.15	(Sutaryo, Ward, and Møller 2014)

to simply degradable complexes. This enzymatic bioprocess can lead to higher COD removal rate in treatment of animal wastewater and enhanced biogas generation. The high price of available commercial enzymes is a major challenge in their application in animal wastewater and manure treatment on a large scale. It is necessary to explore the potential of low-cost substrates for production of enzymes. Agro-waste can be considered as a good option for solid-substrate fermentation.

ACKNOWLEDGMENT

This work was supported by Brain Pool Program through the National Research Foundation of Korea (NRF) funded by the Ministry of Science and ICT (Grant No.: 2019H1D3A1A02071191).

REFERENCES

Alexandre, V M F, A M Valente, Magali C Cammarota, and Denise M G Freire. 2011. "Performance of Anaerobic Bioreactor Treating Fish-Processing Plant Wastewater Pre-Hydrolyzed with a Solid Enzyme Pool." *Renewable Energy* 36 (12): 3439–44. doi: 10.1016/j.renene.2011.05.024.

Bustillo-Lecompte, Ciro Fernando, and Mehrab Mehrvar. 2015. "Slaughterhouse Wastewater Characteristics, Treatment, and Management in the Meat Processing Industry: A Review on Trends and Advances." *Journal of Environmental Management* 161: 287–302. doi: 10.1016/j.jenvman.2015.07.008.

Chang, Mun Yuen, Eng-Seng Chan, and Cher Pin Song. 2021. "Biodiesel Production Catalysed by Low-Cost Liquid Enzyme Eversa® Transform 2.0: Effect of Free Fatty Acid Content on Lipase Methanol Tolerance and Kinetic Model." *Fuel* 283: 119266. doi: 10.1016/j.fuel.2020.119266.

Cheng, Dongle, Yi Liu, Huu Hao Ngo, Wenshan Guo, Soon Woong Chang, Dinh Duc Nguyen, Shicheng Zhang, Gang Luo, and Xuan Thanh Bui. 2021. "Sustainable Enzymatic Technologies in Waste Animal Fat and Protein Management." *Journal of Environmental Management* 284: 112040. doi: 10.1016/j.jenvman.2021.112040.

Gaytán, Itzel, Manuel Burelo, and Herminia Loza-Tavera. 2021. "Current Status on the Biodegradability of Acrylic Polymers: Microorganisms, Enzymes and Metabolic Pathways Involved." *Applied Microbiology and Biotechnology* 105 (3): 991–1006. doi: 10.1007/s00253-020-11073-1.

Kunz, A, M Miele, and R L R Steinmetz. 2009. "Advanced Swine Manure Treatment and Utilization in Brazil." *Bioresource Technology* 100 (22): 5485–89. doi: 10.1016/j. biortech.2008.10.039.

Leno, Naveen, Cheruvelil Rajamma Sudharmaidevi, Gangadharan Byju, Kizhakke Covilakom Manorama Thampatti, Priya Usha Krishnaprasad, Geethu Jacob, and Pratheesh Pradeep Gopinath. 2021. "Thermochemical Digestate Fertilizer from Solid Waste: Characterization, Labile Carbon Dynamics, Dehydrogenase Activity, Water Holding Capacity and Biomass Allocation in Banana." *Waste Management* 123: 1–14. doi: 10.1016/j.wasman.2021.01.002.

Liew, Yuh Xiu, Yi Jing Chan, Sivakumar Manickam, Mei Fong Chong, Siewhui Chong, Timm Joyce Tiong, Jun Wei Lim, and Guan-Ting Pan. 2020. "Enzymatic Pretreatment to Enhance Anaerobic Bioconversion of High Strength Wastewater to Biogas: A Review." *Science of the Total Environment* 713: 136373. doi: 10.1016/j. scitotenv.2019.136373.

Liu, Chong, Xiaohua Li, Shunan Zheng, Zhang Kai, Tuo Jin, Rongguang Shi, Hongkun Huang, and Xiangqun Zheng. 2021. "Effects of Wastewater Treatment and Manure Application on the Dissemination of Antimicrobial Resistance around Swine Feedlots." *Journal of Cleaner Production* 280: 123794. doi: 10.1016/j. jclepro.2020.123794.

Maillard, Émilie, and Denis A Angers. 2014. "Animal Manure Application and Soil Organic Carbon Stocks: A Meta-Analysis."*Global Change Biology* 20 (2): 666–79. doi: 10.1111/gcb.12438.

Manyi-Loh, Christy E, Sampson N Mamphweli, Edson L Meyer, Golden Makaka, Michael Simon, and Anthony I Okoh. 2016. "An Overview of the Control of Bacterial Pathogens in Cattle Manure." *International Journal of Environmental Research and Public Health.* doi: 10.3390/ijerph13090843.

Meng, Ying, Fubo Luan, Hairong Yuan, Xue Chen, and Xiujin Li. 2017. "Enhancing Anaerobic Digestion Performance of Crude Lipid in Food Waste by Enzymatic Pretreatment." *Bioresource Technology* 224: 48–55. doi: 10.1016/j.biortech.2016.10.052.

Mobarak-Qamsari, E., R. Kasra-Kermanshahi, M. Nosrati, and T. Amani. 2012. "Enzymatic Pre-Hydrolysis of High Fat Content Dairy Wastewater as a Pretreatment for Anaerobic Digestion." *International Journal of Environmental Research (IJER)* 6 (2): 475–80.

Ndambi, Oghaiki Asaah, David Everett Pelster, Jesse Omondi Owino, Fridtjof de Buisonjé, and Theun Vellinga. 2019. "Manure Management Practices and Policies in Sub-Saharan Africa: Implications on Manure Quality as a Fertilizer." *Frontiers in Sustainable Food Systems.* https://www.frontiersin.org/article/10.3389/fsufs.2019.00029.

Neshat, Soheil A., Maedeh Mohammadi, Ghasem D. Najafpour, and Pooya Lahijani. 2017. "Anaerobic Co-Digestion of Animal Manures and Lignocellulosic Residues as a Potent Approach for Sustainable Biogas Production." *Renewable and Sustainable Energy Reviews* 79 (May): 308–22. doi: 10.1016/j.rser.2017.05.137.

Parawira, Wilson. 2012. "Enzyme Research and Applications in Biotechnological Intensification of Biogas Production." *Critical Reviews in Biotechnology* 32 (2): 172–86. doi: 10.3109/07388551.2011.595384.

Quiñones, Teresa Suárez, Matthias Plöchl, Jörn Budde, and Monika Heiermann. 2012. "Results of Batch Anaerobic Digestion Test–Effect of Enzyme Addition." *Agricultural Engineering International: CIGR Journal* 14 (1): 38–50.

Rafiee, F, and M Rezaee. 2021. "Different Strategies for the Lipase Immobilization on the Chitosan Based Supports and Their Applications." *International Journal of Biological Macromolecules* 179: 170–95. doi: 10.1016/j.ijbiomac.2021.02.198.

Ramankutty, Navin, Zia Mehrabi, Katharina Waha, Larissa Jarvis, Claire Kremen, Mario Herrero, and Loren H Rieseberg. 2018. "Trends in Global Agricultural Land Use: Implications for Environmental Health and Food Security." *Annual Review of Plant Biology* 69 (1): 789–815. doi: 10.1146/annurev-arplant-042817-040256.

Rigo, Elisandra, Roberta Eletízia Rigoni, Patrícia Lodea, Débora De Oliveira, Denise M G Freire, Helen Treichel, and Marco Di Luccio. 2008. "Comparison of Two Lipases in the Hydrolysis of Oil and Grease in Wastewater of the Swine Meat Industry." *Industrial & Engineering Chemistry Research* 47 (6): 1760–65. doi: 10.1021/ie0708834.

Rosa, Daniela R, Iolanda C S Duarte, N Katia Saavedra, Maria B Varesche, Marcelo Zaiat, Magali C Cammarota, and Denise M G Freire. 2009. "Performance and Molecular Evaluation of an Anaerobic System with Suspended Biomass for Treating Wastewater with High Fat Content after Enzymatic Hydrolysis." *Bioresource Technology* 100 (24): 6170–76. doi: 10.1016/j.biortech.2009.06.089.

Shahid, Muhammad Kashif, Ayesha Kashif, Prangya Ranjan Rout, Muhammad Aslam, Ahmed Fuwad, Younggyun Choi, Rajesh Banu J, Jeong Hoon Park, and Gopalakrishnan Kumar. 2020. "A Brief Review of Anaerobic Membrane Bioreactors Emphasizing Recent Advancements, Fouling Issues and Future Perspectives." *Journal of Environmental Management* 270 (June): 110909. doi: 10.1016/j.jenvman.2020.110909.

Sutaryo, Sutaryo, Alastair James Ward, and Henrik Bjarne Møller. 2014. "The Effect of Mixed-Enzyme Addition in Anaerobic Digestion on Methane Yield of Dairy Cattle Manure." *Environmental Technology* 35 (19): 2476–82. doi: 10.1080/09593330.2014.911356.

Triolo, Jin M, Sven G Sommer, Henrik B Møller, Martin R Weisbjerg, and Xin Y Jiang. 2011. "A New Algorithm to Characterize Biodegradability of Biomass during Anaerobic Digestion: Influence of Lignin Concentration on Methane Production Potential." *Bioresource Technology* 102 (20): 9395–9402. doi: 10.1016/j.biortech.2011.07.026.

Valladão, Alessandra Bormann Garcia, Denise Maria Guimarães Freire, and Magali Christe Cammarota. 2007. "Enzymatic Pre-Hydrolysis Applied to the Anaerobic Treatment of Effluents from Poultry Slaughterhouses." *International Biodeterioration & Biodegradation* 60 (4): 219–25. doi: 10.1016/j.ibiod.2007.03.005.

Wang, Xuemei, Zifu Li, Xiaoqin Zhou, Qiqi Wang, Yanga Wu, Mayiani Saino, and Xue Bai. 2016. "Study on the Bio-Methane Yield and Microbial Community Structure in Enzyme Enhanced Anaerobic Co-Digestion of Cow Manure and Corn Straw." *Bioresource Technology* 219: 150–57. doi: 10.1016/j.biortech.2016.07.116.

Weide, Tobias, Carolina Duque Baquero, Marion Schomaker, Elmar Brügging, and Christof Wetter. 2020. "Effects of Enzyme Addition on Biogas and Methane Yields in the Batch Anaerobic Digestion of Agricultural Waste (Silage, Straw, and Animal Manure)." *Biomass and Bioenergy* 132: 105442. doi: 10.1016/j.biombioe.2019.105442.

13 Application of Membrane Technology for Nutrient Removal/ Recovery from Wastewater

Muhammad Kashif Shahid,[1]
Ahmad Fuwad,[2] *and Younggyun Choi*[3]
[1]Research Institute of Environment &
Biosystem, Chungnam National University,
Daejeon, Republic of Korea
[2]Department of Mechanical Engineering,
Inha University, Incheon, Republic of Korea
[3]Department of Environmental & IT
Engineering, Chungnam National University,
Daejeon, Republic of Korea

CONTENTS

13.1 INTRODUCTION

This industrialization era has tremendous increases in water pollution, especially surface water pollution by excessive waste discharge causing a great threat to fresh water resource sustainability and environmental safety (Gao et al. 2018; Fuwad et al. 2019). Moreover, with an increase in the population, the global demand for fresh water and food increased exponentially (Piñero Eça et al. 2012;

DOI: 10.1201/9781003201076-13

Lee et al. 2021). Hence, in order to achieve more sustainable society, efficient waste removal and resource recovery technologies are necessary (Elser and Bennett 2011). Efficient water purification and nutrients recovery from wastewater leads to a more sustainable society on social, economic, and environmental aspects. Currently, the food security is the major challenge as there is an annual increase of 4% in fertilizer demand to feed increased number of people (Ledezma et al. 2015). Existing nutrients production already at its peak such as ammonia from fossil fuels, phosphorous from mining, and both of these major nutrients production mainly depends on the presence of finite natural resources (Rothausen and Conway 2011). Therefore, nutrient recovery from wastewater resources is an alternative approach to overcome the problems (Fowler et al. 2013).

In recent years, the advancement in technologies improves the nutrient recovery from surface wastewater but still there is high demand for the efficient nutrient recovery from wastewater. Several existing technologies are being employed for nutrient recovery including physiochemical processes (ion exchange, struvite precipitation, steam stripping, and membrane-based separations) and biological nutrient removal (BNR) processes (Liu et al. 2017; Rout et al. 2017; Yin, Wang, and Zhao 2018). Physiochemical processes such as precipitation, sorption and ion exchange, membrane removal are most commonly employed for phosphate removal (Shahid and Choi 2020). Whereas, BNR are most commonly employed for nitrogen and phosphate recovery from different processes like conventional activated sludge process (Khiewwijit et al. 2015), oxidation ditch (OD) (Batstone et al. 2015), biofilm reactor (Etter, Hug, and Udert 2013), single unit packed bed reactor (Rout et al. 2018), sequencing batch reactor (SBR) (Shahid, Kim, and Choi 2019), etc. but most of these processes have several limitation and are not feasible to use on commercial scale such as in ion exchange the removal efficiency decreases with media regeneration (Tarpeh et al. 2018), similarly, steam stripping cause a lot of environmental issues by emitting large amounts of greenhouse gases (Zeng, Mangan, and Li 2006). For struvite precipitation, specific concentration of phosphorous is required in the feed that is 100 mg/L which is difficult to maintain as normal wastewater only contains 5–50 mg/L. Moreover, the presence of heavy metal and toxic compounds also affects the struvite purity (Muhmood et al. 2018).

Conventional membrane-based processes, such as nanofiltration (NF) and reverse osmosis (RO) are most efficient processes for high recovery of nutrients like ammonium, phosphate, potassium, and urea from variety of wastewater (Niewersch et al. 2014; Shahid, Pyo, and Choi 2017a). But these systems are easily prone to fouling due to the presence of different bio-materials in the wastewater, which fouled the membrane under high applied pressure, which increases the process cost by decreasing the membrane lifetime, increasing its maintenance cost and hence decreasing the process productivity (Shahid et al. 2020).

Development of novel membrane processes forward osmosis (FO), membrane distillation (MD), and electrodialysis (ED) increases the membrane-based nutrient recovery productivity due to less prone to membrane fouling, hence low energy consumption and running cost. These technologies can easily achieve high purity of nutrient, their selectivity can be tuned according to the requirement and most importantly they are more energy efficient compared to conventional pressure-driven technologies. A brief overview of each of these three processes is explained in the following sections.

13.2   FORWARD OSMOSIS

Forward osmosis (FO) is the most promising technology in water purification as well as nutrient recovery. It works on the nature's principle that is osmotic pressure difference, the molecules pass through the semipermeable membrane due do concentration difference between two solutions that is feed solution and draw solution. Therefore, the osmotic pressure is a major driving force in FO as compared to the hydraulic pressure in the pressure-driven processes (Roy et al. 2016). This gives an advantage of less fouling of membrane, more specifically, the fouling is irreversible, which decreases the energy consumption as well as the low running and maintenance cost. In nutrients recovery, different kind of source or feed was utilized to study the performance of FO system, such as urine (J. Zhang et al. 2014), municipal wastewater (Xue et al. 2015), sludge (Xie et al. 2014), dairy, and animal wastewater (Wu et al. 2018). Figure 13.1 shows the schematic illustration of the process.

In FO systems, different kinds of draw solutions were used for nutrient recovery including fertilizer drawn FO (FDFO) systems, used to recover fertilizer and can be directly used for irrigation or agricultural purposes. In these FDFO, numerous draw solutions were used including ammonium phosphate monobasic ($NH_4H_2PO_4$) (Jafarinejad 2021), ammonium sulfate (($NH_4)_2SO_4$) (Chekli

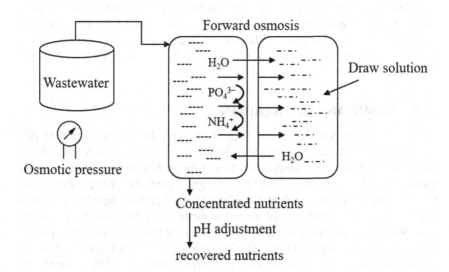

FIGURE 13.1   The schematic illustration of the FO process for nutrient recovery. The nutrients were concentrated in the feed solution, which is then further treated though different processes like chemical precipitation according to the applications or can be directly used.

et al. 2017), ammonium nitrate (NH_4NO_3) (Soler-Cabezas et al. 2018), potassium phosphate monobasic (KH_2PO_4) (Chekli et al. 2017), calcium nitrate ($Ca(NO_3)_2$) (Kim et al. 2017), ammonium chloride (NH_4Cl), sodium nitrate ($NaNO_3$) (Singh et al. 2019), urea (Volpin et al. 2019), seawater (Xue et al. 2015), etc. The experimental and numerical results showed better performance of FO membranes with above solutions. FO membranes in theoretical results, showed over 93% of nutrient recovery whereas, in experimental studies indicated 50–80% of ammonium recovery and over 90% of phosphate recovery.

Although FO membranes have promising future and better performance in pilot scale studies, it has several problems that limit its permeability and selectivity. Other than pH, temperature, and flow rate, the two most important factors that affect the FO membrane performance are (a) membrane material, and (b) draw solution. Ideal FO membrane should have high permeability, high rejection, minimum fouling, minimum concentration polarization, high chemical and mechanical stability (W. Xu, Chen, and Ge 2017; Haupt and Lerch 2018). Different approaches were used to improve the membrane materials such as surface modification (both chemical and physical modifications), and the addition of nanoparticles. But all the improvements come to a final point that is permeability-selectivity trade off (Xie et al. 2016). It is the permeability-selectivity trade off, the membrane with high permeability increase the water flux but on the expense of decreased membrane selectivity. This decline in selectivity increase reverse salt flux by increasing the movement of ions from draw solution and hence, overcoming this factor is crucial for membrane development.

The other major factor which affects the FO membrane performance is the quality of draw solution. Ideally a draw solution should have high osmotic pressure, low viscosity, low reverse salt flux, and high diffusion coefficient (Yan et al. 2018). An optimized draw solution with easy recovery, and non-toxic characteristics is the best for FO process (Jafarinejad 2021). Researchers are working to develop a specialized draw solution for FO processes to increase the efficiency of FO process.

13.3 MEMBRANE DISTILLATION

Unlike FO, membrane distillation (MD) is a thermally driven membrane process that can utilize low-grade heat to perform separation through hydrophobic membrane In MD, a vapor-liquid interface at the surface of the membrane pores allows vapors to pass and retain the rest of liquid. Due to the hydrophobic characteristic of microporous membrane, the liquid cannot pass through the membrane and a partial pressure difference across the membrane causes water vapor transport. Water is transported in vapor form (the purest form) and all the impurities are rejected in the feed solution. This process is considered highly economical as compared with other pressure-driven technologies (Alkhudhiri, Darwish, and Hilal 2012). Furthermore, it has high nutrient recovery as unlikely to FO, it does not depend on the concentration of nutrients in the feed or the osmotic pressure. Hence, the nutrients can be easily concentrated in the feed steam like inorganic nutrients can be separated in the feed and facilitates nutrients precipitation. The vapor pressure in the MD can be created through different methods on the permeate side, based on these factors the MD is divided into four major categories: (i) Direct contact MD: where a cooling water is used in the permeate side; (ii) Vacuum MD: a vacuum is created in the permeate side; (iii) Air gap MD: air is used in the permeate side; and (iv) Sweeping gas MD; an inert gas has been used for permeate side.

MD offers high potential to recover nutrients with both high volatility as well as low volatility. For example, ammonia has high volatility than water and can easily be concentrated in the permeate side. Moreover, MD configuration can be altered (vacuum, sweeping gas, or direct contact) according to the feed like wastewater, urine, swine manure (Thygesen et al. 2014; Zarebska et al. 2014). These processes show over 96% of ammonia recovery and can be used as fertilizer (Ahn, Hwang, and Shin 2011). Moreover, some studies reported enhanced ammonia recovery by capturing the ammonia vapors in the permeate side through stripping sulfuric acid, which increases the ammonia recovery to 99%. The process is illustrated in Figure 13.2.

Although MD offers great performance in nutrient recovery, still there are some challenges need to be addressed for enhancing process efficiency. Feed solution quality affects the MD performance such as volatile organic compounds (fatty acids) present in feed, cause an increase in the vapor pressure and thereby moves along with the water vapors to the permeate side, effecting the purity or selectivity of the nutrient recovery (Van Der Bruggen 2013; Meng et al. 2014). Furthermore, some of the contaminants in the feed (e.g., surfactants) decrease the surface tension and cause wetting of the membrane pores. This wetting of pores leads to the direct flow of the feed solution into permeate side effecting the permeate quality. Moreover the presence of colloids materials and dissolved organic

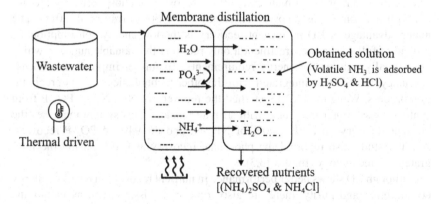

FIGURE 13.2   The schematic of the membrane distillation process. The high volatile ammonia was enriched on the permeate side with the membrane distillation process, for enhanced recovery acid stripping (HCl or H_2SO_4) was used. The resulting solution was then processed again to recover the nutrient.

matter in the feed stream leads to the biofouling of the membrane (Thygesen et al. 2014; Zarebska et al. 2014). To overcome these problems, studies are required on the improvement of the membrane material with high wettability properties such as super hydrophobic or omniphobic.

The superhydrophobic membrane increases the liquid entry pressure, which ultimately decreases the pore wettability and hence, decreases the fouling potential (Liao, Wang, and Fane 2014). On the other hand, omniphobic membranes repel both water as well as the low surface tension liquids such as surfactant (Lin et al. 2014). Different approaches have been reported to increase the MD membranes properties like incorporation of TiO_2 particles in the membrane to increase the liquid entry pressure. Similarly surface modification techniques were used to fabricate an omniphobic membrane by coating different nanoparticles such as silica nanoparticles, graphene, carbon nanotubes on the surface through polymer coating and surface functionalization (Lee et al. 2017). These approaches make it possible to attain new MD membrane with high mass transport and minimum wettability.

13.4   ELECTRODIALYSIS

Electrodialysis (ED) is an emerging electrochemical process for the separation of ions from wastewater. In ED, an alternating series of cation exchange membranes (CEMs) and anion exchange membranes (AEMs) are placed between anode and cathode terminals (T. Xu and Huang 2008). A potential gradient is generated under applied current which results in ion migration across the membranes. Under the influence of direct current, the cations and anions migrate toward the cathode and anode, respectively.

ED has already been utilized in water desalination, concentration of industrial fluids, and removal of minerals from industrial water. Recently, some lab scale studies showed high performance in nutrient recovery of phosphorus and ammonia with the continuous recovery efficiency of 38–50% (Tran et al. 2015). The major advantage of ED in nutrient recovery is the low energy consumption, as instead of full wastewater treatment it only recovers the valuable nutrient, which dramatically reduces the energy requirements. To further improve the process efficiency, the ED was integrated with the other technologies such as precipitation process. Wang et al. (2-13_studied the recovery of NH_4-N and PO_4-P from synthetic wastewater integrated with struvite reactor). This system increases the nutrient recovery up to 96–100% of NH_4-N, and 86–94% of PO_4-P recovery. Another study also reported the phosphate recovery as $CaPO_4$ using the integrated ED technology (Figure 13.3).

Although ED shows better performance in nutrient recovery in terms of energy consumption and purity, there are also some major issues such as membrane scaling or membrane fouling. Membrane fouling occurs due to the buildup of deposition layer during the operation, this decreases the efficiency of the process by increasing the membrane resistance, which leads to the decreased ion migration and selectivity (Shahid and Choi 2017). Fouling in ion exchange membranes

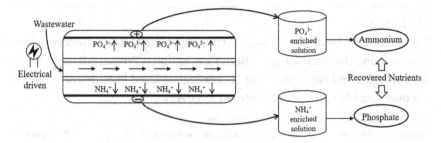

FIGURE 13.3 The schematic illustration of the electrodialysis-based system for nutrient recovery. The PO_4^{3-} ions and NH^{3+} ions can be concentrated at positive and negative terminals, respectively. Both the ions can be recovered with the permeate solution which can be further processed to remove nutrients from the permeate stream.

mainly depends on the presence of charges on the membrane surface, as the opposite charged particles in the solution attached to the membrane surface causing severe fouling. Major methods to avoid the fouling include reversing an electrode polarity after every short interval (Zhang et al. 2012), cleaning membranes with acidic and basic solutions (Shahid, Pyo, and Choi 2017b), optimizing the current density according to the feed, etc.

Membrane-based technologies are better in nutrient recovery as well as water purification compared to other techniques. The emerging membrane technologies show better performances as compared with conventional pressure-based techniques in terms of high nutrient recovery, low energy consumption, low maintenance and running cost, and easy to manage. A brief summary of membrane-based technologies for nutrient recovery is presented in Table 13.1.

13.5 CONCLUSIONS

Emerging membrane technologies discussed above (FO, MD, and ED) have promising future in wastewater treatment, especially for nutrient recovery. Each technology follows a unique principle for mass transfer and nutrient concentration. As in FO, absence of hydraulic pressure increases the membrane lifespan due to decrease in fouling, moreover, its product yield can be maximized by supplementing magnesium ion in a reverse salt flux system. In MD, the process is carried out by the vapor pressure difference, which assists to recover high volatile nutrients such as ammonia at high concentration factor for high quality fertilizer. ED can selectively recover different nutrients such as phosphate due to the use of electric current and specialized ion exchange membrane. Besides multiple advantages, all of these processes have limitations as discussed above. Hence, a substantial research is required to improve the membrane-based nutrient recovery processes for industrial and commercial application. The combination or integration of these processes with other technologies can also improve the process efficiency as well as selectivity. In-depth studies of these processes is necessary to evaluate their energy consumption, operating and maintenance cost, product quality, and dependence

TABLE 13.1

Comparison of Membrane-Based Technologies for Nutrient Recovery by Summarizing the Technologies (Both Conventional Pressure-Based Technologies and New Emerging Technologies) Used in Nutrient Recovery and Feed Used for Nutrient Recovery

Process	Feed Stream	Recovery	Operating Conditions	Remarks	Reference
Hydraulic pressure-driven process	Urine	Ammonium (70%), Phosphate (73%), and potassium (71%)	Hydraulic pressure: 50 bar	The process used RO technology, which shows low nutrient recovery due to membrane fouling and scaling	(Maurer, Pronk, and Larsen 2006)
Osmotic pressure-driven process (FO)	Swine wastewater	Ammonium (>93%), Phosphate (>99%)	Draw solution: 0.5 M $MgCl_2$	High nutrient recovery, reverse fouling occurs in the form of cake layer	(Wu et al. 2018)
Thermal-driven process (MD)	Biogas slurry	Ammonia (>98%)	Feed Temp.: 75°C; vacuum pressure: 10 kPa	The highly volatile NH_3 recovered with high purity; biofouling of membrane occurs due to the presence of solid particles in the feed	(He et al. 2018)
Electrical-driven process (ED)	Synthetic manure	Ammonium (78%), Phosphate (75%)	Current density: 2.7 A/cm^2	High purity nutrients were recovered, energy efficiency was 78%, deposited colloidal particles	(Mondor et al. 2008)

on other factors such as feed characteristics, temperature, and environmental factors including humidity.

DECLARATION OF COMPETING INTERESTS

The authors declare that they have no known competing financial interests or personal relationships that could have appeared to influence the work reported in this article.

ACKNOWLEDGMENT

This work was supported by Brain Pool Program through the National Research Foundation of Korea (NRF) funded by the Ministry of Science and ICT (Grant No.: 2019H1D3A1A02071191) and (NRF-2021R1I1A1A01060319).

REFERENCES

Ahn, Y. T., Y. H. Hwang, and H. S. Shin. 2011. "Application of PTFE Membrane for Ammonia Removal in a Membrane Contactor." *Water Science and Technology* 63 (12): 2944–48. doi:10.2166/wst.2011.141.

Alkhudhiri, Abdullah, Naif Darwish, and Nidal Hilal. 2012. "Membrane Distillation: A Comprehensive Review." *Desalination.* doi:10.1016/j.desal.2011.08.027.

Batstone, D. J., T. Hülsen, C. M. Mehta, and J. Keller. 2015. "Platforms for Energy and Nutrient Recovery from Domestic Wastewater: A Review." *Chemosphere* 140 (December): 2–11. doi:10.1016/j.chemosphere.2014.10.021.

Bruggen, Bart Van Der. 2013. "Integrated Membrane Separation Processes for Recycling of Valuable Wastewater Streams: Nanofiltration, Membrane Distillation, and Membrane Crystallizers Revisited." *ACS Publications* 52 (31): 10335–41. doi:10.1021/ie302880a.

Chekli, Laura, Youngjin Kim, Sherub Phuntsho, Sheng Li, Noreddine Ghaffour, Tor Ove Leiknes, and Ho Kyong Shon. 2017. "Evaluation of Fertilizer-Drawn Forward Osmosis for Sustainable Agriculture and Water Reuse in Arid Regions." *Journal of Environmental Management* 187 (February): 137–45. doi:10.1016/j.jenvman.2016.11.021.

Elser, James, and Elena Bennett. 2011. "Phosphorus Cycle: A Broken Biogeochemical Cycle." *Nature.* doi:10.1038/478029a.

Etter, Bastian, Alexandra Hug, and Kai M Udert. 2013. "Total Nutrient Recovery from Urine-Operation of a Pilot-Scale Nitrification Reactor."

Fowler, David, Mhairi Coyle, Ute Skiba, Mark A. Sutton, J. Neil Cape, Stefan Reis, Lucy J. Sheppard, et al. 2013. "The Global Nitrogen Cycle in the Twenty-first Century." *Philosophical Transactions of the Royal Society B: Biological Sciences* 368 (1621). doi:10.1098/rstb.2013.0164.

Fuwad, A., H. Ryu, N. Malmstadt, S.M. Kim, and T.-J. Jeon. 2019. "Biomimetic Membranes as Potential Tools for Water Purification: Preceding and Future Avenues." *Desalination* 458. doi:10.1016/j.desal.2019.02.003.

Gao, Feng, Yuan Yuan Peng, Chen Li, Wei Cui, Zhao Hui Yang, and Guang Ming Zeng. 2018. "Coupled Nutrient Removal from Secondary Effluent and Algal Biomass Production in Membrane Photobioreactor (MPBR): Effect of HRT and Long-Term Operation." *Chemical Engineering Journal* 335: 169–75. doi:10.1016/j.cej.2017.10.151.

Haupt, Anita, and André Lerch. 2018. "Forward Osmosis Application in Manufacturing Industries: A Short Review." *Membranes* 8 (3): 47. doi:10.3390/membranes8030047.

He, Qingyao, Te Tu, Shuiping Yan, Xing Yang, Mikel Duke, Yanlin Zhang, and Shuaifei Zhao. 2018. "Relating Water Vapor Transfer to Ammonia Recovery from Biogas Slurry by Vacuum Membrane Distillation." *Separation and Purification Technology* 191: 182–91. doi:10.1016/j.seppur.2017.09.030.

Jafarinejad, Shahryar. 2021. "Forward Osmosis Membrane Technology for Nutrient Removal/Recovery from Wastewater: Recent Advances, Proposed Designs, and Future Directions." *Chemosphere.* doi:10.1016/j.chemosphere.2020.128116.

Khiewwijit, Rungnapha, Hardy Temmink, Huub Rijnaarts, and Karel J. Keesman. 2015. "Energy and Nutrient Recovery for Municipal Wastewater Treatment: How to Design a Feasible Plant Layout?" *Environmental Modelling and Software* 68: 156–65. doi:10.1016/j.envsoft.2015.02.011.

Kim, Youngjin, Yun Chul Woo, Sherub Phuntsho, Long D. Nghiem, Ho Kyong Shon, and Seungkwan Hong. 2017. "Evaluation of Fertilizer-Drawn Forward Osmosis for Coal Seam Gas Reverse Osmosis Brine Treatment and Sustainable Agricultural Reuse." *Journal of Membrane Science* 537: 22–31. doi:10.1016/j.memsci.2017.05.032.

Ledezma, Pablo, Philipp Kuntke, Cees J.N. Buisman, Jürg Keller, and Stefano Freguia. 2015. "Source-Separated Urine Opens Golden Opportunities for Microbial Electrochemical Technologies." *Trends in Biotechnology* 33 (4): 214–20. doi:10.1016/j.tibtech.2015.01.007.

Lee, Eui Jong, Alicia Kyoungjin An, Pejman Hadi, Sangho Lee, Yun Chul Woo, and Ho Kyong Shon. 2017. "Advanced Multi-Nozzle Electrospun Functionalized Titanium Dioxide/Polyvinylidene Fluoride-Co-Hexafluoropropylene (TiO$_2$/PVDF-HFP) Composite Membranes for Direct Contact Membrane Distillation." *Journal of Membrane Science* 524: 712–20. doi:10.1016/j.memsci.2016.11.069.

Lee, Eunseok, Prangya Ranjan Rout, and Jaeho Bae. 2021. "The Applicability of Anaerobically Treated Domestic Wastewater as a Nutrient Medium in Hydroponic Lettuce Cultivation: Nitrogen Toxicity and Health Risk Assessment." *Science of the Total Environment* 780: 146482. doi:10.1016/j.scitotenv.2021.146482.

Liao, Yuan, Rong Wang, and Anthony G. Fane. 2014. "Fabrication of Bioinspired Composite Nanofiber Membranes with Robust Superhydrophobicity for Direct Contact Membrane Distillation." *Environmental Science and Technology* 48 (11): 6335–41. doi:10.1021/es405795s.

Lin, Shihong, Siamak Nejati, Chanhee Boo, Yunxia Hu, Chinedum O Osuji, and Menachem Elimelech. 2014. "Omniphobic Membrane for Robust Membrane Distillation." *ACS Publications* 1 (11): 443–47. doi:10.1021/ez500267p.

Liu, Tao, Bin Ma, Xueming Chen, Bing Jie Ni, Yongzhen Peng, and Jianhua Guo. 2017. "Evaluation of Mainstream Nitrogen Removal by Simultaneous Partial Nitrification, Anammox and Denitrification (SNAD) Process in a Granule-Based Reactor." *Chemical Engineering Journal* 327: 973–81. doi:10.1016/j.cej.2017.06.173.

Maurer, M., W. Pronk, and T. A. Larsen. 2006. "Treatment Processes for Source-Separated Urine." *Water Research.* doi:10.1016/j.watres.2006.07.012.

Meng, Suwan, Yun Ye, Jaleh Mansouri, and Vicki Chen. 2014. "Fouling and Crystallisation Behaviour of Superhydrophobic Nano-Composite PVDF Membranes in Direct Contact Membrane Distillation." *Journal of Membrane Science* 463: 102–12. doi:10.1016/j.memsci.2014.03.027.

Mondor, M., L. Masse, D. Ippersiel, F. Lamarche, and D. I. Massé. 2008. "Use of Electrodialysis and Reverse Osmosis for the Recovery and Concentration of Ammonia from Swine Manure." *Bioresource Technology* 99 (15): 7363–68. doi:10.1016/j.biortech.2006.12.039.

Muhmood, Atif, Shubiao Wu, Jiaxin Lu, Zeeshan Ajmal, Hongzhen Luo, and Renjie Dong. 2018. "Nutrient Recovery from Anaerobically Digested Chicken Slurry via Struvite: Performance Optimization and Interactions with Heavy Metals and Pathogens." *Science of the Total Environment* 635: 1–9. doi:10.1016/j.scitotenv.2018.04.129.

Niewersch, Claudia, A. L. Battaglia Bloch, Süleyman Yüce, Thomas Melin, and Matthias Wessling. 2014. "Nanofiltration for the Recovery of Phosphorus – Development of a Mass Transport Model." *Desalination* 346: 70–78. doi:10.1016/j.desal.2014.05.011.

Piñero Eça, Lilian, Débora Galdino Pinto, Ariene Murari Soares de Pinho, Marcelo Paulo Vaccari Mazzetti, and Marina Emiko YagimaOdo. 2012. "Autologous Fibroblast Culture in the Repair of Aging Skin." *Dermatologic Surgery* 38 (2 Part 1): 180–84. doi:10.1111/j.1524-4725.2011.02192.x.

Rothausen, Sabrina G.S.A., and Declan Conway. 2011. "Greenhouse-Gas Emissions from Energy Use in the Water Sector." *Nature Climate Change.* doi:10.1038/nclimate1147.

Rout, Prangya Ranjan, Puspendu Bhunia, and Rajesh Roshan Dash. 2017. "Simultaneous Removal of Nitrogen and Phosphorous from Domestic Wastewater Using *Bacillus cereus* GS-5 Strain Exhibiting Heterotrophic Nitrification, Aerobic Denitrification and Denitrifying Phosphorous Removal." *Bioresource Technology* 244 (2017): 484–495. doi:10.1016/j.biortech.2017.07.186.

Rout, Prangya Ranjan, Rajesh Roshan Dash, Puspendu Bhunia, and Surampalli Rao. 2018. "Role of *Bacillus cereus* GS-5 Strain on Simultaneous Nitrogen and Phosphorous Removal from Domestic Wastewater in an Inventive Single Unit Multi-layer Packed Bed Bioreactor." *Bioresource Technology* 262 (2018): 251–260. doi:10.1016/j. biortech.2018.04.087.

Roy, Dany, Mohamed Rahni, Pascale Pierre, and Viviane Yargeau. 2016. "Forward Osmosis for the Concentration and Reuse of Process Saline Wastewater." *Chemical Engineering Journal* 287 (March): 277–84. doi:10.1016/j.cej.2015.11.012.

Shahid, Muhammad Kashif, and Young-Gyun Choi. 2017. "The Comparative Study for Scale Inhibition on Surface of RO Membranes in Wastewater Reclamation: CO_2 Purging versus Three Different Antiscalants." *Journal of Membrane Science* 546: 61–69. doi:10.1016/j.memsci.2017.09.087.

Shahid, Muhammad Kashif, and Younggyun Choi. 2020. "Characterization and Application of Magnetite Particles, Synthesized by Reverse Coprecipitation Method in Open Air from Mill Scale." *Journal of Magnetism and Magnetic Materials* 495 (August 2019): 165823. doi:10.1016/j.jmmm.2019.165823.

Shahid, Muhammad Kashif, Ayesha Kashif, Prangya Ranjan Rout, Muhammad Aslam, Ahmed Fuwad, Younggyun Choi, Rajesh Banu J, Jeong Hoon Park, and Gopalakrishnan Kumar. 2020. "A Brief Review of Anaerobic Membrane Bioreactors Emphasizing Recent Advancements, Fouling Issues and Future Perspectives." *Journal of Environmental Management* 270 (June): 110909. doi:10.1016/j. jenvman.2020.110909.

Shahid, Muhammad Kashif, Yunjung Kim, and Young-Gyun Choi. 2019. "Magnetite Synthesis Using Iron Oxide Waste and Its Application for Phosphate Adsorption with Column and Batch Reactors." *Chemical Engineering Research and Design* 148: 169–79. doi:10.1016/j.cherd.2019.06.001.

Shahid, Muhammad Kashif, Minsu Pyo, and Young-gyun Choi. 2017a. "Carbonate Scale Reduction in Reverse Osmosis Membrane by CO_2 in Wastewater Reclamation." *Membrane Water Treatment* 8 (2): 125–36. doi:10.12989/mwt.2017.8.2.125.

Shahid, Muhammad Kashif, Minsu Pyo, and Young-Gyun Choi. 2017b. "The Operation of Reverse Osmosis System with CO_2 as a Scale Inhibitor: A Study on Operational Behavior and Membrane Morphology." *Desalination* 426: 11–20. doi:10.1016/j. desal.2017.10.020.

Singh, N., S. Dhiman, S. Basu, M. Balakrishnan, I. Petrinic, and C. Helix-Nielsen. 2019. "Dewatering of Sewage for Nutrients and Water Recovery by Forward Osmosis (FO) Using Divalent Draw Solution." *Journal of Water Process Engineering* 31: 100853. doi:10.1016/j.jwpe.2019.100853.

Soler-Cabezas, J. L., J. A. Mendoza-Roca, M. C. Vincent-Vela, M. J. Luján-Facundo, and L. Pastor-Alcañiz. 2018. "Simultaneous Concentration of Nutrients from Anaerobically Digested Sludge Centrate and Pre-Treatment of Industrial Effluents by Forward Osmosis." *Separation and Purification Technology* 193: 289–96. doi:10.1016/j.seppur.2017.10.058.

Tarpeh, William A., Ileana Wald, Maja Wiprächtiger, and Kara L. Nelson. 2018. "Effects of Operating and Design Parameters on Ion Exchange Columns for Nutrient Recovery from Urine." *Environmental Science: Water Research and Technology* 4 (6): 828–38. doi:10.1039/c7ew00478h.

Thygesen, Ole, Martin A.B. Hedegaard, Agata Zarebska, Claudia Beleites, and Christoph Krafft. 2014. "Membrane Fouling from Ammonia Recovery Analyzed by ATR-FTIR Imaging." *Vibrational Spectroscopy* 72: 119–23. doi:10.1016/j.vibspec.2014.03.004.

Tran, Ahn T.K., Yang Zhang, JiuYang Li, Priyanka Mondal, Wenyuan Ye, Boudewijns Meesschaert, Luc Pinoy, and Bart Van der Bruggena. 2015. "Phosphate Pre-Concentration from Municipal Wastewater by Selectrodialysis: Effect of Competing Components." *Separation and Purification Technology* 141: 38–47. doi:10.1016/j.seppur.2014.11.017.

Volpin, Federico, Huijin Heo, Md Abu Hasan Johir, Jaeweon Cho, Sherub Phuntsho, and Ho Kyong Shon. 2019. "Techno-Economic Feasibility of Recovering Phosphorus, Nitrogen and Water from Dilute Human Urine via Forward Osmosis." *Water Research* 150: 47–55. doi:10.1016/j.watres.2018.11.056.

Wang, Xiaolin, Yaoming Wang, Xu Zhang, Hongyan Feng, Chuanrun Li, and Tongwen Xu. 2013. "Phosphate Recovery from Excess Sludge by Conventional Electrodialysis (CED) and Electrodialysis with Bipolar Membranes (EDBM)." *Industrial and Engineering Chemistry Research* 52 (45): 15896–904. doi:10.1021/ie4014088.

Wu, Zhenyu, Shiqiang Zou, Bo Zhang, Lijun Wang, and Zhen He. 2018. "Forward Osmosis Promoted In-Situ Formation of Struvite with Simultaneous Water Recovery from Digested Swine Wastewater." *Chemical Engineering Journal* 342: 274–80. doi:10.1016/j.cej.2018.02.082.

Xie, Ming, Long D. Nghiem, William E. Price, and Menachem Elimelech. 2014. "Toward Resource Recovery from Wastewater: Extraction of Phosphorus from Digested Sludge Using a Hybrid Forward Osmosis-Membrane Distillation Process." *Environmental Science and Technology Letters* 1 (2): 191–95. doi:10.1021/ez400189z.

Xie, Ming, Ho Kyong Shon, Stephen R. Gray, and Menachem Elimelech. 2016. "Membrane-Based Processes for Wastewater Nutrient Recovery: Technology, Challenges, and Future Direction." *Water Research* 89: 210–21. doi:10.1016/j.watres.2015.11.045.

Xu, Tongwen, and Chuanhui Huang. 2008. "Electrodialysis-Based Separation Technologies: A Critical Review." *American Institute of Chemical Engineers AIChE J* 54 (12): 3147–59. doi:10.1002/aic.11643.

Xu, Wenxuan, Qiaozhen Chen, and Qingchun Ge. 2017. "Recent Advances in Forward Osmosis (FO) Membrane: Chemical Modifications on Membranes for FO Processes." *Desalination.* doi:10.1016/j.desal.2017.06.007.

Xue, Wenchao, Tomohiro Tobino, Fumiyuki Nakajima, and Kazuo Yamamoto. 2015. "Seawater-Driven Forward Osmosis for Enriching Nitrogen and Phosphorous in Treated Municipal Wastewater: Effect of Membrane Properties and Feed Solution Chemistry." *Water Research* 69: 120–30. doi:10.1016/j.watres.2014.11.007.

Yan, Tao, Yuanyao Ye, Hongmin Ma, Yong Zhang, Wenshan Guo, Bin Du, Qin Wei, Dong Wei, and Huu Hao Ngo. 2018. "A Critical Review on Membrane Hybrid System for Nutrient Recovery from Wastewater." *Chemical Engineering Journal.* doi:10.1016/j.cej.2018.04.166.

Yin, Qianqian, Ruikun Wang, and Zhenghui Zhao. 2018. "Application of Mg–Al-Modified Biochar for Simultaneous Removal of Ammonium, Nitrate, and Phosphate from Eutrophic Water." *Journal of Cleaner Production* 176: 230–40. doi:10.1016/j.jclepro.2017.12.117.

Zarebska, A., D. Romero Nieto, K. V. Christensen, and B. Norddahl. 2014. "Ammonia Recovery from Agricultural Wastes by Membrane Distillation: Fouling Characterization and Mechanism." *Water Research* 56: 1–10. doi:10.1016/j.watres.2014.02.037.

Zeng, L., C. Mangan, and X. Li. 2006. "Ammonia Recovery from Anaerobically Digested Cattle Manure by Steam Stripping." In *Water Science and Technology*, 54:137–45. doi:10.2166/wst.2006.852.

Zhang, Jiefeng, Qianhong She, Victor W.C. Chang, Chuyang Y. Tang, and Richard D. Webster. 2014. "Mining Nutrients (N, K, P) from Urban Source-Separated Urine by Forward Osmosis Dewatering." *Environmental Science and Technology* 48 (6): 3386–94. doi:10.1021/es405266d.

Zhang, Yang, Simon Paepen, Luc Pinoy, Boudewijn Meesschaert, and Bart Van der Bruggen. 2012. "Selectrodialysis: Fractionation of Divalent Ions from Monovalent Ions in a Novel Electrodialysis Stack." *Separation and Purification Technology* 88: 191–201. doi:10.1016/j.seppur.2011.12.017.

14 Photocatalytic Membrane Reactors (PMRs) for Wastewater Treatment
Photodegradation Mechanism, Types, and Optimized Factors

Rizwan Ahmad,[1] Ali Ehsan,[1] Imran Ullah Khan,[1] Muhammad Aslam,[2] and Prangya Ranjan Rout[3]
[1]Department of Chemical and Energy Engineering, Pak-Austria Fachhochschule: Institute of Applied Sciences & Technology (PAF-IAST), Haripur, Pakistan
[2]Department of Chemical Engineering, COMSATS University Islamabad (CUI), Lahore Campus, Lahore, Pakistan
[3]School of Energy and Environment, Thapar Institute of Engineering and Technology, Patiala, Punjab, India

CONTENTS

DOI: 10.1201/9781003201076-14

14.1 INTRODUCTION

The heterogeneous photocatalysis based upon the application of light photons and a semiconductor photocatalyst is an advanced oxidation process which is capable to generate reducing/oxidizing species. It has been the focus of numerous investigations and extensively researched in recent years after Fujishima and Honda firstly discovered the photocatalytic splitting of water on TiO_2 electrodes (Fujishima and Honda, 1972). Indeed when photocatalyst is excited by photons with energy equal to or higher than their band gap energy level, valence electrons are promoted from valence band to conduction band, resulting electrons and holes migrate to the surface of photocatalyst; and reduce and oxidize the adsorbed substrate, respectively (Ahmad et al., 2016). The key advantage of this process is that photocatalytic oxidation can be carried out under ambient temperature and organic pollutants are mineralized to CO_2, H_2O, and inorganic constituents (Athanasekou et al., 2014).

Among many semiconductor photocatalysts, there is a common consensus among researchers that titanium dioxide (TiO_2) is more superior because of its high activity, easily availability, inexpensive, non-toxic, and large stability to light illumination (Choi et al., 2007). However, heterogeneous photocatalysis is limited to large-scale applications because of key challenges such as high energy consumption and recovery of photocatalyst due to the difficulty of the infinitesimally small photocatalyst particles after photodegradation (Mozia, 2010). In order to overcome this issue, the hybrid processes integrating photocatalysis and membrane separation offer an exciting technology because both methods outperform and overcome the challenges of others. Currently, photocatalysis can be combined with membrane separation in two configurations: the reactors that use the immobilized/embedded photocatalyst membrane and a reactor in which catalyst is suspended in reaction mixture (Molinari et al., 2017). The photocatalytic membrane combined with built-in photocatalytic functionality has captivated the scientific attention during recent years (Ahmad et al., 2017d; Goei and Lim, 2014b). The membrane offers the *in situ* degradation of organic contaminants with enhanced membrane permeability flux and separation in one step thus minimizing the environmental and economic impact without compromising green chemistry principles. Furthermore, the photocatalyst immobilized membranes outperform the suspended type configuration, where catalyst recovery and fouling caused by the photocatalyst itself is a serious issue (Ahmad et al., 2017d).

The performance of these novel photocatalytic membranes largely depends upon the effective distribution of catalytic sites and morphological control of photocatalyst film. The homogenous distribution of photocatalyst sites has potential to enhance the activity and alleviate membrane fouling more effectively (Ahmad et al., 2017c,d). Based upon the type of the membrane fabrication and substrate pore size, varieties of membrane surface morphologies have been achieved. The efficient photocatalyst distribution on membrane surface as well as within the membrane pores has found to be more attractive than only surface immobilized photocatalytic membranes (Athanasekou et al., 2012; Horovitz et al., 2016). Membrane performance not only depends of morphological control, but the operational condition can also play an important role in defining the kinetics and the activity of the system (Mozia, 2010). Herein, we discussed the dependence of reaction kinetics on the operational condition of the reactor. The types of photocatalytic membranes along with their photodegradation mechanism were included. Finally, the most important factors affecting the performance were described.

14.2   MECHANISM AND KINETICS OF PHOTODEGRADATION

Figure 14.1 provides the conceptual mechanism of photocatalytic degradation associated with photocatalytic membranes. The light illumination to membranes caused the generation of electron and hole pair in conduction and valance band, respectively. The charge carriers can initiate redox reaction with the particles being adsorbed at the surface of photocatalyst. The study of the kinetics associated with the degradation of pollutants plays a key role while considering the effectiveness of operation. In the view of photocatalytic degradation, reaction kinetics provides a basic knowledge to optimize the conditions that would enhance the overall efficiency of configuration. Langmuir-Hinshelwood (L-H) model could be successfully applied for the degradation kinetics of many organic compounds. It assumes that adsorption plays a vital role on degradation efficiency and it also assumes that there is a saturation point beyond which adsorption become ineffective, so initial concentration has a key role while deciding the effectiveness of operation. The effect of photodegradation rate and the organic concentration as given by L-H model could be given as (Mendret et al., 2013):

$$r = -\frac{dC}{dt} = \frac{kKC}{1+KC} \tag{14.1}$$

where "r" denotes the initial decomposition rate (mg/L min), "t" is the time (min) for the illumination of UV light, "C" represents the concentration of reactants (mg/L),

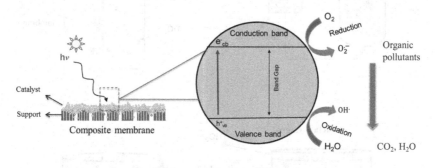

e⁻$_{cb}$ = Electron, h⁺$_{vb}$ = Hole

FIGURE 14.1   Illustration for photocatalytic activity on membrane surfaces.

"K" is the adsorption coefficient of the reactants (L/min), and "k" denotes reaction rate constant (mg/L min). The degradation kinetics for organic dye is considered to follow pseudo first-order kinetics. The rate equation for their degradation may be given below as (Choi et al., 2006):

$$\ln\left(\frac{C_0}{C_t}\right) = K_{app}t \tag{14.2}$$

where C_0 is the initial concentration of dye, C_t is the concentration at time t (min), while K_{aap} (min^{-1}) is the first-order rate constant.

Many researchers have reported that the photocatalytic degradation associated with photocatalytic membranes follows first-order kinetics (Mendret et al., 2013; Mozia, 2010), however contradictory finding has also been reported. Wang et al. (2008) investigated the degradation of Acid Red 4 and found that the kinetic order followed the order of 0.3. The authors concluded that light intensity and system configuration are important parameters in determining the rate constant for photocatalytic degradation (Wang et al., 2008). The rate of pollutant degradation depends upon adsorption of pollutant on catalyst surface, operational parameters such as temperature, pH, light intensity, and the presence ionic species. The intrinsic properties of photocatalyst also affects the kinetics of photocatalytic reactions (Mozia, 2010). The optimization of all the system variables is extensively required which has led to the successful application of photocatalytic membranes, which is promising.

14.3   TYPES OF PHOTOCATALYTIC MEMBRANE REACTORS

The photocatalytic membrane reactors (PMRs) are mainly categorized in different types such as suspended and immobilized type reactors. Figure 14.2 provides the

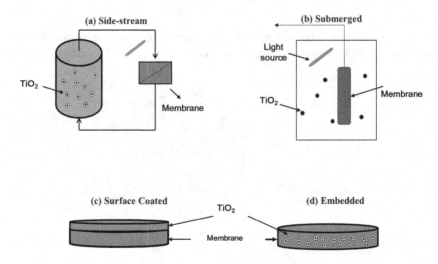

FIGURE 14.2   Types of photocatalytic membranes used in reactors.

schematic illustration of each configuration. The suspended type of configuration of the photodegradation and catalyst recovery occurs in a single chamber (Choo et al., 2008a,b). The suspended type of configuration can be further classified into two types, i.e., side-stream (Figure 14.2a) and submerged type (Figure 14.2b) membrane reactors (Choo, 2018). The catalyst suspension is maintained by stirring, cross-flow, or aeration, which hinder catalyst from deposition onto the membrane (Ryu et al., 2005). This configuration provides the benefits such as high interaction of pollutants with catalyst by providing high surface area. Furthermore, the catalyst concentration can be varied within the reactor. These features made this configuration promising for the application in large-scale applications. Nevertheless, despite the outstanding feature, the issues such as the membrane fouling by catalyst deposition on membrane, particularly in the presence of foulants, are serious challenges (Chong et al., 2010). The damage to membrane especially polymeric membranes by light irradiation is also a bottleneck in the full-scale development of this type. Furthermore, few studies have reported damage to the membrane during long-term operation of the reactor due to the abrasive effect of catalyst suspension (Mozia, 2010).

For immobilized reactors, two types of photocatalytic membranes have been widely utilized in reactors. Figure 14.2c,d presents the schematic illustration of membrane configurations. In surface immobilized types of the membranes (Figure 14.2c), the catalytic activity predominantly occurs on the surface. Particularly, when thick catalytic film with less porosity is deposited on membrane, the activity is only limited to the membrane surface (Ahmad et al., 2016). It has been revealed that the thin film with high porosity imparts advantageous attributes by providing high surface reactivity thus high anti-fouling characteristics. As the availability of high photocatalyst on membrane surface can improve catalytic activity and many researchers introduced multi-layer coatings on membrane surface. However, it has been suggested that high number of coatings can cause surface defects which pose detrimental effect on membrane performance, therefore, optimal number of coatings should be considered (Mendret et al., 2013). The deterioration in performance of membrane has been attributed to the less interaction of photocatalyst active layer with support membrane which leads to the failure of separation function (Choi et al., 2007). Therefore, the fabrication of defect-free surface coated membrane is of key importance. Choi et al. (2007) fabricated various PMs by changing the number of coatings on alumina substrate and found that the optimal number of coating should exist for better performance. Likewise, Mendret et al. (2013) also investigated the effect of multi-layer coating on the morphology and anti-fouling properties of the membrane. It was concluded that the six layers of coating was needed to have efficient membrane. The increase in coating from 6 to 8, 10 and 12 caused poor morphological control.

In embedded types (Figure 14.2d) of PMs, the photocatalyst is mostly immobilized in the membrane pore matrix. This type of configuration is used when polymeric membranes are used, and UV light photons penetrate the membrane matrix to cause in-pore photocatalytic activity. One of the main challenges in utilizing these membranes is the stability of membrane under UV light and concomitant attack of strong radicals with membrane materials.

14.4 TYPES OF MEMBRANE MATERIALS AND CONFIGURATIONS IN REACTORS

There are mainly two types of membrane materials, i.e., polymeric and ceramic membranes used in the reactors. The polymeric materials offer a cheap source for the development of photocatalytic membrane. Furthermore, the application of stable polymeric materials such as polyvinyl chloride (PVDF) provides benefit to use membrane in harsh environment where ultraviolet (UV) light irradiation and concomitant attack of the strong radicals can be omitted during short-term operation (Fischer et al., 2015). Other than PVDF, the materials such as polyethersulfone (PES) are another potential candidate for the preparation of catalytic membrane due to their hydrophilic nature (Xu et al., 2016). Nevertheless, many studies have suggested that the membrane can be damaged due to shearing effect of TiO$_2$ particles in suspension. Furthermore, it has been suggested that the aging of polymeric membrane materials can occur under long-term operation of the reactor under UV light irradiation (Mozia et al., 2014).

The ceramic membranes are expected to outperform polymeric membranes because of their excellent chemical and thermal resistance, high hydrophilicity, longer life, better fouling resistance, and ability to withstand the interaction of strong radicals as a result of photocatalytic activity (Hatat-Fraile et al., 2017). However, ceramic membranes are often expensive than the polymeric membrane, therefore, low-cost membrane should be developed (Ahmad et al., 2018). Currently, alumina (Al$_2$O$_3$) membranes have been widely used in PMRs. Both ceramic as well as the polymeric membranes have been utilized in PMRs in as flat sheet, tubular, and hollow fiber membranes. Figure 14.3 provides the schematic illustrations of different membrane configuration used in PMRs. Various configuration such as film coated (Figure 14.3a), nanotubes coated (Figure 14.3b), nanowires coated (Figure 14.3c), tubular coated (Figure 14.3d), and hollow fiber (Figure 14.3e) have been used in PMRs. Although, the film coated have been

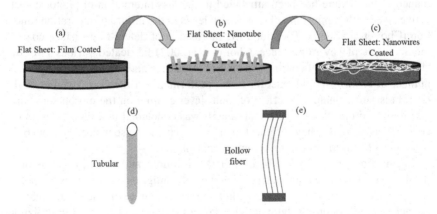

FIGURE 14.3 Various configurations of catalytic membranes; flat sheet (a) film coated; (b) nanotube coated; (c) nanowires coated; (d) tubular; and (e) hollow fiber membranes.

extensively investigated by many researchers, however, the hierarchical structures as shown in Figure 14.3b,c have received considerable attention during recent years (Ahmad et al., 2019).

14.5 FACTORS AFFECTING THE PERFORMANCE OF PMRS

There are various factors which have significant effect on operation and efficiency of photocatalytic membrane hybrid reactors. Sometimes many variables are dependent on each other, so change in one may lead to variation among others and it is necessary to understand the influence of each parameter on the performance of PMRs. The study of optimum conditions is necessary that governs the degradation kinetics of numerous kinds of pollutants. The factors that can contribute toward the variation in PMR performance are given below.

14.5.1 EFFECT OF pH CHANGE

The magnitude and nature of charge on membrane surface depends upon the pH of the solution to be filtered. The interaction between the surface group and ionic spices in solution decides the rejection or passage of the specie through membrane, the unlike charge may contribute to the high rejection while it can simultaneously cause high flux decline and irreversible fouling. Along with various parameters the strength of ion on membrane surface and pH are the governing parameter to be considered. The positive aspects of UV irradiation can be pronounced at acidic pH when the water contains anionic pollutants (Mendret et al., 2013) but acidic pH offers TiO_2 particles agglomeration thus decreasing surface area as well as the adsorption of pollutants on catalyst surface (Paz, 2006). Low pH (pH ~5) may also cause increase in flux of membrane by decreasing the turbidity within tank containing suspended photocatalyst (Damodar et al., 2012). Every photocatalyst have certain pH range where its point of zero charge (PZC) exists on its surface. The PZC of TiO_2 is pH 6.3. The catalyst surface occupies positive charge at pH lower than PZC and negative at higher pH and thus acting as source of electrostatic interaction for the pollutant species. While higher pH facilitates photocatalytic process by enhancing the amount of hydroxyl radicals. Transient behavior of zeta potential at different solution pH plays a vital role because electrostatic interactions between pollutant and photocatalyst can improve photocatalytic efficiency and membrane fouling (Mozia, 2010). In PMR due to photocatalytic reactions, large varieties of intermediates could be produced which has potential to adsorb on catalyst surface (Sleiman et al., 2007) and possess own characteristics therefore its behavior toward the membrane surface charge is totally different. Sometime a compound shows significant change in its physio-chemical properties while some has no effect toward pH change. The degradation kinetics of an organic compound is also dependent upon the surface charge of TiO_2, so extraordinary attention is required while dealing with wastewater that differs with respect to the pollutants present in it.

14.5.2   Light Intensity

Light intensity is an important factor in PMR's operations because light source intensity decides the quantity of photons that would be ejected by that source and thus controlling the rate of photocatalytic degradation. The rate of degradation increases by increasing the intensity of light source, however, much high intensity may result in loss of photons and making the operation less cost-effective. Higher light intensity has no effect on the rate of degradation but for intermediate light intensities, the rate of degradation is directly proportional to the square root of light intensity whereas in the case of low light intensity the rate could be considered to follow first-order kinetics (Mozia, 2010; Zhao & Zou, 2011; Zhao et al., 2013). Lower light intensity is incapable to excite the electron in valence shell thus limiting the activity due to low generation of electron-hole pair. However, the recombination phenomenon could be yielded due to high light intensities, which are considered to faster than the oxidation process and for this case the rate constant as well as reaction rate become constant regardless of increase in light intensity (Wang et al., 2008). Based upon the configuration of PMR, different light intensities have own characteristics, so it is never easy to conclude the optimum value of light intensity.

14.5.3   Residence Time

In PMRs, residence time affects the performance of operation at various concentrations of reactants. It has close relation with the photocatalytic activity as well as membrane separation process. Flux is the governing parameter that controls residence time. There exists an inverse relationship between filtration flux and residence time. Lower filtration flux allows high removal efficiency of pollutants because its facilities the greater contact of pollutants with photocatalyst surface and thus increasing the photocatalytic activity. In the case of cross-flow filtration, the low flux value (high residence time) also helps to maintain low TMP rise and decrease the retaining of photocatalyst on membrane surface (Chin et al., 2007).

14.5.4   Effect of Aeration

Aeration plays a vital role in submerged membrane photocatalytic reactor systems where it helps in maintaining homogenous suspension of photocatalyst, creating turbulent condition, and increasing the interaction between pollutant and photocatalyst. Air flow through the system has beneficial effect in terms of photocatalytic activity, as oxygen can act as an oxidant which can decrease the electron-hole recombination. However, the aeration rate beyond 0.5 L/min has no beneficial effect in terms of enhancing photocatalytic degradation; it is due to the fact that the transmission of UV light may be impaired due to the presence of bubble cloud (Chin et al., 2007).

Fu et al. (2006) also investigated that photodegradation of fulvic acid is more effective in acidic conditions at airflow of 0.06 m^3/h. Due to the shearing effect caused by the air flow the pollutants layer formed on the surface of membrane can be removed and this property enables system to run at high washing cycles (Hua et al., 2007). The mixing affects also facilities in minimizing concentration polarization. Aeration rate can also help in controlling the size of photocatalyst. In suspension photocatalyst particles tend to agglomerate and thus increasing the particle size. At higher rate aggregated particles are disintegrated as aeration rate has inverse relation with the particle size (Chin et al., 2007). Higher aeration rate not only controls particle size in fact the film mass transfer coefficient around the aggregate particles can also be increased.

14.5.5 EFFECT OF TRANSMEMBRANE PRESSURE

Transmembrane pressure (TMP) plays an important role during the operation of PMR. TMP acts as a driving force which compels feed solution to filter through the membrane. Rising TMP favors high permeate flux value, while photocatalytic activity may be decreased due to the acceleration of more solution through membrane, more contaminants are adsorbed on membrane and the effect of TMP become less prominent during later stages of filtration (Mozia, 2010). This is due to the fact that concentration polarization layer provides resistance to the solvent molecules and thus offering resistance to the flow of feed solution. Increased TMP also generates thick solute layer on membrane surface which facilitates convective transport, thus flux become independent of enhanced TMP and solely depends on mass transfer region. More severe flux decline could be observed, as enhanced TMP may form a compact cake layer on membrane surface and leading to high cake and fouling resistance (Mozia et al., 2014). As far as the photocatalytic activity is concerned, photocatalytic activity may also be decreased at higher pressure because the available active sites of photocatalyst and UV light photon availability could be reduced (Mozia, 2010). The optimum value of TMP may vary from one process to other however TMP should be maintained as low as possible.

14.6 MEMBRANE FOULING CONTROL AND TREATMENT EFFICIENCY

The inherent anti-fouling ability with high permeate flux is of highly importance in membranes for effective water treatment. The photocatalytic degradation of foulants on membrane surface has shown promising potential to reduce membrane fouling and permeate quality of PMs. The catalytic membrane surface offers extraordinary feature to selectively degrade the contaminants without compromising the anti-fouling properties of membrane. Nevertheless, the foulants accumulation on membrane surface attenuates the access of UV light photos and as pose detrimental effect of the performance of PMs. The fouling tendency is more pronounced when catalyst interacts with foulants to produce aggregate in suspended type configuration of PMRs (Erdei et al., 2008; Shon et al., 2008). Table 14.1 compares the treatment of different wastewaters in PMRs. Cake layer formation has been found the main reason for the deterioration of membrane permeate flux. Cross-flow filtration resulted in better control toward membrane fouling, however it has been observed that the degradation efficiency with dead-end filtration was higher as compared to cross-flow filtration (Wang et al., 2008). Horovitz et al. (2016) proposed that the indirect UV illumination system during dead end filtration

TABLE 14.1

Comparison of Removal Efficiency and Membrane Fouling during the Treatment of Various Wastewaters

PMRs Type	Types of Wastewater	Feed Composition	Membrane	Fouling Mechanism	Treatment Efficiency (%)	References
Suspended	Secondary treated effluent	Humic acid, beef extract, peptone, and tannic acid	Hollow fiber membrane	Aggregation on membrane	92	Erdei et al. (2008)
	Turbid water	Dihydroxybenzoic acid	Flat sheet membrane PAN 650, Flat sheet membrane PAI8 (LGC), Cellophane hollow fiber membranes	Deposition of pollutant on membrane	NA	Azrague et al. (2007)
	Pharmaceutical wastewater	Tamoxifen and gemfibrozil	Polyethersulfone polymer	Catalyst deposition and adsorption of pollutants	92	Erdei et al. (2008)
	Biologically treated sewage effluent	Beef extract, peptone, humic acid, and tannic acid	Flat-sheet PVC	Photocatalyst particles cake layer	90–97	Shon et al. (2008)
	Greywater	Shower gel, shampoo, bathroom cleaner, conditioner, hand soap, and bubble bath	Polyethylene	Pore blocking by large foulants-TiO_2 aggregates	80	Pidou et al. (2009)
	Biologically treated sewage effluent	Organic pollutants	Polyethylene	Organic fouling	80	Shon et al. (2008)
	Toxic organic wastewater	Trichloroethylene, $CaCl_2$, $NaHCO_3$, and HA	Polysulfone hollow fiber	Deposition of humic acids and TiO_2 particles	60	Choo et al. (2008a)
	Oily wastewater	Palm oil	Commercial nanofiltration membranes	Irreversible fouling	99.49 (COD)	Sidik et al. (2019)

(*Continued*)

TABLE 14.1   (Continued)
Comparison of Removal Efficiency and Membrane Fouling during the Treatment of Various Wastewaters

PMRs Type	Types of Wastewater	Feed Composition	Membrane	Fouling Mechanism	Treatment Efficiency (%)	References
Immobilized	Textile wastewater	Congo Red	Flat sheet α-Al$_2$O$_3$ ceramic membranes	Cake layer	95	Ahmad et al. (2017d)
	Oily wastewater	Oil contents	TiO$_2$/γ-Al$_2$O$_3$-modified ceramic membrane	Concentration polarization	90 (TOC)	Golshenas et al. (2020)
	Dye wastewater	Rhodamine B	Polyvinylidene fluoride (PVDF) membrane	Mainly irreversible fouling	53.29 (TOC)	Liu et al. (2020)
	Textile wastewater	Acid Orange 7 and Congo Red	Flat-sheet α-Al$_2$O$_3$ ceramic membranes	Cake layer	93.5	Ahmad et al. (2019)
	Dye wastewater	Methyl Orange	α-Alumina NF membrane tubes	No significant fouling	94–98	Romanos et al. (2012)
	Micro-polluted water	Humic acid	Alumina membranes	Cake layer and pore blocking	95.9	Zhang et al. (2008)
	Dyestuff effluent	Rhodamine B	TiO$_2$-coated membrane	Adsorption fouling of pollutant	95	Goei et al. (2013)
	Organic wastewater	Acid orange 7	Flat-sheet a-Al$_2$O$_3$ ceramic membranes	Accumulation of acid orange 7 on membrane	95	Mendret et al. (2013)
	Micro-polluted water	Humic acids	CNTs–TiO$_2$ composite membrane	Humic acid adsorption	85	Zhao et al. (2013)
	Campus sewage	Sewage	TiO$_2$ hollow fiber	Cake layer and pore blocking	90.2	Zhang et al. (2014)
	Organic wastewater	Phenol	Polypropylene macroporous membrane grafted poly (acrylic acid) and TiO$_2$	Cake layer of foulants and TiO$_2$	32.5	Yang et al. (2011)
	Textile wastewater	Congo Red	Flat sheet α-Al$_2$O$_3$ ceramic membranes	Cake layer	~90	Ahmad et al. (2020)

operation in which the permeate side of membrane was coated with photocatalyst and UV light was irradiated on the permeate side (Horovitz et al., 2016). This configuration allows the benefit such as the photocatalytic layer is less exposed to surface fouling and only smaller molecules than the pore size of membrane can reach the permeate side and consequently subjected to photodegradation (Horovitz et al., 2016). In other study, the light illumination on both sides of PM was investigated. High degradation efficiency was observed using this configuration, which can enhance the anti-fouling property of membrane. The authors concluded that the beneficial effect of external photocatalyst layer acting as a pre-treatment stage to the subsequent degradation on the internal pore surface during UV illumination on both sides of membrane (Athanasekou et al., 2012).

Recently, the application of different surface directing agents to control the structure of photocatalyst on membrane has been widely investigated (Ahmad et al., 2017a,b,d; Goei and Lim, 2014a; Zhao et al., 2014). The modified and photocatalyst decorated membranes prepared using sacrificial polymeric pore forming agent showed better performance than conventional photocatalytic membranes. The surface modification produce more rebinding cites and complementary cavities, which increase the adsorption of contaminants and their simultaneous degradation (Goei and Lim, 2014b; Zhao et al., 2014). The efficient photodegradation not only on membrane surface but also on membrane pores has further added the positive effect in enhancing anti-fouling and degradation efficiency. Despite of recent interest, the effect of different surface directing agents for the development of more efficient membranes need further investigations. It can be observed in Table 14.1 that most of the wastewaters considered in studies are synthetic. Nevertheless, with the use of unique reactor configurations, the researchers have successfully treated real wastewater.

14.7 CONCLUSIONS

The PMRs especially employing photocatalytic membranes have been used with interesting results in the photodegradation of organic pollutants as well as the anti-fouling characteristics. This chapter reviews the dependence of reaction kinetics on the operational condition of the reactor. A significant development has been reported to investigate the mechanism and kinetics of photodegradation associated with photocatalytic membranes to overcome all challenges with inexpensive alternatives. It has been attempted to solve engineering issues and provide a great flexibility for modifying the membrane configuration. In particular, type of membrane materials and their performance. It is an attempted to provide critical consideration for all the features which optimize the performance of photocatalytic membrane. A detail comparison of removal efficiency and membrane fouling during the treatment of various wastewaters has been described. However, more efforts are required to investigate the factors which affect the removal performance. Based on the discussion, it still requires many improvements before it can be routinely applied in industrial applications. Therefore, more efforts are needed to channel the knowledge gap between such new methods and large-scale operations.

REFERENCES

Ahmad, R., Ahmad, Z., Khan, A.U., Mastoi, N.R., Aslam, M. and Kim, J. 2016. Photocatalytic systems as an advanced environmental remediation: Recent developments, limitations and new avenues for applications: *Journal of Environmental Chemical Engineering* 4(4), 4143–4164.

Ahmad, R., Aslam, M., Park, E., Chang, S., Kwon, D., and Kim, J. 2018. Submerged low-cost pyrophyllite ceramic membrane filtration combined with GAC as fluidized particles for industrial wastewater treatment: *Chemosphere*, 206, 784–792.

Ahmad, R., Kim, J. K., Kim, J. H., and Kim, J. 2019. Diethylene glycol-assisted organized TiO_2 nanostructures for photocatalytic wastewater treatment ceramic membranes: *Water*, 11(4), 750.

Ahmad, R., Kim, J.K., Kim, J.H. and Kim, J. 2017a. Effect of polymer template on structure and membrane fouling of TiO_2/Al_2O_3 composite membranes for wastewater treatment: *Journal of Industrial and Engineering Chemistry* 57, 55–63.

Ahmad, R., Kim, J.K., Kim, J.H. and Kim, J. 2017b. In-situ TiO_2 formation and performance on ceramic membranes in photocatalytic membrane reactor: *Membrane Journal* 27(4), 328–335.

Ahmad, R., Kim, J.K., Kim, J.H. and Kim, J. 2017c Nanostructured ceramic photocatalytic membrane modified with a polymer template for textile wastewater treatment: *Applied Sciences* 7(12), 1284.

Ahmad, R., Kim, J.K., Kim, J.H. and Kim, J. 2017d. Well-organized, mesoporous nanocrystalline TiO_2 on alumina membranes with hierarchical architecture: *Antifouling and Photocatalytic Activities. Catalysis Today* 282(Part 1), 2–12.

Ahmad, R., Lee, C. S., Kim, J. H., & Kim, J. 2020. Partially coated TiO_2 on Al_2O_3 membrane for high water flux and photodegradation by novel filtration strategy in photocatalytic membrane reactors: *Chemical Engineering Research and Design*, 163, 138–148.

Alem, A., Sarpoolaky, H. and Keshmiri, M. 2009. Titania ultrafiltration membrane: Preparation, characterization and photocatalytic activity: *Journal of the European Ceramic Society* 29(4), 629–635.

Athanasekou, C.P., Morales-Torres, S., Likodimos, V., Romanos, G.E., Pastrana-Martinez, L.M., Falaras, P., Dionysiou, D.D., Faria, J.L., Figueiredo, J.L. and Silva, A.M.T. 2014. Prototype composite membranes of partially reduced graphene oxide/TiO_2 for photocatalytic ultrafiltration water treatment under visible light: *Applied Catalysis B: Environmental* 158(Supplement C), 361–372.

Athanasekou, C., Romanos, G., Katsaros, F., Kordatos, K., Likodimos, V. and Falaras, P. 2012. Very efficient composite titania membranes in hybrid ultrafiltration/photocatalysis water treatment processes: *Journal of Membrane Science* 392, 192–203.

Azrague, K., Aimar, P., Benoit-Marquie, F. and Maurette, M.T. 2007. A new combination of a membrane and a photocatalytic reactor for the depollution of turbid water: *Applied Catalysis B: Environmental*, 72(3–4), 197–204.

Banerjee, S., Pillai, S.C., Falaras, P., O'shea, K.E., Byrne, J.A. and Dionysiou, D.D. 2014. New insights into the mechanism of visible light photocatalysis: *The Journal of Physical Chemistry Letters*, 5(15), 2543–2554.

Chin, S. S., Lim, T. M., Chiang, K., and Fane, A. G. 2007. Factors affecting the performance of a low-pressure submerged membrane photocatalytic reactor: *Chemical Engineering Journal*, 130(1), 53–63.

Choi, H., Stathatos, E. and Dionysiou, D.D. 2006. Sol–gel preparation of mesoporous photocatalytic TiO_2 films and TiO_2/Al_2O_3 composite membranes for environmental applications: *Applied Catalysis B: Environmental* 63(1), 60–67.

Choi, H., Stathatos, E. and Dionysiou, D.D. 2007. Photocatalytic TiO$_2$ films and membranes for the development of efficient wastewater treatment and reuse systems: *Desalination* 202(1), 199–206.

Chong, M.N., Jin, B., Chow, C.W. and Saint, C. 2010 Recent developments in photocatalytic water treatment technology: A review. *Water Research* 44, 2997–3027.

Choo, Kwang-Ho. 2018. Modeling photocatalytic membrane reactors: In *Current Trends and Future Developments on (Bio-)Membranes*, 297–316. Elsevier.

Choo, K.H., Chang, D.I., Park, K.W. and Kim, M.H. 2008a. Use of an integrated photocatalysis/hollow fiber microfiltration system for the removal of trichloroethylene in water: *Journal of Hazardous Materials*, 152(1), 183–190.

Choo, K.H., Tao, R. and Kim, M.J. 2008b. Use of a photocatalytic membrane reactor for the removal of natural organic matter in water: Effect of photoinduced desorption and ferrihydrite adsorption: *Journal of Membrane Science* 322(2), 368–374.

Damodar, R. A., You, S. J., and Chiou, G. W. 2012. Investigation on the conditions mitigating membrane fouling caused by TiO$_2$ deposition in a membrane photocatalytic reactor (MPR) used for dye wastewater treatment: *Journal of Hazardous Materials*, 203, 348–356.

Damodar, R.A., You, S.J. and Ou, S.H. 2010. Coupling of membrane separation with photocatalytic slurry reactor for advanced dye wastewater treatment: *Journal of Separation and Purification Technology*, 76(1), 64–71.

Dominguez, S., Ribao, P., Rivero, M.J. and Ortiz, I. 2015. Influence of radiation and TiO$_2$ concentration on the hydroxyl radicals generation in a photocatalytic LED reactor. Application to dodecylbenzenesulfonate degradation: *Journal of Applied Catalysis B: Environmental*, 178, 165–169.

Erdei, L., Arecrachakul, N. and Vigneswaran, S. 2008. A combined photocatalytic slurry reactor–immersed membrane module system for advanced wastewater treatment: *Journal of Separation and Purification Technology*, 62(2), 382–388.

Fischer, K., Grimm, M., Meyers, J., Dietrich, C., Gläser, R., and Schulze, A. 2015. Photoactive microfiltration membranes via directed synthesis of TiO$_2$ nanoparticles on the polymer surface for removal of drugs from water. *Journal of Membrane Science*, 478, 49–57.

Fu, J., Ji, M., Wang, Z., Jin, L., and An, D. 2006. A new submerged membrane photocatalysis reactor (SMPR) for fulvic acid removal using a nano-structured photocatalyst: *Journal of Hazardous Materials*, 131(1–3), 238–242.

Fujishima, A. and Honda, K. 1972. Electrochemical photolysis of water at a semiconductor electrode: *Nature* 238(5358), 37–38.

Gao, Y., Hu, M. and Mi, B. 2014. Membrane surface modification with TiO$_2$–graphene oxide for enhanced photocatalytic performance: *Journal of Membrane Science* 455(Supplement C), 349–356.

Goei, R., Dong, Z. and Lim, T.T. 2013. High-permeability pluronic-based TiO$_2$ hybrid photocatalytic membrane with hierarchical porosity: Fabrication, characterizations and performances: *Chemical Engineering Journal*, 228, 1030–1039.

Goei, R. and Lim, T.-T. 2014a. Ag-decorated TiO$_2$ photocatalytic membrane with hierarchical architecture: photocatalytic and anti-bacterial activities: *Water Research* 59, 207–218.

Goei, R. and Lim, T.-T. 2014b. Asymmetric TiO$_2$ hybrid photocatalytic ceramic membrane with porosity gradient: Effect of structure directing agent on the resulting membranes architecture and performances: *Ceramics International*, 40(5), 6747–6757.

Golshenas, A., Sadeghian, Z., and Ashrafizadeh, S. N. 2020. Performance evaluation of a ceramic-based photocatalytic membrane reactor for treatment of oily wastewater: *Journal of Water Process Engineering*, 36, 101186.

Guo, B., Pasco, E.V., Xagoraraki, I. and Tarabara, V.V. 2015. Virus removal and inactivation in a hybrid microfiltration–UV process with a photocatalytic membrane: *Separation and Purification Technology* 149, 245–254.

Hatat-Fraile, M., Liang, R., Arlos, M. J., He, R. X., Peng, P., Servos, M. R., and Zhou, Y. N. 2017. Concurrent photocatalytic and filtration processes using doped TiO$_2$ coated quartz fiber membranes in a photocatalytic membrane reactor: *Chemical Engineering Journal*, 330, 531–540.

Horovitz, I., Avisar, D., Baker, M.A., Grilli, R., et al. 2016 Carbamazepine degradation using a N-doped TiO$_2$ coated photocatalytic membrane reactor: Influence of physical parameters: *Journal of Hazardous Materials* 310, 98–107.

Hua, F. L., Tsang, Y. F., Wang, Y. J., Chan, S. Y., Chua, H., and Sin, S. N. 2007. Performance study of ceramic microfiltration membrane for oily wastewater treatment: *Chemical Engineering Journal*, 128(2–3), 169–175.

Ismail, A.A., and Bahnemann, D.W. 2014. Photochemical splitting of water for hydrogen production by photocatalysis: A review. *Journal of Solar Energy Materials and Solar Cells*, 128, 85–101.

Jiang, H., Zhang, G., Huang, T., Chen, J., Wang, Q. and Meng, Q. 2010. Photocatalytic membrane reactor for degradation of acid red B wastewater: *Journal of Chemical Engineering*, 156(3), 571–577.

Kim, J. and Van der Bruggen, B, 2010. The use of nanoparticles in polymeric and ceramic membrane structures: review of manufacturing procedures and performance improvement for water treatment: *Environmental Pollution* 158(7), 2335–2349.

Lin, C.J., Yu, Y.H. and Liou, Y.H. 2009. Free-standing TiO$_2$ nanotube array films sensitized with CdS as highly active solar light-driven photocatalysts: *Journal of Applied Catalysis B: Environmental*, 93(1–2), 119–125.

Liu, T., Wang, L., Liu, X., Sun, C., Lv, Y., Miao, R., and Wang, X. 2020. Dynamic photocatalytic membrane coated with ZnIn$_2$S$_4$ for enhanced photocatalytic performance and antifouling property: *Chemical Engineering Journal*, 379, 122379.

Loddo, V., Augugliaro, V. and Palmisano, L. 2009. Photocatalytic membrane reactors: Case studies and perspectives: *Asia-Pacific Journal of Chemical Engineering*, 4(3), 380–384

Mendret, J., Hatat-Fraile, M., Rivallin, M. and Brosillon, S. 2013. Hydrophilic composite membranes for simultaneous separation and photocatalytic degradation of organic pollutants: *Separation and Purification Technology* 111(Supplement C), 9–19.

Molinari, R., Grande, C., Drioli, E., Palmisano, L. and Schiavello, M. 2001. Photocatalytic membrane reactors for degradation of organic pollutants in water: *Journal of Catalysis Today*, 67(1–3), 273–279.

Molinari, R., Lavorato, C. and Argurio, P. 2017. Recent progress of photocatalytic membrane reactors in water treatment and in synthesis of organic compounds: A review. *Catalysis Today* 281, 144–164.

Moustakas, N., Katsaros, F., Kontos, A., Romanos, G.E., Dionysiou, D. and Falaras, P. 2014. Visible light active TiO$_2$ photocatalytic filtration membranes with improved permeability and low energy consumption: *Catalysis Today* 224, 56–69.

Mozia, S. 2010. Photocatalytic membrane reactors (PMRs) in water and wastewater treatment: A review. *Separation and Purification Technology* 73(2), 71–91.

Mozia, S., Darowna, D., Orecki, A., Wróbel, R., Wilpiszewska, K., and Morawski, A. W. 2014. Microscopic studies on TiO$_2$ fouling of MF/UF polyethersulfone membranes in a photocatalytic membrane reactor: *Journal of Membrane Science*, 470, 356–368.

Paz, Y. 2006. Preferential photodegradation – Why and how? *Comptes Rendus Chimie*, 9(5–6), 774–787.

Pidou, M., Parsons, S.A., Raymond, G., Jeffrey, P., Stephenson, T. and Jefferson, B. 2009. Fouling control of a membrane coupled photocatalytic process treating greywater. *Water Research*, 43(16), 3932–3939.

Romanos, G.E., Athanasekou, C.P., Katsaros, F.K., Kanellopoulos, N.K., Dionysiou, D.D., Likodimos, V. and Falaras, P. 2012. Double-side active TiO_2-modified nanofiltration membranes in continuous flow photocatalytic reactors for effective water purification: *Journal of Hazardous Materials* 211–212(Supplement C), 304–316.

Ryu, J., Choi, W., Choo, K. 2005. A pilot-scale photocatalyst-membrane hybrid reactor: Performance and characterization: *Water Science and Technology*, 51(6e7), 491e497.

Shon, H.K., Phuntsho, S. and Vigneswaran, S. 2008. Effect of photocatalysis on the membrane hybrid system for wastewater treatment. *Desalination*, 225(1–3), 235–248.

Sidik, D. A. B., Hairom, N. H. H., and Mohammad, A. W. 2019. Performance and fouling assessment of different membrane types in a hybrid photocatalytic membrane reactor (PMR) for palm oil mill secondary effluent (POMSE) treatment: *Process Safety and Environmental Protection*, 130, 265–274.

Sleiman, M., Vildozo, D., Ferronato, C., and Chovelon, J. M. 2007. Photocatalytic degradation of azo dye Metanil Yellow: Optimization and kinetic modeling using a chemometric approach: *Applied Catalysis B: Environmental*, 77(1–2), 1–11.

Wang, W.Y., Irawan, A. and Ku, Y. 2008. Photocatalytic degradation of Acid Red 4 using a titanium dioxide membrane supported on a porous ceramic tube: *Water Research* 42(19), 4725–4732.

Xu, Z., Ye, S., Zhang, G., Li, W., Gao, C., Shen, C., and Meng, Q. 2016. Antimicrobial polysulfone blended ultrafiltration membranes prepared with Ag/Cu_2O hybrid nanowires: *Journal of Membrane Science*, 509, 83–93.

Yang, S., Gu, J.S., Yu, H.Y., Zhou, J., Li, S.F., Wu, X.M. and Wang, L. 2011. Polypropylene membrane surface modification by RAFT grafting polymerization and TiO_2 photocatalysts immobilization for phenol decomposition in a photocatalytic membrane reactor: *Separation and Purification Technology*, 83, 157–165.

Zhang, X., Wang, D.K., Lopez, D.R.S. and da Costa, J.C.D. 2014. Fabrication of nanostructured TiO_2 hollow fiber photocatalytic membrane and application for wastewater treatment: *Chemical Engineering Journal*, 236, 314–322.

Zhao, H., Li, H., Yu, H., Chang, H., Quan, X. and Chen, S. 2013. $CNTs–TiO_2/Al_2O_3$ composite membrane with a photocatalytic function: Fabrication and energetic performance in water treatment: *Separation and Purification Technology*, 116, 360–365.

Zhao, K., Feng, L., Lin, H., Fu, Y., Lin, B., Cui, W., Li, S. and Wei, J. 2014. Adsorption and photocatalytic degradation of methyl orange imprinted composite membranes using TiO_2/calcium alginate hydrogel as matrix: *Catalysis Today* 236 (Part A), 127–134.

Zhao, S., and Zou, L. 2011. Effects of working temperature on separation performance, membrane scaling and cleaning in forward osmosis desalination: *Desalination*, 278(1–3), 157–164.

Zakersalehi, A., Choi, H., Andersen, J. and Dionysiou, D.D. 2013. Photocatalytic ceramic membranes: *Journal of Membrane Science and Technology*, 1–22.

Zhang, X., Du, A.J., Lee, P., Sun, D.D. and Leckie, J.O. 2008. Grafted multifunctional titanium dioxide nanotube membrane: separation and photodegradation of aquatic pollutant: *Applied Catalysis B: Environmental*, 84(1–2), 262–267.

Index

Printed in the United States
by Baker & Taylor Publisher Services